Scanning Electron Microscopy in Biology

R. G. Kessel   C. Y. Shih

Scanning Electron Microscopy in

BIOLOGY

A Students' Atlas on Biological Organization

With 22 Figures and 132 Plates

Springer-Verlag
Berlin Heidelberg New York 1974

Professor Richard G. Kessel
The University of Iowa, Department of Zoology, Iowa City, IA 52242/USA

Dr. Ching Y. Shih
The University of Iowa, Scanning Electron Microscope Laboratory,
Iowa City, IA 52242/USA

The picture on the front cover is an enlargement of figure 2 on page 39; those on the back cover are taken from pages 139, 153, 303 and 339.

Library of Congress Cataloging in Publication Data

Kessel, Richard Glen, 1931—
    Scanning electron microscopy in biology.

    Includes bibliographies.
    1. Scanning electron microscopy. I. Shih, Ching Yuan, 1934—   joint author.
    II. Title.
[DNLM: 1. Biology. 2. Microscopy, Electron, Scanning. QH212.S3 K42s]
QH212.S3K47   578′.4′5   74-11164

ISBN 3-540-06724-8 Springer-Verlag Berlin · Heidelberg · New York
ISBN 0-387-06724-8 Springer-Verlag New York · Heidelberg · Berlin

Preface

In the continuing quest to explore structure and to relate structural organization to functional significance, the scientist has developed a vast array of microscopes. The scanning electron microscope (SEM) represents a recent and important advance in the development of useful tools for investigating the structural organization of matter. Recent progress in both technology and methodology has resulted in numerous biological publications in which the SEM has been utilized exclusively or in connection with other types of microscopes to reveal surface as well as intracellular details in plant and animal tissues and organs. Because of the resolution and depth of focus presented in the SEM photograph when compared, for example, with that in the light microscope photographs, images recorded with the SEM have widely circulated in newspapers, periodicals and scientific journals in recent times.

Considering the utility and present status of scanning electron microscopy, it seemed to us to be a particularly appropriate time to assemble a text-atlas dealing with biological applications of scanning electron microscopy so that such information might be presented to the student and to others not yet familiar with its capabilities in teaching and research. The major goal of this book, therefore, has been to assemble material that would be useful to those students beginning their study of botany or zoology, as well as to beginning medical students and students in advanced biology courses.

It should be emphasized that the SEM is only one tool capable of providing a particular kind of useful information, just as, for example, the light microscope or the transmission electron microscope. While a limited number of light photomicrographs and transmission electron micrographs have been included for clarity in certain cases, the emphasis is on the view obtainable with the SEM, since this represents new information in some cases or has provided additional clarity in other cases. The student, therefore, should consult freely the large number of books in which there is information on the finer intracellular and extracellular details such as can be obtained with the transmission electron microscope.

A phylogenetic and systematic approach is utilized in the organization of this book and is intended to serve students as an adjunct to their textbooks and laboratory manuals. The format of this book was designed for maximum utility on the part of the student. In most sections we have attempted to cover one subject in each plate, placing the corresponding descriptive information close at hand so that it can be quickly and easily related to the illustrations. Present-day textbooks in biology tend to extensively utilize drawings to illustrate structural details and interrelationships as well as processes or concepts. The student of science can well appreciate the significance of the old adages " one look is worth a thousand words " and " seeing is believing." In this connection, the three-dimensional quality of the SEM image as well as the excellent resolution and large surface area in the SEM picture serve a particularly valuable and realistic teaching function.

In our desire to provide rather broad coverage of both plants and animals, many of the preparations were made specifically for this book. In a few cases where information was already available we have used illustrations kindly provided by other investigators. In addition to our desire to select and present information scientifically useful to the student, we have attempted to present images that were esthetically pleasing and interesting as well.

The bibliographic citations in this book have been held to a minimum since a complete and excellent bibliography dealing with research papers involving scanning electron microscopy is published each year as part of the Proceedings of the Annual Scanning Electron Microscope Symposium sponsored by the Illinois Institute of Technology Research Institute and organized by Dr. OM JOHARI.

It is not difficult to predict that the SEM will be increasingly used in the biological, medical, and physical sciences. The number of instruments in use and the number of users continue to increase yearly. Furthermore, scientists continue their efforts to improve the operation of the instrument, the methods of specimen preparation, and the instrumental accessories that can enhance the information output. We are hopeful that this book will stimulate additional interest and progress in this field.

Iowa, Summer 1974   R.G. KESSEL
                                                    C.Y. SHIH

Acknowledgments

We deeply appreciate the efforts of many individuals who have contributed in various ways to the completion of this book. Special gratitude is extended to Professor Emeritus H. W. BEAMS for providing abundant material and suggestions for use in this book. Without his assistance and stimulation this book would have been impossible.

Special thanks are extended to CYNTHIA ASMUSSEN WATSON for her excellent technical assistance in the extensive photographic work required for this book.

The authors are very much indebted to those scientists who kindly provided illustrations for use in this book. They include Drs. S.L. ERLANDSEN, L.E. DeBAULT, D.E. HILLMAN, A.J. WHITE, H.J. BUCHSBAUM, and B. LARSEN of the University of Iowa College of Medicine; Dr. J. SCHABILION, Botany Department, University of Iowa; Dr. T. TANAKA, Okayama University Medical School; Drs. M.J. TEGNER and D. EPEL, Scripps Institution of Oceanography; Dr. G.J. STEWART, Temple University Health Science Center; Dr. M.D. SCHNEIDER, Illinois Institute of Technology Research Institute; Dr. J.H. ANDERSON, M.D. Anderson Hospital and Tumor Institute; Dr. E.R. DIRKSEN, University of California, Berkeley; Dr. H.D. SYBERS, University of California, San Diego; Dr. E.R. LEWIS, University of California, Berkeley; and Drs. T.M. CHEN and C.C. BLACK, University of Georgia.

We should like to acknowledge the invaluable assistance of several graduate students in the Departments of Zoology and Botany at the University of Iowa who made contributions to the preparation of this book. These individuals are F. FEUCHTER, S. EVANS, M. WITTE, A. ENDRESS, K. JENSEN, H. LING, R. LUTZ, R. KARDON, and R. MASON. The authors also appreciate the cooperation of those at the University of Iowa who provided specimen for our use. They include Drs. G. CAIN, M. SOLURSH, G. GUSSIN, D. SOLL, R. CRUDEN, H. DEAN, R. HULBARY, R. MUIR, M. ROSINSKY, R.W. EMBREE and S. SURZYCKI. Drs. H.W. BEAMS, R.D. SJOLUND, C.S. HUANG, and J. FRANKEL kindly read portions or all of the text material and provided helpful suggestions.

Finally, we thank JESSIE INGSTAD, VICKI MOOTHART, and BARBARA O'DONNELL for typing the manuscript.

Contents

Chapter 1
Introduction

Chapter 1   Introduction

A number of scanning electron microscopes are now commercially available, and one such instrument is illustrated in Fig. 1. The control console of a scanning electron microscope consists of two major assemblies. The electron-optical column with the specimen chamber is located at the left of Fig. 1 and is mounted on a plinth containing the vacuum pumping unit. The assembly illustrated in the right portion of Fig. 1 includes display racks containing most of the operator controls, display cathode ray tube (CRT) screens, processing units, meters, etc.

Comparison of the Scanning Electron Microscope with other Microscopes

Magnification

The maximum effective magnification of the light microscope (LM) is about 1 200 diameters, whereas the effective magnification of the best transmission electron microscope (TEM) approaches 1 000 000 diameters. The scanning electron microscope (SEM) can provide a range of magnification varying from about 15 diameters to about 50 000 diameters, depending on the nature and form of the material examined. The value of expanding the low magnification range of the SEM from about $2\times$ to $50\times$ has been emphasized (MEAKIN and FALLON, 1973). Useful pictures in this range can be obtained by increasing the working distances to reduce distortion.

Resolution

The maximum resolution (the minimum distance two objects may be separated and still observed as distinct) of the LM is about 2 000 to 3 000 Å (200 to 300 nm). The best TEM's have a resolving power of 2 to 5 Å (0.2 to 0.5 nm). Most SEM's have a resolution of 100 to 200 Å (10 to 20 nm), depending upon the nature and conditions of the sample as well as on the operating voltage of the instrument and final aperture size. A resolution of 100 Å at 30 KV is common.

Depth of Field

The outstanding feature of SEM micrographs is the remarkable three-dimensional quality. In contrast, the typical micrograph obtained with the LM or the TEM is a two-dimensional image. The LM can be focused in only one plane; therefore, its depth of field is severely limited. The LM is useful for revealing the shape of specimens only at low magnification since the depth of field of a LM decreases with increasing magnification. For internal detail the LM is best used with thin or flat samples, and even thinner sections are required for the TEM. Useful sections for study in the LM commonly range from 8 to 15 µm, but sections to be studied in the TEM generally range from 300 to 700 Å in thickness and must be cut on a special microtome. Because the specimens examined in the TEM must be so thin, the depth of field is limited and an essentially two-dimensional image is obtained. No such limits are found with the SEM. In contrast to the LM or the TEM, in which the illuminating source (light or electrons) must pass through the specimen, in the SEM the electrons that are recorded do not pass through the specimen, but rather secondary electrons are collected from the surface of the specimen and are used to form an image. These secondary electrons need not be focused, only collected. Therefore, in the SEM there is virtually no constraint on the size of the specimen to be studied, the size being limited only by the capacity of the specimen stage.

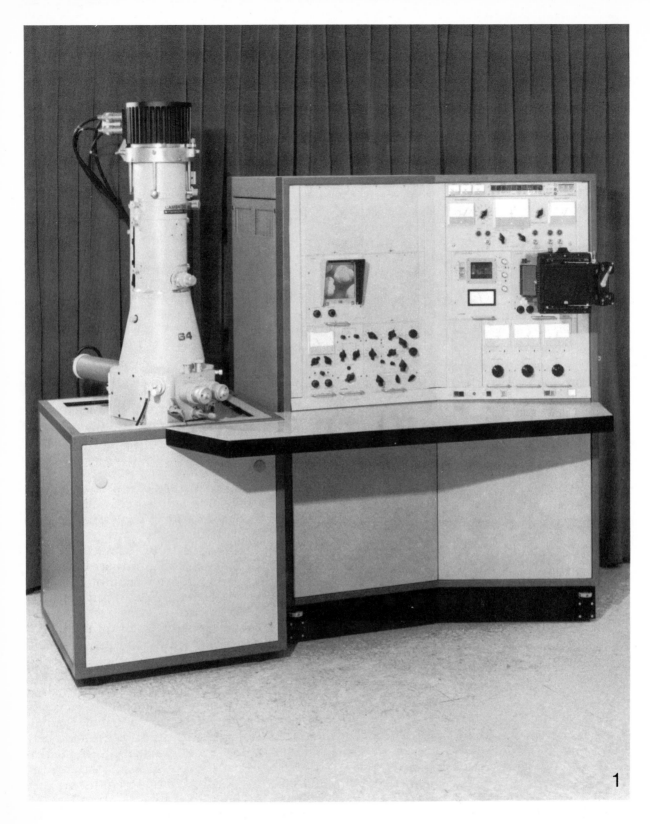

Fig. 1. A Cambridge Stereoscan S4 Scanning Electron Microscope (photograph courtesy of Kent Cambridge Scientific Co.)

2

In terms of both magnification and resolution the SEM has a range generally intermediate between that of the LM and the SEM. Truly unique features of the SEM include the large specimen size that can be observed and the tremendous depth of field. The depth of field in the SEM, compared with the LM, is better at comparable magnifications by a factor of at least 300.

Operating Modes

The SEM can also obtain other kinds of information not found with the LM or the TEM. The major modes of imaging in the SEM include (1) emissive, (2) reflective, (3) absorptive, (4) transmission, (5) cathodoluminescence, (6) x-ray, and (7) beam-induced conductivity.

In summary, the overall advantages of the SEM include (1) a tremendous depth of field and the resulting three-dimensional image, (2) the capability of viewing a large size specimen, (3) a broad magnification range, (4) resolution, (5) ease of varying magnification without changing the focal length so that the depth of field remains constant, and (6) various kinds of information that may be obtained from the specimen. Extensive work is presently underway to improve various technological features of the SEM. Among the important efforts being made are those concerned with providing a better electron beam source and a better vacuum system. This effort is reflected in the development of a field emission electron gun, an ion pump-lanthanum hexaboride gun, and a scanning-transmission electron microscope (cf. CREWE and WALL, 1970; AHMED and NIXON, 1973; KUYPERS et al., 1973; SWANN and KYNASTON, 1973).

Basic Theory and Operation of the Scanning Electron Microscope

A schematic comparison of the light optical and a transmission electron optical system with the scanning electron microscope system is illustrated in Fig. 2. The SEM can be simply characterized as a closed-circuit television system in which the object observed is illuminated by a constantly moving spot of electrons. The electrons in a SEM are emitted by the heating of a tungsten filament (cathode) located within the gun in the upper portion of the microscope column. The cathode is generally held at 20000 V (but is variable over 1 to 30 K V range) below ground potential of the anode.

The electron gun thus produces the electron source (a crossover of high current density). Below the gun are three prealigned electromagnetic condenser lenses which serve to accelerate the electrons and focus them to a small point. The condenser lenses also serve to progressively demagnify the electron beam into a probe which is focused on the surface of the specimen. The electron probe thus formed may range from 10 to 50 nm in diameter. The probe of electrons is deflected by scanning coils which drive the beam over the specimen surface in a square raster comparable to the situation in a television screen. This electron beam or probe is synchronized with the electron beam of a cathode ray picture tube. The scanning coil assembly is located in the bore of the final or lowest condenser lens, which is sometimes referred to as an objective lens. The final lens is fitted with three interchangeable apertures. The specimen chamber is situated at the base of the microscope column in line with the electron beam. The final aperture limits the divergence of the electron beam at the specimen surface. Usually three different sizes of apertures can be manually selected by a control external to the specimen chamber (Fig. 1). This control (isolation valve) is also used to seal off the specimen chamber from the column when samples are changed. It is important that the final aperture, which is located just above the specimen, be clean and in alignment during high-resolution work so as to avoid astigmatism and charging artifacts. Otherwise the image will shift during focusing or other lens adjustments. Larger aperture sizes are used at higher magnifications, but they limit the depth of field obtainable in a focused picture.

Separate diffusion pumps are provided for the column and the specimen chamber and are backed by a single rotary pump. Column vacuum can be maintained during specimen changes and during replacement of the filament (electron source). Both the column and specimen chamber must be under adequate vacuum (less than 10^{-4} torr) for operation of the instrument. The quality of the vacuum in the specimen chamber, very important in high-resolution operation, minimizes contamination. With an inadequate vacuum, contamination of the system occurs rapidly as the hydrocarbons undergo cracking in the electron beam. This, in turn, decreases the signal amplitude and results in poorer resolution. Some SEM's utilize vacuum systems with liquid nitrogen cold traps to reduce the possibility of oil from back-streaming into the chamber from diffusion pumps.

Movements of the specimen in the x-, y-, and z-directions, as well as tilting and rotation, are possible

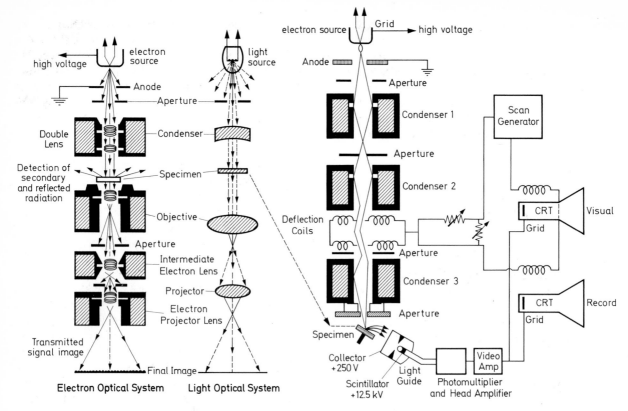

Fig. 2. Diagram illustrating details of a transmission electron optical system, light optical system and SEM (From P.E. MEE: "Microscopy and its contributions to computer technology." *Microstructures*, Oct./Nov. 1972)

Fig. 3. Stereopair, oral view of *Hydra*. ×46

by means of controls outside the specimen chamber. If a specimen is tilted 8 to 10° between successive exposures, stereo pairs can be obtained and viewed with a stereoviewer. The stereo pairs illustrated in Fig. 3 are oral views of the coelenterate *Hydra oligactis*, showing the mouth and tentacles. A simple stereoviewer placed in proper position over this micrograph should illustrate a three-dimensional depth.

Modes of Operation of the Scanning Electron Microscope

As the electron beam is passed over or scanned across the specimen surface, the interaction of electrons with matter (the specimen) results in the production of a variety of electron types as well as x-rays and cathodoluminescense. This results in the possibility of several major modes of imaging in the SEM (Fig. 4).

Emissive Mode

The most important and widely used operating mode is the emissive. When the high-energy electrons in the probe strike the specimen, secondary electrons are excited so that a signal is formed from these low-energy secondary electrons (energies of about 4 eV) emitted from the top 50 to 100 Å of the specimen. Secondary electrons can provide three kinds of specimen information: topographical detail, superimposed voltage contrast, and superimposed crystallographic orientation patterns. The number of secondary electrons produced depends on the surface topography as well as the composition of the specimen. It is generally necessary to evaporate a metal of high atomic number (usually gold-palladium alloy) over the specimen, particularly those that do not conduct well or those that are poor emitters of secondary electrons. The secondary electrons emitted from the specimen pass to the collector (+charge), which accelerates the electrons to strike the scintillator (+charge, but higher voltage). Here photons are produced which enter and pass through a photomultiplier, which serves to produce large numbers of additional electrons (100 000 to 50 million). The electrons leaving the photomultiplier pass to display and record cathode ray tubes. Each of the cathode ray tube (CRT) displays in an instrument such as illustrated in Fig. 1 is 100 mm square. One of these magnetically focused displays is for visual observations

and has a long-persistence phosphor coating (about 600 lines resolution), whereas the other CRT has a short-persistence phosphor coating for high-resolution photographic recording (about 1 000 lines center resolution). The magnification in a SEM is the ratio of the scan length on the visual display cathode ray tube (constant) to the scan length on the specimen (variable). The magnification can thus be increased by decreasing the length of the specimen scan.

In summary, in the emissive mode, low-energy secondary electrons emitted from the specimen by primary electron-probe excitation are drawn usually in a curved trajectory to a positively biased scintillator—light-guide—photomultiplier collector system (Fig. 2). After amplification, this is the signal that modulates the CRT beam.

Applications of the emissive mode in the study of surface topography are virtually unlimited. In addition, large single-crystal specimens examined at low magnifications illustrate patterns of crystallographic orientation similar in appearance to Kikuchi patterns. Finally, it also is possible to display "voltage contrast" with this mode. Generally, negative potentials appear bright and positive potentials appear dark. This provides a useful way of studying semiconductor devices and integrated circuits.

Reflective Mode

In this mode of operation backscattered electrons with a higher energy than secondary electrons are collected. Since they have their origin from depths of up to several microns in the specimen surface, these backscattered images have lower resolution than the secondary images. Collected backscattered electrons can be used to obtain information regarding the kinds of atoms in a specimen. The resolution limit is about 500 to 2 000 Å. The resolution limit for the other modes indicated below is typically 0.1 to 1 μm.

Absorptive Mode

In this condition the signal is collected by monitoring the current flowing to earth from the specimen.

Transmission Mode

In this mode electrons pass through a specimen (as with a section of biological material), and the electrons which have penetrated the specimen are collected. Here

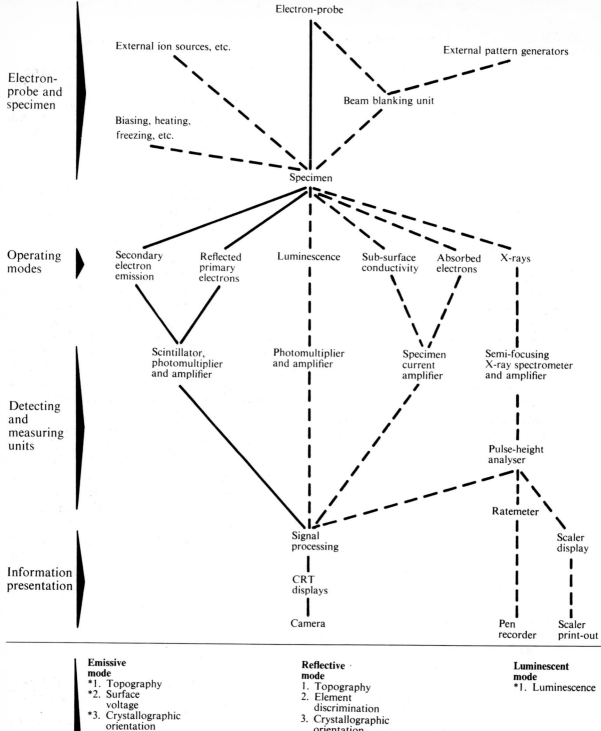

Electron-probe and specimen

Operating modes

Detecting and measuring units

Information presentation

Electron-probe

External ion sources, etc.

External pattern generators

Beam blanking unit

Biasing, heating, freezing, etc.

Specimen

Secondary electron emission

Reflected primary electrons

Luminescence

Sub-surface conductivity

Absorbed electrons

X-rays

Scintillator, photomultiplier and amplifier

Photomultiplier and amplifier

Specimen current amplifier

Semi-focusing X-ray spectrometer and amplifier

Pulse-height analyser

Ratemeter

Signal processing

CRT displays

Camera

Scaler display

Pen recorder

Scaler print-out

Information provided

Emissive mode
*1. Topography
*2. Surface voltage
*3. Crystallographic orientation

Conductive mode
*1. Sub-surface conductivity

Reflective mode
1. Topography
2. Element discrimination
3. Crystallographic orientation

Absorptive mode
1. Topography
2. Crystallographic orientation
3. Surface voltage
4. Element discrimination

Luminescent mode
*1. Luminescence

X-ray mode
*1. Quantitative microanalysis
*2. Qualitative microanalysis

*Preferred method

it is possible to examine thicker specimens than can be viewed with a conventional transmission electron microscope.

X-Ray Mode

In the x-ray mode the x-rays generated when the electron beam reacts with the specimen are collected, analyzed, and used to form a signal. The ability to perform x-ray microanalysis of elements in the SEM is one of its most valuable assets. It is accomplished by adding to the SEM crystal diffraction spectrometers, energy dispersion spectrometers, or a combination of both. Crystal spectrometers can operate at high resolution and have good linearity at high counts. Generally, they have been preferred for *quantitative* analysis with high-probe currents. Energy-dispersive spectrometers are useful for rapid *qualitative* analysis (sometimes also with quantitative analysis), and they have a high collection efficiency but poorer energy resolution (LIFSHIN and CICCARELLI, 1973).

Cathodoluminescence Mode

When a high-energy electron beam interacts with a solid specimen, one measurable phenomenon that can be displayed is cathodoluminescence. When a substance that is bombarded with electrons emits light, this phenomenon can provide useful information, such as structural details, that would otherwise not be visible. It is also possible to examine the behavior of cathodoluminescence material. The effect produced is somewhat similar to optical fluorescence microscopy. The cathodoluminescence mode of the SEM has not as yet been extensively used (YOFFE *et al.,* 1973). An initial application of this mode to biological material was made by PEASE and HAYES (1966), who used the microscope in the luminescent mode to observe sections of spinach leaves treated with Thioflavin T. This dye appears to selectively bind with the cell walls to make these structures clearly visible. This mode is also useful in semiconductor work.

Beam-Induced Conductivity Mode

In this condition electrical connections are made to the specimen and an external voltage is applied. The current flow that results is monitored to provide a signal. When coupled with the changes possible with the incident beam itself, the technique is of great value in the study of junction field regions in semiconductor devices.

Thus, we have seen that a scanning electron microscope system can detect, display, and in some cases measure information derived from the action of a very fine beam of electrons scanning a specimen surface in a square raster. In general terms the information derived includes surface detail (microscopy), x-ray microanalysis, surface voltage contrast, sub-surface current, crystallographic orientation, and luminescence. The specimen can be manipulated and its state changed during investigation. Further, the specimen can be moved linearly, rotated, and tilted. With the addition of suitable modules, the specimen can also be heated, cooled, bombarded with ions, etc. With the addition of suitable external pattern generators, the SEM can also be used for electron-beam processing on an experimental basis.

Further, there are many factors contributing to the acquisition of high-quality pictures of biological material in the SEM. These include factors of (1) specimen fixation and drying, which are dealt with subsequently, (2) suitable coating of the specimen to reduce charging artifacts, (3) probe size and signal-to-noise ratio (the diameter of the probe is dependent to a large extent on the accelerating voltage used as well as lens aberration and settings), (4) properly aligned gun, (5) accelerating voltage of electrons, (6) final aperture size and its cleanliness, and (7) quality of the vacuum.

References

AHMED, H., and NIXON, W.C.: "Boride guns for high signal level SEM", In Scanning Electron Microscopy/1973, O. JOHARI and I. CORVIN, eds. IITRI **6**, 217—224 (1973).

CREWE, A.V., and WALL, J.: "A scanning microscope with 5 Å resolution." J. Mol. Biol. **48**, 375—393 (1970).

EVERHART, T.E., and HAYES, T.L.: "The scanning electron microscope." Sci. Am. **226**, 55—69 (1972).

HEARLE, J.W., SPARROW, J., and CROSS, P.M.: Use of the Scanning Electron Microscope, New York: Pergamon Press (1972).

JOY, D.: "The scanning electron microscope—principles and applications," In Scanning Electron Microscopy/1973, O. JOHARI and I. CORVIN, eds. IITRI **6**, 743—750 (1973).

KUYPERS, W., THOMPSON, M.N., and ANDERSON, W.H.J.: "A scanning transmission electron microscope," In Scanning Electron Microscopy/1973, O. JOHARI and I. CORVIN, eds. IITRI **6**, 9—16 (1973).

LIFSHIN, E., and CICCARELLI, M.F.: "Present trends in x-ray microanalysis with the SEM", In Scanning Electron Microscopy/1973, O. JOHARI and I. CORVIN, eds. IITRI **6**, 89—96 (1973).

MEAKIN, J.D., and FALLON, L.M.: "Low magnification scanning electron microscopy", In Scanning Electron Microscopy/1973, O. JOHARI and I. CORVIN, eds. IITRI **6**, 145–150 (1973).

OATLEY, C.W.: "The Scanning Electron Microscope": Part 1, The Instrument, Cambridge Monographs in Physics, Cambridge, England: Cambridge University Press (1972).

PEASE, R.F.W., and HAYES, T.L.: "Scanning electron microscopy of biological material", Nature **210**, 1049 (1966).

SWANN, D.J., and KYNASTON, D.: "The development of a field emission SEM", In Scanning Electron Microscopy/1973, O. JOHARI and I. CORVIN, eds. IITRI **6**, 57—64 (1973).

THORNTON, P.R.: Scanning Electron Microscopy, London: Chapman and Hall (1967).

YOFFE, A.D., HOWLETT, K.J., and WILLIAMS, P.M.: "Cathodoluminescence studies in the SEM", In Scanning Electron Microscopy/1973, O. JOHARI and I. CORVIN, eds. IITRI **6**, 309—316 (1973).

Methods of Specimen Preparation in Scanning Electron Microscopy

It is only within the past several years that the SEM has been used appreciably in the study of cells and soft tissues. This is somewhat surprising, especially in view of the fact that this technique can provide a picture of whole cell surfaces that markedly improves on the images of the light microscope (even Nomarski optics) yet avoids the limitations of replication techniques. The depth of field and the increased resolution afforded by the scanning electron microscope when compared with the light microscope permit the visualization of a completely new realm of biological structure.

The reasons for the limited use of scanning electron microscopy in biology in the past are doubtless several. Not only has the instrument been of limited availability, but perhaps more important, techniques available for the preparation of soft, hydrated tissues have only recently been developed at a level providing useful images. Hard objects such as insect mouthparts, pollen grains, and teeth lend themselves well to examination by scanning electron microscopy because they do not require specialized methods to preserve their natural morphology. Until recently, soft hydrated tissues have not been prepared with sufficient care so as to avoid

drying artifacts. As a consequence published images in many cases have been only slightly representative of the original cells. However, it is now quite possible with the variety of preparative procedures available and now in use to observe structures associated with cell surfaces and with cellular interrelationships in a manner impossible to obtain by any other method.

Fresh Material

Initial studies of biological material in the scanning electron microscope (BOYDE and STEWART, 1962) involved hard tissues such as teeth, hair, bone, and nails after treatment to remove debris such as dust or mucus. Such specimens retain their original shape in the dried state and can be examined without a metal coating in the SEM by using low accelerating voltages and low magnification. Hard objects such as rocks, shells, and small insects, which have a hard exoskeleton and plant structures including pollen grains can still be usefully studied in the SEM without fixation. However, before viewing, the specimens are sometimes coated with a thin layer of carbon and gold-palladium to reduce charging effects in the electron beam.

Specimens in an electron microscope are exposed to a vacuum pressure generally in the range of 10^{-4} to 10^{-6} torr. If a wet specimen is introduced to such vacuums, its water content will evaporate and the specimen will freeze. The surface tension forces generated during evaporation and the resulting ice crystal formation and growth is suffciently harmful so as to render the resultant images almost useless.

Air-Drying

The problem of drying artifacts became critical when cells and soft tissues were to be studied in the SEM. In the past, specimens were often fixed, rinsed with distilled water, dried overnight in a vacuum dessicator, and examined directly in the SEM. Initial preparations were air-dried from such solvents as water, alcohol, or acetone. Air-drying of cells and tissues is now known to produce a number of drying artifacts in which surface structures may be obliterated because of their collapse during the drying process. There is, however, less distortion and artifact production if soft objects are air-dried from a solvent that has a low surface tension (e.g., acetone or propylene oxide).

The ileum of the rat or mouse is commonly infected with a long bacterium, *Streptobacillus moniliformis*.

This condition is useful to illustrate the effects of various drying methods after fixation. A low magnification view of the villi in the rat ileum is illustrated in Fig. 5 and a higher magnification in Fig. 6. In both cases the material was air-dried from water. At low magnification the bacteria are not visible, but at higher magnification it can be seen that they have collapsed onto the surface of the villi during the air-drying process. The surface of epithelial and goblet cells are not distinct under these conditions. In Figs. 7 and 8 portions of intestinal villi and bacteria from the mouse are illustrated as dried from absolute ethanol and propylene oxide, respectively. Under these conditions, the bacteria are more apparent and not as collapsed onto the surface of the villi. Further, the cell boundaries of epithelial cells and goblet cells are more apparent. A low magnification view of the rat ileum dried by the critical point method from carbon dioxide is illustrated in Fig. 9. At this magnification the bacteria are quite apparent and many of them stand erect and project into or through the intervillous spaces. This condition is also illustrated at higher magnification in Fig. 10, which includes a portion of two villi and the intervening space. The bacteria are erect and well preserved. The effect of various drying methods on cultured cells has been studied by PORTER et al. (1972).

Chemical Fixation of Cells, Tissues, and Organs

A wide variety of chemical fixatives have been employed in scanning electron microscopy of biological tissues. In many cases the fixatives are similar to those used in preparating tissues for examination by transmission electron microscopy. Fixatives commonly employed include

(1) 1 or 2% osmium tetroxide in 0.5 or 1.0 M sodium phosphate or sodium cacodylate buffer (pH 7.2 to 7.4).

(2) 1.5 to 3% glutaraldehyde prepared in 0.5 or 1.0 M sodium phosphate or sodium cacodylate buffer (pH 7.2 to 7.4).

(3) KARNOVSKY's (1965) paraformaldehyde-glutaraldehyde fixative prepared using one of the buffers already indicated. This fixative has also been used at dilutions of one-half to one-fourth of the full strength mixture depending upon the osmolality of the particular tissue to be studied.

(4) A "cocktail" fixative containing both glutaraldehyde and osmium tetroxide has been employed to good advantage to preserve structural details in biological specimens. The mixture frequently used consists of cold 1.25% glutaraldehyde and 2% osmium tetroxide in 0.1 M sodium cacodylate buffer (pH 7.2 to 7.4). In this case the glutaraldehyde and osmium tetroxide are combined from stock solution just prior to use. Fixation is carried out for 1 to 3 hours at 4° C.

Commonly, the tissues are initially fixed in glutaraldehyde (primary fixative) followed by a buffer rinse and secondary fixation in osmium tetroxide. Thus, a dual fixation consisting of an aldehyde (reducing fixative) and osmium tetroxide (oxidizing fixative) is frequently employed. Both glutaraldehyde and osmium tetroxide are noncoagulating fixatives (cf. HAYAT, 1970). In some cases the tissues are fixed only in osmium tetroxide.

Both glutaraldehyde and osmium tetroxide penetrate tissue blocks slowly. Therefore, care should be exercised so that relatively small tissue blocks are exposed to the fixative to insure good fixation. Further, fixation in glutaraldehyde is often for 24 to 48 hours or more during which time the fixative is changed twice.

Many individual cells such as protozoa, blood cells, and sperm are extremely sensitive to the osmolality of the fixative (BESSIS and WEED, 1972). For example, if mammalian erythrocytes are placed in a hypertonic fixative, they become covered with short spine-like processes (crenated) due to shrinkage of cells (Fig. 11). In hypotonic fixatives the erythrocytes tend to swell and become spherical rather than biconcave. In these cases especially, it is important that the concentration of the fixative and buffer be adjusted to approximate that of the tissues being fixed. In some cases it is useful to add sucrose to the fixative to increase its osmolality.

Fixation in osmium tetroxide is especially useful for those specimens covered by a thin mucus film since osmium tetroxide tends to remove this potentially obscuring layer. In those structures where mucus is more abundant (e.g., stomach, intestine, trachea, etc.) the tissues must be carefully rinsed with buffer or Ringer's solution prior to fixation. In some cases it is helpful to remove mucus by rinsing such tissue blocks prior to fixation with a 1% solution of sodium carbonate (prepared with the appropriate Ringer's solution and adjusted to pH 8 with hydrochloric acid) for about one minute.

In general, primary fixation of biological material for SEM observation involves an aldehyde and occurs over a period of several hours to several days. The tissue is then rinsed several times with buffer, saline, or distilled water (for about 30 to 60 minutes) and then 1 or 2% osmium tetroxide is used as a secondary

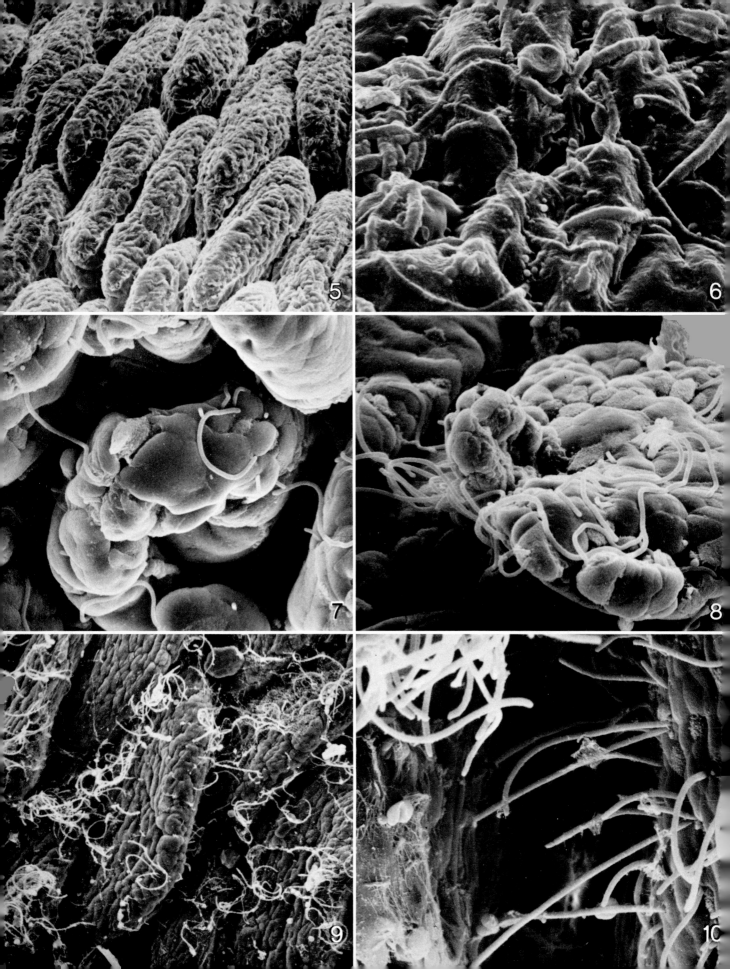

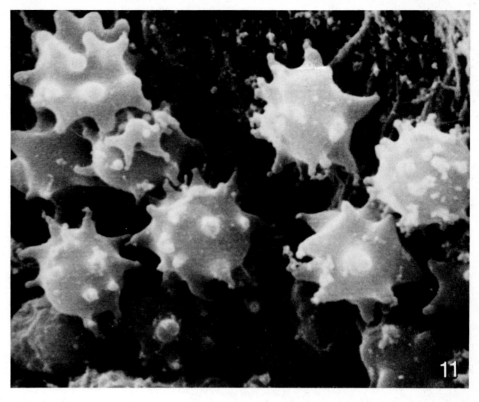

Fig. 11. Erythrocytes (crenated) from rat. × 8625

fixative. Fixation with osmium tetroxide is usually carried out for 1 to 3 hours at 4° C.

The methods of preserving biological specimens for SEM are indeed extensive, and the literature should be consulted for additional details and variations. More extensive information regarding methods for preparing tissue culture cells and organs for SEM study is provided in subsequent sections.

Fig. 5. Rat small intestine, dried from water. × 270. (Kindly provided by Drs. S.L. ERLANDSEN and M. COAN)

Fig. 6. Rat small intestine, dried from water. × 2625. (Kindly provided by Drs. S.L. ERLANDSEN and M. COAN)

Fig. 7. Rat small intestine, dried from alcohol. × 1340

Fig. 8. Rat small intestine, dried from propylene oxide. × 1390

Fig. 9. Rat small intestine, critical point drying. × 275. (Kindly provided by Drs. S.L. ERLANDSEN and M. COAN)

Fig. 10. Rat small intestine, critical point drying. × 1545. (Kindly provided by Drs. S.L. ERLANDSEN and M. COAN)

Dehydration

Following the appropriate fixation method, cells, tissue, or organ blocks are commonly dehydrated through a graded series of either ethanol of acetone (e.g., 30%, 50%, 70%, 85%, 95% for 5 to 10 minutes each, and then to absolute ethanol or acetone with several changes over 30 to 60 minutes). Blocks of tissues or organs are easily transferred during dehydration but individual cells (e.g., protozoa) can frequently be conveniently manipulated either by attaching them to a small coverglass or by carrying them through dehydration in a centrifuge tube. The cells can then be collected at the bottom of the tube prior to each solution change. As indicated previously, in the past cells were examined after being air-dried from water, alcohol, or acetone. The common drying procedure now, however, is the critical point method using liquid carbon dioxide (ANDERSON, 1951). The dehydrated specimen is transferred from absolute ethanol or acetone to a critical point drying apparatus. In some cases the tissue is transferred to isoamyl acetate prior to critical point drying (CPD).

Critical Point Drying

This technique was introduced to electron microscopy by ANDERSON (1951). The method takes advantage of the fact that at its critical point a fluid passes imperceptibly from a liquid to a gas with no evident boundary and no associated distortional forces. A number of commercially available devices for critical point drying exist, but a homemade instrument can be devised for this procedure. The major part of such an instrument includes a specimen drying chamber in which the specimens are placed during replacement of the acetone or alcohol by carbon dioxide (Fig. 12). The specimen drying chamber is provided with a screw cap cover. Another major portion of the instrument is a system containing two needle pressure valves, one for admitting the liquid carbon dioxide to the specimen chamber, the second for permitting the escape of carbon dioxide from the specimen chamber. The container illustrated in Fig. 12 is filled with hot tap water (about 40 to 45° C) and is used to warm the specimen chamber in order to achieve the critical point of liquid carbon dioxide (1072 p.s.i. and 31° C). Several different devices may hold the specimens while they are drying in the specimen chamber, and these are illustrated in Fig. 13. Metal containers with a cap provided with holes can be used, as can plastic capsules that have been punctured a number of times with a needle. The holes in such containers provide for easy evaporation and removal of the dehydrating fluid. When cells are attached to cover glasses or to pieces of plastic petri dishes, it is often convenient to stack them on a device such as illustrated in Fig. 13 to maximize the number of specimens being dried. These containers are then placed within a wire basket that fits inside the specimen holder of the critical point drying apparatus. Specimens are transferred to the drying chamber with just enough alcohol or acetone (absolute) to prevent air drying prior to critical point drying. The cover of the specimen chamber is secured and liquid carbon dioxide is admitted to the chamber from a tank through a siphon. When the inlet valve is opened, the pressure within the chamber rises but eventually stabilizes at about 900 p.s.i. With the inlet valve opened, the exit valve is gradually opened so that carbon dioxide can flow slowly through the chamber for 2 to 3 minutes. Then, both valves are closed and the

Fig. 12. Major components of a critical point drying apparatus

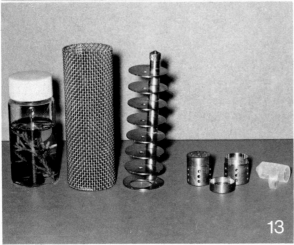

Fig. 13. Several types of specimen holders for critical point drying

specimens remain exposed to the carbon dioxide for about 5 minutes. A fresh flow of carbon dioxide is then provided for about 1 or 2 minutes; then the valves are closed for about 5 minutes. This procedure is repeated five or six times and the two needle valves are eventually closed. At this time a container of hot tap water is placed around the specimen chamber, heating it to about 40° C. This temperature is sufficient to achieve the critical point of carbon dioxide, which is 31° C. During the heating process, the pressure within the chamber rises to about 1500 p.s.i. With the specimen chamber still warm, the exit valve is slowly opened and the chamber is depressurized over a period of 10 minutes so that chances of decompression damage to the specimen are minimized or prevented. It has been shown by PORTER *et al.* (1972) that Freon 13 as well as nitrous oxide can be used in place of carbon dioxide for good results in the critical point drying method.

Processing Specimens after Drying

Specimens dried by the critical point method are transferred to specimen stubs (Fig. 14) which insert into the stage of the SEM (Fig. 15). The blocks can be attached to these specimen stubs after the stubs are coated with a thin layer of silver-conducting paint, or they can be attached to the stubs by a sticky copper-conducting tape (by 3M Co.). The specimens are then transferred to a vacuum evaporator (Denton DV 502) equipped with an omni rotating-tilting device as well as a carbon rod and a metal, evaporating unit (Fig. 16). The specimens are placed about 10 cm from the evaporation source, which consists of a tungsten wire basket in which is placed a tightly coiled 10 to 14 cm length of 8 mil diameter gold-palladium wire. Initially, a thin layer of carbon is evaporated onto the specimens from a carbon electrode source within the evaporator. Then, a 10 to 30 nm layer of gold-palladium alloy is evaporated. During evaporation, the specimen stubs rotate and tilt continuously to ensure a uniformly thin layer. All specimens to be illustrated were studied in a Cambridge Stereoscan S4 scanning electron microscope operating at accelerating voltages ranging from 5 to 20 KV, and images were recorded photographically on Polaroid type 55 P/N film.

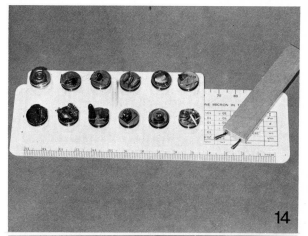

Fig. 14. Specimens mounted on stubs

Fig. 15. Standard stage with specimen holder (Cambridge Stereoscan S4)

Fig. 16. Specimen stubs on tilting omni-stage for evaporation with carbon and gold-palladium

13

Physical Methods of Tissue Preparation

Freeze Method

In this technique tissues or organs are frozen and then immediately examined in the SEM. Although this method permits rapid examination of specimens under conditions approaching the living, some artifacts due to ice may be introduced if sufficient care is not exercised. Further, the specimen must be examined within a short time, at low magnifications, and at a reduced resolution (e.g., 50 to 60 nm). A radish root tip with root hairs is shown preserved by the freeze method in Fig. 17 and can be compared to a similar specimen in Fig. 18 that was fixed and dried by the critical point method. In the frozen state the root hairs are preserved in an extended condition.

More recently, a special apparatus and a modified procedure for freezing samples to be studied in the cold stage of a SEM have been reported by ECHLIN and MARETON (1973). The reader is referred to this article for specific details in the use of this method. Essentially, unfixed and noncryoprotected specimens are quench-frozen on aluminum stubs that have demountable caps. The specimen is sealed under liquid nitrogen and the closed unit is placed in a vacuum evaporator precooled to —150°C and equipped with a rotating cold table. The cap on the specimen stub is removed (at a pressure of about 10^{-7} torr) and the specimen is coated with carbon and metal alloy. The cap is then replaced under vacuum and the entire unit transferred to a precooled cold stage (—150°C) in the SEM. The cap is removed just before examination. This procedure permits bulk specimens to remain

Fig. 17. Raddish root tip and root hairs (frozen). × 44

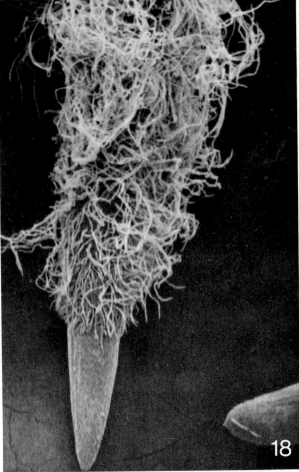

Fig. 18. Raddish root tip and root hairs after fixation and critical point drying. × 39

below —130°C throughout preparation and observation, and frost formation on the specimen is prevented. Further, better resolution of detail is achieved by this method.

Freeze-Dry Method

This technique involves the drying of a specimen as it is frozen. Freeze-dry procedures have been used in the past for cultured cells to be studied in the SEM. After fixation, the cells, which are covered with a small amount of water, are instantly frozen to —158°C with Freon 12 cooled with liquid nitrogen. Freeze drying may also be conducted from other solvents that are suitably volatile when frozen. The specimens are transferred to a freeze-dry unit at a temperature of —80°C. After sublimation of water for several hours, the specimen is warmed to —40°C, where drying is completed from 2 to 6 hours. A liquid nitrogen cold finger is used to trap water removed during the drying process. The specimens are then coated as described previously and examined in the SEM.

Although this procedure can be used in some cases, several artifactual conditions not present in critical point drying have been reported, such as the development of micro-ridges on the normally smooth cell surface and alterations in the diameter as well as distortion of the microvilli, filopodia, and lamellipodia (PORTER et al., 1972). Critical point drying of cells and soft tissues appears to represent a more favorable method. However, improved methods for rapid freezing and handling of specimens have been described by BOYDE and ECHLIN (1973) and should be consulted for details.

Correlation of Biological Organization in the Scanning Electron Microscope, the Transmission Electron Microscope, and the Light Microscope

Advances in SEM methodology are being reported continually. One such recent and useful procedure permits specimens prepared for and examined initially in the TEM to be examined in the SEM after removal of the embedding material (Epon 812). Thus, it is now possible to study tissue blocks in the SEM that were originally fixed, dehydrated, and embedded in resins for sectioning and observation in the TEM. For example, tissue embedded in Epon 812 may be sectioned 1 μm thick on an ultramicrotome, affixed to glass slides, and stained in one of variety of ways for obser-

vation in the light microscope. Adjacent thin sections can then be cut and viewed in the TEM. The embedding material may then be removed and the tissue block examined in the SEM. In this technique the excess Epon 812 is trimmed from the block and the remaining resin dissolved with an epoxy solvent. This solvent is prepared by aging 1 to 2% sodium hydroxide (or potassium hydroxide) in absolute methanol (or absolute ethanol) for 3 to 4 days. The aging process results in the production of an active sodium methoxide. Blocks may be placed for 3 to 18 hours in clean glass vials containing 4 to 5 ml of the solvent. However, to minimize the contamination of the tissue, a relatively simple and inexpensive continuous flow apparatus can be made in which the tissue blocks are perfused with the epoxy solvent. For the details of this apparatus, the reader is referred to the article by ERLANDSEN et al. (1973). After the Epon has been completely removed from the tissue block (as determined by monitoring with the light microscope), the tissue is thoroughly washed with absolute methanol, followed by treatment with amyl acetate and then critical point drying. The tissue block, after study in the SEM, may be transferred to propylene oxide and re-embedded in Epon for sectioning and TEM observation once again. This procedure is, therefore, extremely useful in the correlation of information provided by the light microscope, the transmission electron microscope, or the scanning electron microscope.

The usefulness of correlating information obtained from the light microscope with that of the scanning electron microscope is further emphasized by a technique recently developed by WETZEL et al. (1973). This method permits the identification of Giemsa-stained leukocytes in the light microscope, yet preserves the delicate surface features so that the same cells can be studied in the SEM (WETZEL et al., 1973). In this method leukocytes are fixed to coverslips previously scribed with an asymmetric design so that individual cells can be mapped and identified by both light microscopy and SEM. The Giemsa staining is performed on wet preparations which are then examined and photographed. The wet but stained preparations are then rinsed with distilled water, dehydrated in a graded series of ethanols into amyl acetate, dried by the carbon dioxide critical point procedure, and examined in the SEM after carbon and metal evaporation. This technique is especially useful in such a heterogeneous cell population as blood leukocytes for it permits positive identification of cells studied with respect to their surface features in the SEM. The usefulness of this technique is illustrated in Figs. 19 through 22. In the photo-

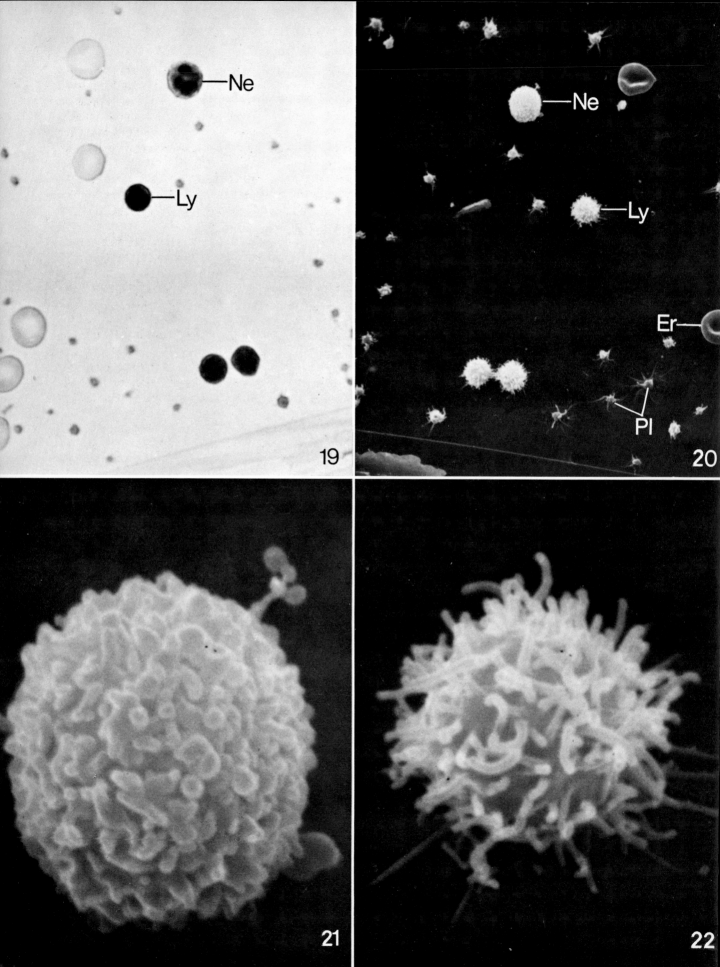

micrograph (Fig. 19), a neutrophil (Ne) and a lymphocyte (Ly) can be specifically identified. The same neutrophil (Ne) and lymphocyte (Ly) are illustrated by SEM in Fig. 20, which also shows erythrocytes (Er) and platelets (Pl). An enlarged view of the neutrophil is illustrated in Fig. 21, while the lymphocyte is enlarged in Fig. 22.

Special Techniques

Observation of Conventionally Prepared Tissue

Many biologists have extensive collections of tissue blocks embedded in paraffin or collodion for light microscope studies. These tissue blocks can also be studied in the SEM, using relatively simple procedures (*cf.* HODGKIN, 1972). Paraffin is removed by two successive washings in benzene. The tissue is then transferred to 100% ethanol or acetone, dried by the critical point method, metalized, and examined under the SEM. Collodion-embedded material can be immersed in acetone to remove the collodion and then processed as indicated.

If a correlation between the light microscopic image and the SEM image is desired, a 15 to 20 µm ribbon of sections can be cut and affixed to a cover glass. Paraffin can be removed and the section stained appropriately, but the cover slip should be mounted face down in buffer, with nail polish sealing the edges to prevent drying. After a photomicrograph is obtained, the cover slip can be removed, and the tissue section dehydrated and critical point dried for scanning electron microscopy. Thus, additional correlative information can be obtained.

Replication

Replication is an indirect method of examining living specimens such as skin or teeth by looking at their negative or positive casts or impressions without removing the tissue. A silicone rubber/epoxy-cast is useful for a skin replica where hair preservation is desired. In this technique silicone rubber is pressed on the skin and a positive cast is made by pouring epoxy resin into the cured silicone rubber. The resulting replica is then coated with metal and observed in the SEM. For the palm skin a celluloid impression can be made. With this method a celluloid plate is softened by a drop of amyl acetate and pressed for a few minutes on the skin. The negative replica can be directly observed with the SEM by using reverse imaging to obtain a positive picture. This skin replica has been shown by FUJITA *et al.* (1961) to be free from deformation and shrinkage and is considered superior to using the skin itself.

Replication of hard dental structures such as enamel, dentin, and dental restorative materials can be accomplished in two steps. The original sample is copied by means of an acetate tape softened on one side with acetone. The wet side is placed over the object to be copied and the tape is allowed to harden under moderate pressure. After separation the negative is filled with a silicone rubber, and when the polymerization is completed, the positive replica is ready for coating and examination. For soft tissue and fragile structures a hydrocolloid material can be used for impressions (PAMEIJER and STALLARD, 1973).

Cryofracture

Cryofracture of the specimen enables one to observe extracellular and intracellular surfaces. With this method it is possible to correlate intercellular arrangement to the surface structure. The procedure for obtaining a fractured surface can be as simple as taking a piece of tissue, freezing it in liquid nitrogen, and then cracking it in half with a razor blade, either with the tissue submerged or else on a block equilibrated with liquid nitrogen. To control the fracture direction, a V-notch can be made with a razor blade on the tissue before the specimen is frozen. Cryofracture can be performed at any stage during specimen preparation. If it is carried out after fixation, a presoaking of the tissue in glycerine is necessary (NEMANIC, 1972). When it is carried out in 70% alcohol, the specimen will crack by itself in the liquid nitrogen (LIM, 1971). This self-cracking is very useful for studying delicate biological organs such as the inner ear. Specimens in 100% alcohol can also be cryofractured. These

---

Fig. 19. LM photomicrograph human blood smear, Giemsa stain. × 1590

Fig. 20. SEM of same blood cells. × 1590

Fig. 21. SEM of human neutrophil. × 15900

Fig. 22. SEM of human lymphocyte. × 15900

results will be shown in plates of tobacco leaf and the lung. Critical point dried tissue can also be frozen in liquid nitrogen and cracked the same way.

References

ANDERSON, T. F.: "Techniques for the preservation of three-dimensional structure in preparing specimens for the electron microscope." Trans. N. Y. Acad. Sci. Ser. II, **13**, 130—134 (1951).

BESSIS, M., and WEED, R. I.: "Preparation of red blood cells (RBC) for SEM. A survey of various artifacts", In Scanning Electron Microscopy/1972, Part II, O. JOHARI and I. COVIN, eds. IITRI **5**, 289—296 (1972).

BOYDE, A., and STEWART, A. D. G.: "SEM of the surface of developing mammalian dental enamel." J. Ultrastruct. Res. **7**, 159—172 (1962).

BOYDE, A., and ECHLIN, P.: "Freeze and freeze-drying—a preparative technique for SEM", In Scanning Electron Microscopy/1973, O. JOHARI and I. CORVIN, eds. IITRI **6**, 759—766 (1973).

ECHLIN, P., and MORETON, R.: "The preparation, coating and examination of frozen biological materials in the SEM", In Scanning Electron Microscopy/1973, Part III, O. JOHARI and I. CORVIN, eds. IITRI **6**, 325—332 (1973).

ERLANDSEN, S. L., THOMAS, A., and WENDELSCHAFER, F.: "A simple technique for correlating SEM with TEM on biological tissue originally embedded in epoxy resin for TEM", In Scanning Electron Microscopy/1973, Part III, O. JOHARI and I. CORVIN, eds. IITRI **6**, 349—356 (1973).

FUJITA, T., TOKUNAWA, J., and IONUE, H.: "Scanning electron microscopy of the skin using celluloid impressions." Arch. Histol. Jap. **30**, 321—326 (1969).

HAYAT, M. A.: Principles and Techniques of Electron Microscopy: Biological Applications. Vol. I., New York: Van Nostrand Reinhold Co. (1970).

HODGKIN, N. M.: "Electron scanning microscopy of biological material, comparative technique." Microstructures **3**, 17—22 (1972).

KARNOVSKY, M. J.: "A formaldehyde-glutaraldehyde fixative of high osmolality for use in electron microscopy." J. Cell Biol. **27**, 137A (1965).

LIM, D. J.: "Scanning electron microscopic observation on non-mechanically cryofractured biological tissue", In Scanning Electron Microscopy/1971, O. JOHARI and I. CORVIN, eds. IITRI **4**, 257—264 (1971).

NEMANIC, M. K.: "Critical point drying, cryofracture, and serial sectioning", In Scanning Electron Microscopy/1972, O. JOHARI and I. CORVIN, eds. IITRI **5**, 297—304 (1972).

PAMEIJER, C. H., and STALLARD, R. E.: "Three replica techniques for biological specimen preparations", In Scanning Electron Microscopy/1973, O. JOHARI and I. CORVIN, eds. IITRI **6**, 357—364 (1973).

PORTER, K. R., KELLEY, D., and ANDREWS, P. M.: "The preparation of cultured cells and soft tissues for scanning electron microscopy", In Proceedings of the Fifth Annual Stereoscan Scanning Electron Microscopy Colloquium, Kent Cambridge Scientific, Inc., Morton Grove, Illinois (1972).

WETZEL, B., ERICKSON, B., and LEVIS, W.: "The need for positive identification of leukocytes examined by SEM." In Scanning Electron Microscopy/1973, O. JOHARI and I. CORVIN, eds. IITRI **6**, 535 (1973).

Chapter 2
One-Celled Organisms

Chapter 2 One-Celled Organisms

The Protozoa comprise a large group of organisms that exhibit considerable diversity in organization and complexity. Some of the Protozoa exhibit affinities to both the animal and the plant kingdom. Further, many Protozoa are single and cellular, others are more organismic in organization, and still others may be multiple or colonial in form. For some time the Protozoa were referred to as acellular, but considerable controversy exists regarding their unicellular or acellular nature. While the organelles of Protozoa and metazoan cells are quite similar in most cases, there are also cellular structures as well as complexities in life cycles that are not represented in metazoan cells. Alternatively, the question arises as to the logic for making such distinctions. There are advantages from a cytological standpoint in equating Protozoa to cells of multicellular organisms, while in dealing with details in their behavior, systematics, evolution, and ecology, it may be more useful to deal with Protozoa as organisms. Thus, Protozoa have features common to both metazoan cells and metazoan organisms. The term "acellular" has been abandoned by some (GRIMSTON, 1961).

The TEM has demonstrated considerable structural complexity in both the cytoplasm and the nucleus of Protozoa. The study and characterization of surface details of Protozoa, however, have proved to be a difficult and time-consuming task with this technique, since the examination of many sections of different planes are required before a three-dimensional reconstruction of the entire surface can be obtained and visualized. The scanning electron microscope can serve, as demonstrated by the examples in this section, as a useful technique for visualizing the surface details of many of the surface complex Protozoa, such as the ciliates. It is convenient at this point to indicate that cilia may form several different kinds of specializations, which are illustrated in the scanning electron micrographs in this section. Cilia may form membranelles that are generally composed of two or three rows of cilia coordinated in such a way that the organelle tends to beat as a paddle. Cirri consist of bundles of cilia that closely adhere together and taper at their ends. An undulating membrane generally consists of a single row of cilia, which functions overall as a membrane. In this section representatives of three major groups of Protozoa are illustrated: the ciliates, the flagellates, and the amoebae.

Reference

GRIMSTON, A. V.: "Fine structure and morphorgnesis of Protozoa." Biol. Rev. Cambridge Philos. Soc. **36**, 97—150 (1961).

Ciliate Protozoa

Phylum Protozoa—Subphylum Ciliophora

Class Ciliatea—Subclass Holotrichia—
Order Hymenostomatida

Paramecium multimicronucleatum

The advantages of the scanning electron microscope in the study of Protozoa are well illustrated by the well-known holotrich ciliate, *Paramecium*. Observations of these slipper-shaped organisms in the scanning electron microscope illustrate the number and distribution of body or somatic cilia (SC), which are approximately 6 to 7 µm long (Figs. 1 through 3). The *Paramecium* illustrated in Fig. 1 is approximately 200 µm long and about 50 µm at its widest point. The anterior tip of the organism is more blunt than the posterior end. When viewed from the oral side (Fig. 2), a depression termed the oral groove (OG), or vestibulum, is observed to extend from the right side of the anterior end of the organism diagonally toward the left side of the animal, where an aperture, termed a buccal overture (BO), is present at the entrance to a closed buccal cavity. Food, such as bacteria and yeast, is directed posteriorly by the beat of cilia in the oral groove to the buccal overture and into the narrow and tubular buccal cavity. Food particles (phagotrophic feeding) become incorporated into large blebs of the surface membrane at the base of the buccal cavity, and as these blebs detach from the remainder of the surface membrane, the food particles surrounded by the membrane of the bleb become internalized into the cytoplasm as a food vacuole.

The outer surface of *Paramecium* consists of a trilayered structure termed the pellicle (periplast). The slipper-shaped body is maintained by the semirigid pellicle, which has a wide range of plasticity or elasticity. The ridges and furrows in the pellicle form a hexagonal lattice arrangement, and cilia emerge from the cell in the depressions of the pellicle (arrows in Fig. 3). To demonstrate the comparative information provided by the transmission and scanning electron microscopes, a transmission electron micrograph of a portion of the surface of *Paramecium* is shown in Fig. 4. This section passes at a right angle to the surface and illustrates the ridged pellicle (Pe), one cilium in longitudinal view (CL), and a basal body (BB), or kinetosome. Two mitochondria (Mi) are closely apposed to the kinetosome of the cilium. A transverse section of a single cilium illustrates the typical nine peripheral doublet microtubules and the central pair of microtubules (Fig. 4, inset). A transverse section of a trichocyst (Tr) is illustrated in Fig. 4. Trichocysts are located in the ectoplasm and are discharged as long threads through pores in the pellicle ridges.

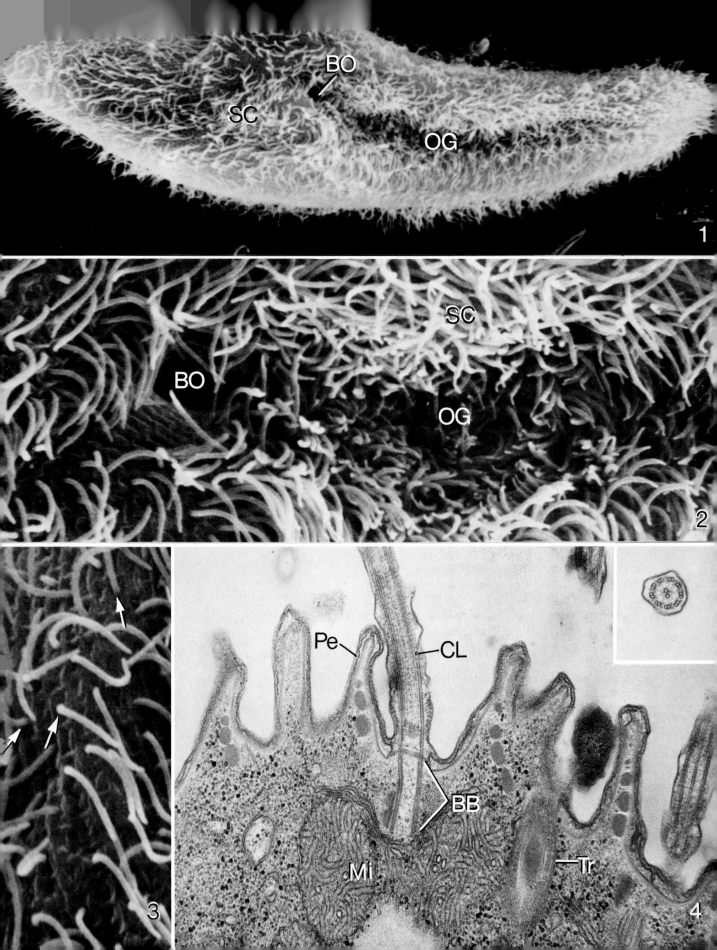

Trichocysts of Paramecium

DC   Dense core (of unexploded trichocyst)
Pe   Pellicle
Sh   Sheath (of unexploded trichocyst)
TB   Trichocyst body (unexploded)
TH   Trichocyst head (exploded)
Th   Threads (of exploded trichocysts)

Fig. 1, ×1010; Fig. 2, ×50000; Fig. 3, ×15900; Fig. 4, ×15900.

These interesting structures represent a secretory product of *Paramecium*, but the secretory product is not exported from the cell until the appropriate stimulus is applied. A section through an unexploded trichocyst of *Paramecium multimicronucleatum* is illustrated in the transmission electron micrograph (Fig. 2). The pointed end of this elongated structure is closely applied to the ridged portion of the pellicle (Pe in Fig. 2). The tapered end of the trichocyst consists of a long, dense core (DC) surrounded by a less dense, granular sheath (Sh). The trichocyst body (TB) measures about 1.5 μm long by 1 μm at its widest diameter. In *P. aurelia* it has been estimated that about 1500 trichocysts are present within a single organism. The tip of the trichocyst is about 1.6 μm long, and the central core frequently illustrates a crystalline organization. The organism illustrated in Fig. 1 was exposed to Parker's washable black ink, which stimulates discharge of the trichocysts; the organism was then fixed in 95% ethanol. The discharged trichocysts (Figs. 3 and 4) have a distal, or terminal, cap shaped much like an arrowhead (TH). The exploded threads (Th) may be 20 μm or more in length and are about 0.38 μm in width. The thread has a banded appearance, and each of the banded units is about 0.2 μm long.

The functional significance of trichocysts is not clear. It has been assumed that the discharge of trichocysts serves to deter aggressors, but there is little or no evidence for this. The trichocysts of *Paramecium* do not deter *Didinium* from engulfing it. Another suggestion is that the trichocysts are adhesive and serve to anchor the organism while feeding, but organisms can swim and feed at the same time. Once trichocysts have been discharged, new trichocysts must be resynthesized to replace them.

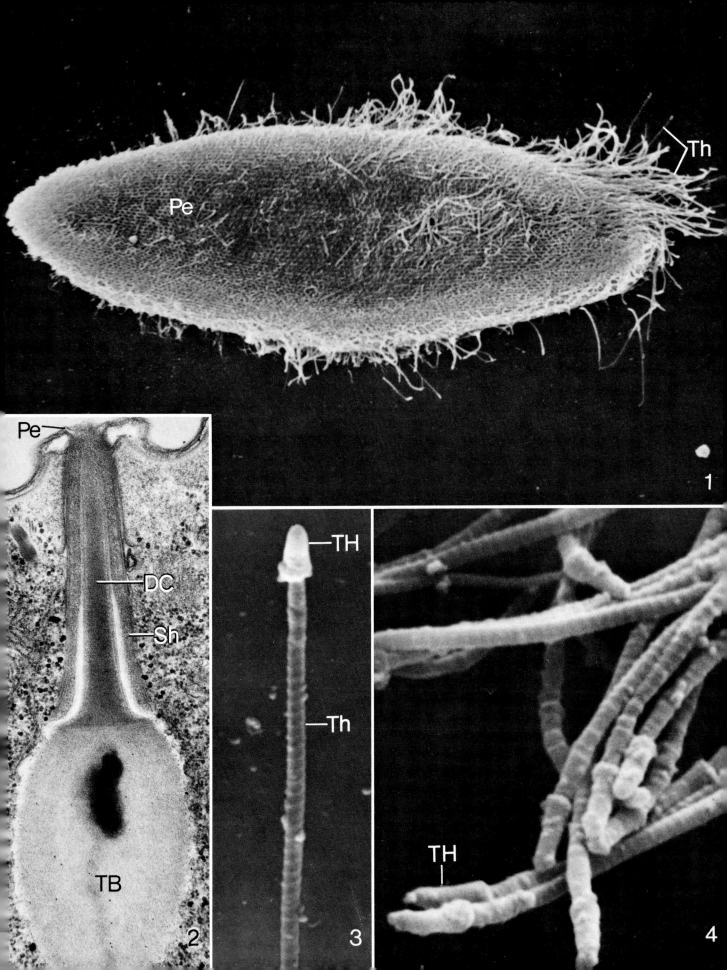

Feeding
in Paramecium multimicronucleatum

Ch   *Chilomonas*
ET   Exploded trichocysts (of *Paramecium*)
Fl   Flagella (of *Chilomonas*)
OG Oral groove (of *Paramecium*)
SC   Somatic cilia (of *Paramecium*)

Fig. 1, ×910; Fig. 2, ×4650; Fig. 3, ×4815; Fig. 4, ×7290.

Chilomonas paramecium constitutes the major food source for *Paramecium multimicronucleatum*. *Chilomonas* is a colorless cryptomonad flagellate that has a vestibule slightly lateral to the anterior end of the organism. From this region also, two flagella (Fl in Fig. 2) emerge. Details in the feeding process of *Paramecium* on *Chilomonas* are illustrated in Figs. 1 through 4. The presence of *Chilomonas* (Ch) in the region of the oral groove (OG) of *Paramecium* is shown in Fig. 1. The somatic cilia (SC) of *Paramecium* are evident in Fig. 2. Note the numerous and long threads that tend to form an investing meshwork around *Chilomonas* in Figs. 3 and 4. These threads are exploded trichocysts (ET in Figs. 3 and 4) and are considerably thinner than the somatic cilia. Particularly noteworthy is the fact that the trichocysts of *Paramecium* are discharged only from that region of the body pellicle to which *Chilomonas* adheres. The details illustrated indicate that the trichocysts of *Paramecium* are used in the capture of *Chilomonas* prior to its injestion.

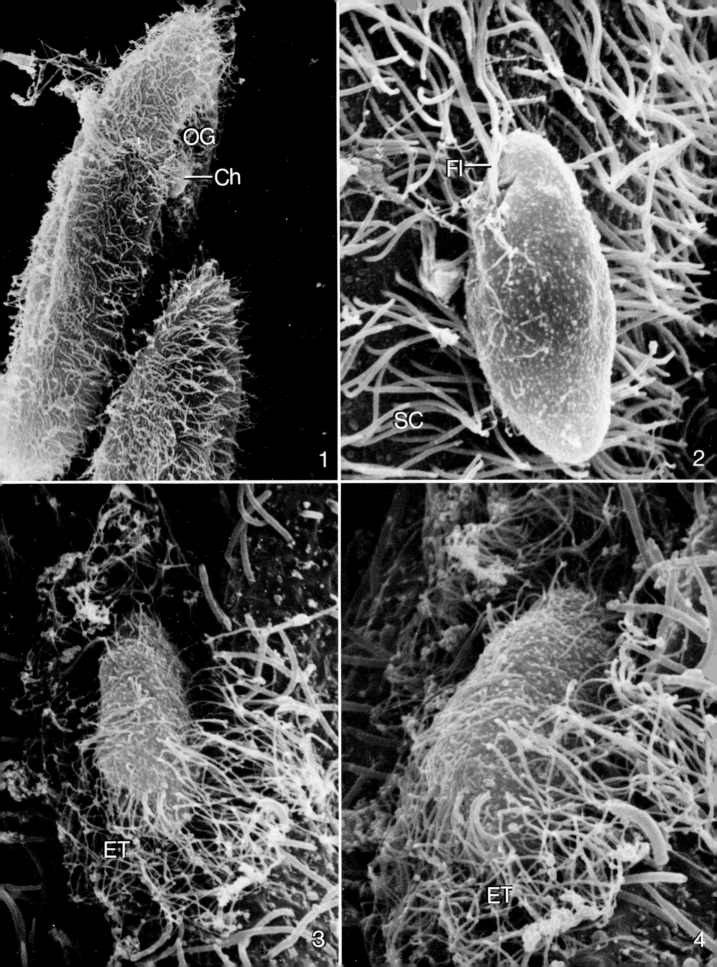

Reproduction in Paramecium (Binary Fission)

CF   Constriction furrow

Fig. 1 (fission in *P. multimicronucleatum*), ×895; Fig. 2 (conjugation in *P. multimicronucleatum*), ×765; Fig. 3 (fission furrow), ×2395; Fig. 4 (conjugation), ×660.

Most Protozoa and some multicellular animals can increase their number by binary fission. In this process in *Paramecium multimicronucleatum* a transverse constriction furrow (CF in Figs. 1 and 3) extends inward from the equator so that one organism is eventually divided into two daughter animals, each of which eventually achieves the volume of the parent. Immediately after the fission process the daughter cells are shorter and more rounded than the parent. The daughter *Paramecium* derived from the anterior half of the parent is called the proter and must differentiate a new posterior end. That daughter cell derived from the posterior half of the parent is called the opisthe. The opisthe must, therefore, regenerate a new anterior end. In *Paramecium* the micronuclei undergo mitosis shortly before fission, whereas the macronucleus is cleaved during fission by the advancing cleavage or fission furrow. If autogamy does not occur and after a number of binary fissions, the resulting organisms are genetically identical and called a clone. The gullet of the proter is derived from the parent, whereas a new gullet develops in the opisthe a short time before the fission furrow is apparent (*cf.* JURAND and SELMAN, 1969, for review of additional details). During the formation of a new gullet (stomatogenesis), new basal bodies appear, they become organized into an anarchic field of basal granules, and later they become arranged in rows and develop cilia. The fission furrow cuts between the two gullets.

Sex in Paramecium (Conjugation)

Under the appropriate environmental conditions, conjugation of complementary mating types can be induced in *Paramecium*. Prior to this process the surfaces of the organisms become "sticky" and the animals tend to attach to each other at their anterior ventral tips. Initially several organisms may stick together, but eventually only pairs are effectively attached. The initial attachment at the anterior ventral regions of two organisms is followed by a union in the region of the gullet (Figs. 2 and 4). In this region the somatic cilia disappear and the pellicle ridges of adjacent organisms become tightly apposed. Eventually, several cytoplasmic bridges or pores develop in the apposed pellicle ridges. While the attachment region lacks cilia and trichocysts, basal granules remain.

During attachment three divisions occur in the micronucleus, after which eight haploid nuclei (or pronuclei) remain in each of the conjugants. Typically seven of these nuclei degenerate, and the remaining pronucleus undergoes an equatorial division. One of these nuclei then migrates or is pushed into the mate through the cytoplasmic bridge. This nucleus is highly folded and becomes constricted as it passes through the cytoplasmic bridge. There is, thus, a reciprocal cross-fertilization during conjugation, at which time the macronucleus fragments. After genetic exchange, the conjugants become separated. The two pronuclei in each animal fuse to form a zygote nucleus. Then two mitotic divisions occur in the zygote nucleus of each animal so that four diploid nuclei are formed. Two of these nuclei enlarge and eventually form a macronucleus, whereas the remaining two nuclei become micronuclei. In autogamy, essentially the same process occurs as in conjugation except that individuals do not fuse and there is thus no genetic exchange between organisms.

Reference

JURAND, A., and SELMAN, G. G.: The Anatomy of *Paramecium aurelia*, London: Macmillan and Co., Ltd (1969).

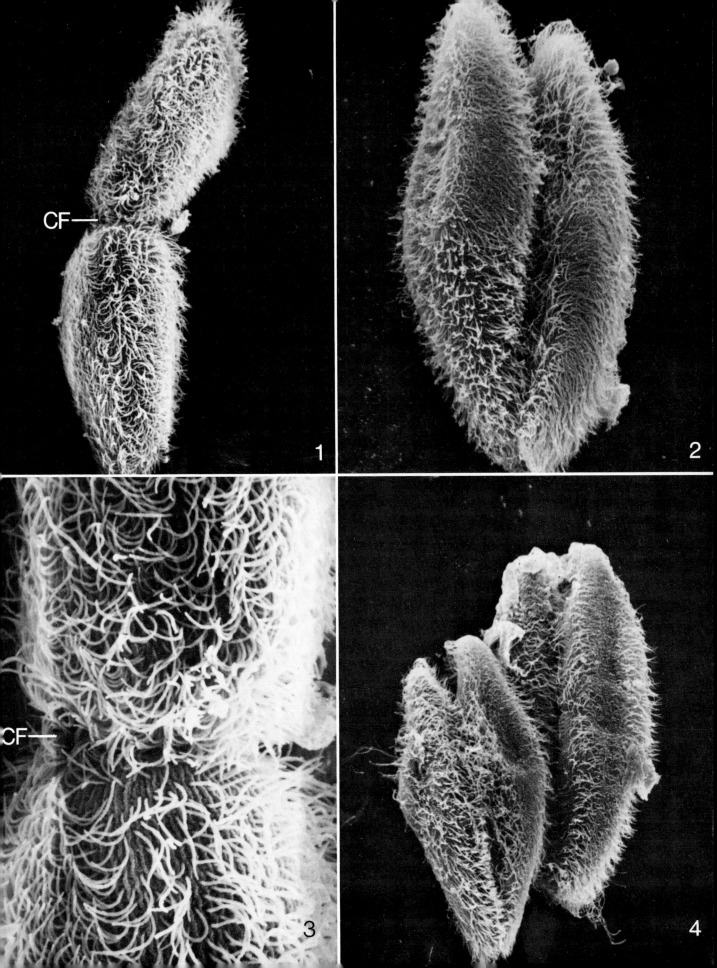

Phylum Protozoa—Subphylum Ciliophora

Class Ciliatea—Subclass Holotrichia—
Order Hymenostomatida

Paramecium bursaria (Ciliary Beat)

The overall shape of *Paramecium bursaria* is illustrated in Figs. 1 and 2. A small opening, called a buccal overture (BO), is present on the ventral side of the organism (Fig. 1). The ciliary rows on the dorsal surface are illustrated in Fig. 2. In Figs. 1 through 4 the anterior end is at the right side of each figure. The anterior end (A), the posterior end (P), the right side (R), and the left side (L) of the organism are indicated in Fig. 4, which is a side view.

It has been demonstrated that suitable fixation of Protozoa (*Opalina* and *Paramecium multimicronucleatum*) coupled with critical point drying can accurately preserve the pattern of ciliary coordination and the form of ciliary beat for scanning electron microscopy. Thus, it becomes possible to easily view and analyze the ciliary pattern in relation to locomotion over an entire organism (TAMM and HORRIDGE, 1970; TAMM, 1972). Such a condition is illustrated here for *Paramecium bursaria*. The organisms were washed in a medium containing 1 mM $CaCl_2$, 2 mM KCl, and 5 mM Tris buffer (pH 7.0) as described by NAITOH (1966). While swimming in this equilibration medium, the cells are fixed instantaneously by the addition of large quantities of 2.5% osmium tetroxide and 2.3% $HgCl_2$ (PARDUCZ, 1967). Following fixation for 30 to 60 minutes, the cells were dehydrated in ethyl alcohol and dried by the critical point method. In the forward-moving animal the somatic cilia, which are aligned in rows somewhat oblique to the anterior-posterior axis, beat metachronally (i.e., with regular phase differences). In contrast, the beat cycle of ciliary rows perpendicular to this direction is in synchronism. An individual cilium "successively assumes the position of its posterior rather than its anterior neighbor" (PARDUCZ, 1967), so that metachronal waves of ciliary movement travel from the left posterior to the right anterior surface of the cells, as can be seen with the SEM in those cells fixed during the forward motion (MW in Fig. 4). The effective stroke of all cilia occurs from a left anterior to a right posterior direction out of the plane of the micrograph (ES in Fig. 4). Therefore, by examining successive somatic cilia in the SEM micrographs from an anterior to a posterior direction over a single metachronal wave, it becomes possible to determine form changes (i.e., bending) of a somatic cilium during a complete beat cycle (TAMM, 1972). In Fig. 4 the cilia that are at the end of an effective power stroke are identified at (ES). These cilia lie close to the body pellicle and extend posteriorly to the right in a position nearly parallel to the wave fronts. Cilia are also shown close to the pellicle surface that wave counterclockwise and have an S-shaped configuration (extending obliquely forward to the left). These cilia represent those at the end of the recovery stroke (eRS) and are prepared to initiate the next effective stroke. The straighter cilia (ES) represent those fixed in the process of performing an effective stroke and are located between cilia ready to initiate an effective stroke and those that have just completed an effective stroke. The effective stroke proceeds from a left anterior direction to a right posterior direction in a plane nearly parallel with the metachronal wave front, which moves anteriorly. The recovery phase of the ciliary beat cycle occurs close to the cell surface so that it does not occur in the plane of the effective stroke. The complete ciliary beat cycle in *Paramecium bursaria* is thus an asymmetric movement in three dimensions similar to the condition described and illustrated by TAMM (1972) for *Paramecium multimicronucleatum*.

A diagram illustrating the form of the beat of one somatic cilium is illustrated in the line drawing. The positions of cilia 1 to 5 are correspondingly labeled 1 to 5 in Fig. 4.

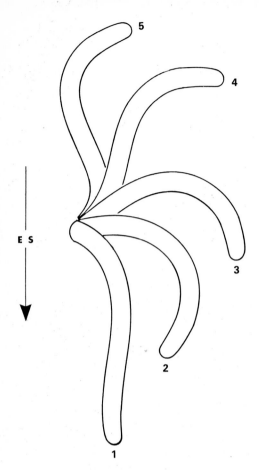

A   Anterior end
BO   Buccal overture
ES   Cilia in effective stroke
eES Cilia at end of an effective (power) stroke
eRS Cilia at end of a recovery stroke
L   Left side
MW Direction of metachronal waves
P   Posterior end
R   Right side
RS   Cilia in a recovery stroke

Fig. 1, ×900; Fig. 2, ×1130; Fig. 3, ×2750; Fig. 4, ×2250.

Diagram illustrating form of beat of one somatic cilium as viewed from above. Numbers refer to consecutive stages in the beat cycle: (1) at end of effective stroke, (2 through 4) counterclockwise rotation during the recovery stroke, (5) at end of the recovery stroke and preparatory to the effective stroke. The effective stroke (ES arrow) takes place from position five to position one, out of the plane of the paper. (Reproduced from TAMM, J. Cell Biol. **55**, 254, 1972, by permission of the Rockefeller University Press)

References

NAITOH, Y.: "Reversal response elicited in nonbeating cilia of *Paramecium* by membrane depolarization", Science (Wash. D.C.), **154**, 660—662 (1966).

PARDUCZ, B.: "Ciliary movement and coordination in ciliates", Intern. Rev. Cytol. **21**, 91—128 (1967).

TAMM, S.L., and HORRIDGE, G.A.: "The relation between the orientation of the central fibrils and the direction of beat in cilia of *Opalina*", Proc. Roy. Soc. Lond. B. Biol. Sci. **175**, 219—233 (1970).

TAMM, S.L.: "Ciliary motion in *Paramecium*", J. Cell Biol. **55**, 250—255 (1972).

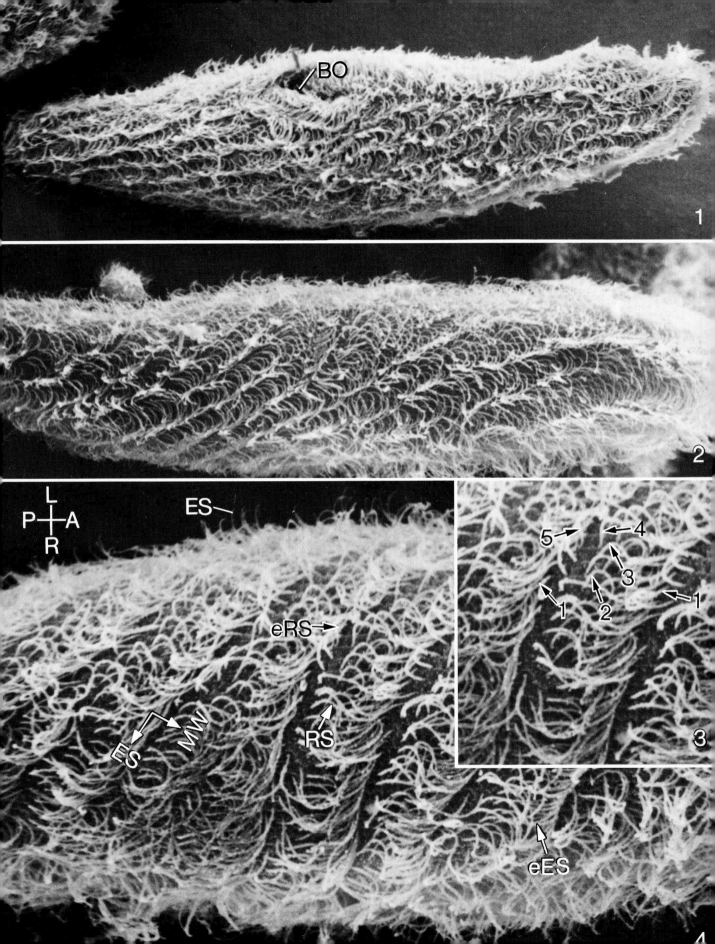

Phylum Protozoa—Subphylum Ciliophora

Class Ciliatea—Subclass Holotrichia—
Order Hymenostomatida

Tetrahymena pyriformis

This holotrich ciliate is a ciliary feeder in which the cilia associated with feeding are rather primitively developed. The surface structures of *Tetrahymena* as viewed with the SEM are illustrated in Fig. 2. The organism is ovoid and the anterior end is more pointed than the posterior end (PE). Single somatic cilia extend around the organism in rows (generally 16 to 20) arranged roughly parallel to the long axis of the cell (SC in Fig. 2). The buccal or oral apparatus is located on the ventral side of the organism somewhat posterior to the anterior end. The shallow depression associated with the mouth is called the buccal cavity (BC in Fig. 2) and has associated with it four morphologically distinct ciliary specializations. A row of closely apposed cilia comprising an undulating membrane (UM in Fig. 2) lies on the right side of the buccal cavity. On the left side of the buccal cavity are three diagonally disposed groups of cilia (each group composed of three rows each) generally called membranelles (Me in Fig. 2). A silver-stained preparation of a fixed *Tetrahymena* is illustrated in Fig. 1. Such a preparation clearly demarcates the rows of kinetosomes (arrow) associated with the somatic cilia. In addition, the kinetosomes of the specialized cilia associated with the buccal cavity and feeding are illustrated in the oral region. The position of the undulating membrane (UM) and three rows of membranelles (Me) is indicated in Fig. 1.

Silver-stained preparations of *Tetrahymena* during stages of binary fission are shown in Figs. 3 and 4. In Fig. 3 a new oral apparatus (OA) has already been formed in the daughter cells and a cleavage, or fission furrow (FF), is just beginning to form. A slightly later stage in the progression of the fission furrow (FF) is illustrated in Fig. 4, and the oral apparatus (OA) of the daughter cells is also indicated. A comparable stage in binary fission to that shown in Fig. 4 is included in the SEM micrograph of Fig. 5. The fission furrow (FF) between the daughter cells is quite deep and the components of the oral apparatus, including the undulating membrane (UM) and the cilia comprising the three membranellar bands (Me) are identified. (Figs. 2 and 5 by permission of Dr. John J. Ruffolo, Jr.)

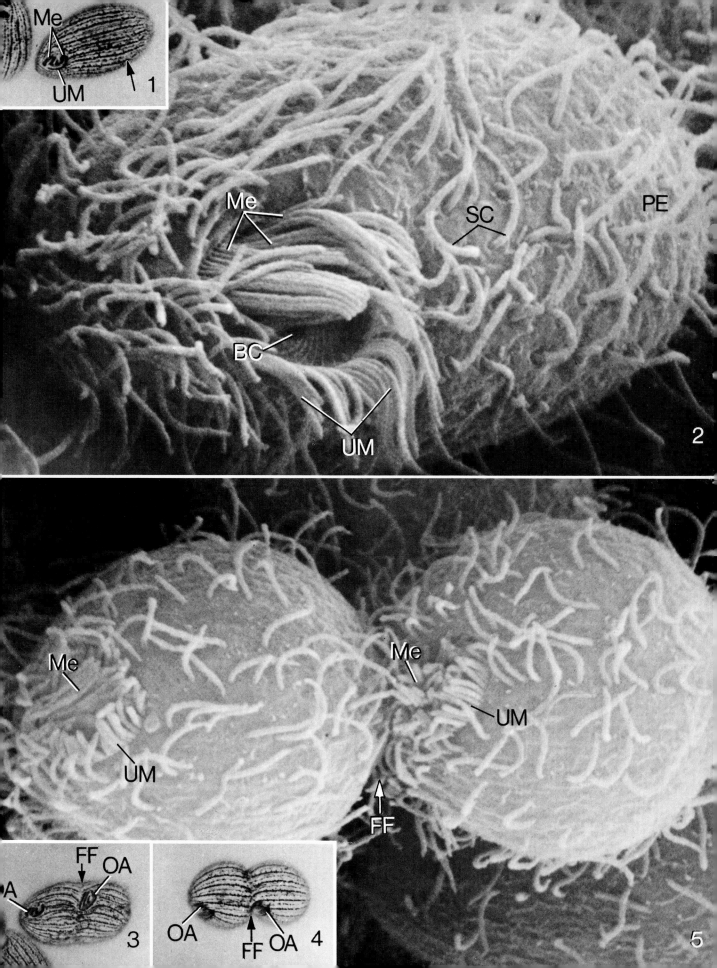

Phylum Protozoa—Subphylum Ciliophora

Class Ciliatea—Subclass Holotrichia—
Order Gymnostomatida

Didinium nasutum

CM	Closed mouth
Pe	Pectinelles
Pr	Proboscis
SM	Probable secretory material

Arrows: Surface projections

Fig. 1, × 1700; Fig. 2, × 3250; Fig. 3, × 10500.

This ciliate has a barrel-shaped body and is a raptorial feeder, eating other ciliates such as *Paramecium*. The anterior tip of the organism is modified into a tube-shaped proboscis (Pr in Fig. 1). The mouth (CM in Fig. 2), which is located at the tip of the proboscis, is closed in the specimens illustrated, but the mouth can be opened to a diameter nearly as great as the body of the organism when it is feeding on very large prey. The proboscis has been described as being made up of parallel trichocyst-like bodies that give the structure a hyaline appearance.

Two separate rings of cilia, called pectinelles (Pe), encircle the body (Fig. 1). Thus, the distribution of cilia on *Didinium* is quite different from that of *Paramecium*. The cilia comprising the pectinelles may be nearly 15 μm in length. One of the ciliary rings is located at the base of the proboscis and the other is located more posteriorly. Small but numerous longitudinal ridges and furrows are arranged along the anterior-posterior axis of the body. Rounded masses of apparent secretory material (SM) can be observed oriented in linear array in association with some of these longitudinal striations (Fig. 3). These bodies may represent the product of recently discharged mucocysts. Mucocysts occur immediately under the plasma membrane in the ectoplasm and are oriented with their longitudinal axes perpendicular to the surface (WESSENBERG and ANTIPA, 1968). These structures are observed to discharge with increased frequency during initial stages of encystment in *Didinium* (HOLT and CHAPMAN, 1971). Numerous but small projections from the body surface (arrows) are apparent in Fig. 3.

References

HOLT, P. A., and CHAPMAN, G. B.: "The fine structure of the cyst wall of the ciliated protozoan *Didinium nasutum.*" J. Protozool. **18**, 604—614 (1971).

WESSENBERG, H., and ANTIPA, G.: "Studies on *Didinium nasutum.* I. Structure and ultrastructure." Protistologica 4, 427—447 (1968).

WESSENBERG, H., and ANTIPA, G.: "Capture and injection of *Paramecium* by *Didinium nasutum.*" J. Protozool. **17**, 250—270 (1970).

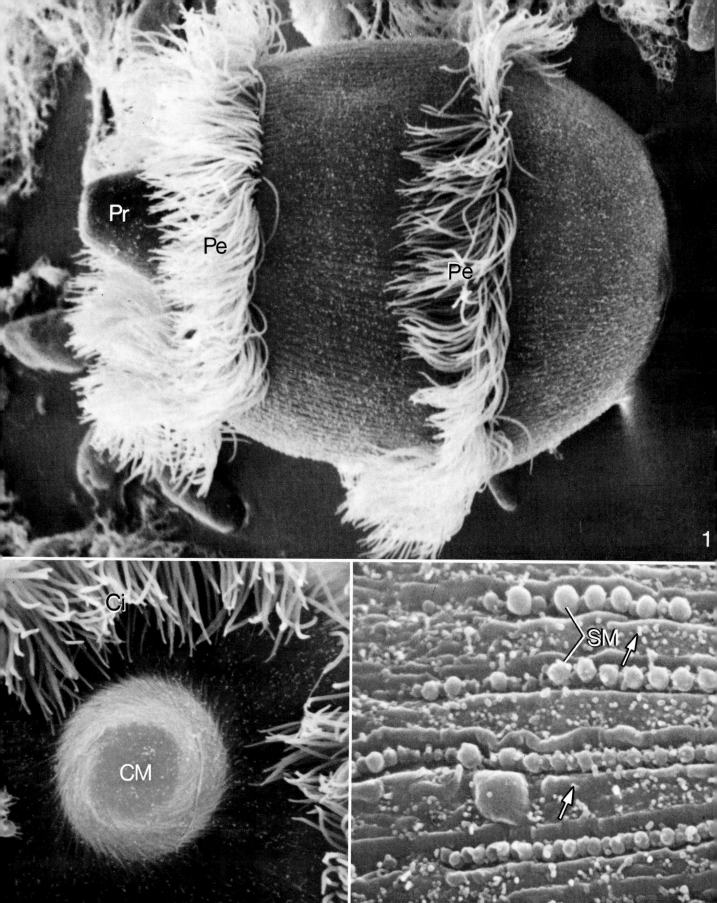

Phylum Protozoa—Subphylum Ciliophora

Class Ciliatea—Subclass Holotrichia—
Order Gymnostomatida

Dileptus anser

BCi	Body or somatic cilia
CP	Caudal process
Cy	Cytostome
FG	Feeding groove (on proboscis)
OC	Oral cilia
PCi	Cilia on proboscis
Pr	Proboscis

Arrows: Surface projections of toxicysts

Fig. 1, $\times 910$; Fig. 2, $\times 1020$; Fig. 3, $\times 1410$; Fig. 4, $\times 4400$; Fig. 5, $\times 2200$.

Dileptus anser is a multinucleate ciliate measuring approximately 200 to 400 μm in length and 30 to 40 μm in width (Figs. 1 and 2). This elongate holotrichous ciliate is characterized by a highly attenuated anterior portion called a proboscis (Pr), or snout. The rostral appendage is movable and serves as an exploratory structure used in feeding. The proboscis possesses about 10 to 15 rows of cilia (PCi in Fig. 3), which are, however, absent along the mid-ventral surface, which is called a feeding groove (FG in Figs. 3 and 4). The long feeding cilia located at the margins of the feeding groove probably assist in the transport of food to the mouth. The ventral surface of the proboscis has a number of short, rounded projections, which denote the position of toxicysts positioned in the cortical cytoplasm immediately beneath the pellicle (arrows in Fig. 4).

At the base of the proboscis on the ventral side of the organism is a circular cytostome (Cy), or mouth (Figs. 1 and 5). Although cilia are located around the mouth (OC in Figs. 3 and 5), they are not arranged into membranelles and undulating membranes as in many other ciliates. The body of *Dileptus* has 27 to 30 rows of cilia (BCi in Figs. 2 and 5) and the posterior end of the animal tapers into a caudal process (CP in Fig. 2). About 20 contractile vacuoles are located inside *Dileptus*. The depression apparent in the body of the animal illustrated in Fig. 2 may represent the discharge of one of these contractile vacuoles.

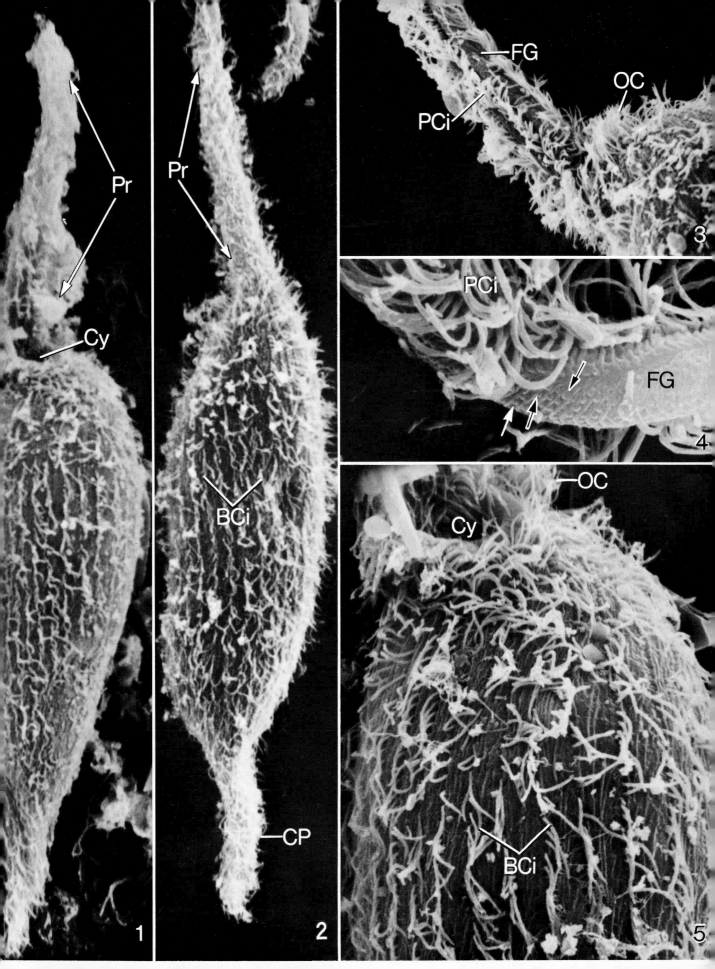

Phylum Protozoa—Subphylum Ciliophora

Class Ciliatea—Subclass Holotrichia—
Order Peritrichida

Vorticella

AS	Attachment of stalk to substratum	OC	Outer ciliary row
Ba	Bacteria	PB	Peristomial border
BC	Buccal cilia	PD	Peristomial disc (buccal cavity)
Bo	Body		
CS	Coiled spring form of stalk	PS	Pellicle striations of body
FP	Folded pellicle of stalk	SmP	Smooth pellicle of stalk
IC	Inner ciliary rows	St	Stalk

Arrows: Folds in pellicle of stalk

Fig. 1, ×720; Fig. 2, ×1375; Fig. 3, ×2080; Fig. 4, ×2025.

Vorticella is a peritrichous ciliate with a body (Bo) which is often bell-shaped and a long, and a slender stalk (St), by which the organism attaches to the substratum (Fig. 1). The animal is, however, capable of varying both its size and shape. The body cilia have completely disappeared in the adult form of this specialized ciliate, but a prominent buccal ciliature is present and restricted in position to a peristomial region. The flat, apical region of the body is organized into a wide peristomial disc (PD in Fig. 4), or buccal cavity, and a marginal peristomial border (PB in Figs. 3 and 4). The buccal ciliature (BC in Fig. 3) is arranged in a band that is probably homologous to the adoral zone of membranelles of other ciliates. The cilia are attached basally, but separated distally. The buccal cilia consist of three rows of cilia that are located in a peristomial groove between the peristomial border and peristomial disc. As viewed from the apical end, the band of buccal cilia spirals in a counterclockwise direction around the disc margin unlike the condition in *Stentor*. The inner row of cilia (IC in Fig. 4) is double, and the cilia are erect and in constant motion. The single outer ciliary row (OC in Fig. 4) tends to direct food toward a small buccal overture which extends internally from the peristomial disc. The buccal overture is not visible in these preparations.

The pellicle covering the body of this sessile ciliate has distinct circular striations (PS in Fig. 3) which extend from the attachment region between the stalk and body to the margin of the peristomial disc. *Vorticella* is capable of alternately contracting and extending the stalk. This is accomplished by contractile filaments (myoneme) which extend from the body into the stalk. The stalk associated with the *Vorticella* in Fig. 1 is only partially contracted, but note the regional differences in the degree of coiling. When the myoneme is completely contracted, the stalk has the appearance of a coiled spring (CS in Fig. 2) and small folds occur in the pellicle (arrows in Fig. 2). A more relaxed or extended stalk is also shown in Fig. 2. Note that in some regions of the stalk the covering pellicle is smooth (SmP) and that these areas alternate with those in which the pellicle covering the stalk is folded (FP). The stalk serves also to attach the organisms to a substratum. The region of stalk attachment (AS) to the stem surface of timothy hay is illustrated in Fig. 2. Bacteria have been described living commensally on the stalk of *Vorticella*. In these preparations bacteria (Ba in Fig. 2) were often observed adhering to the surface of the hay, but rarely were they observed attached to *Vorticella*.

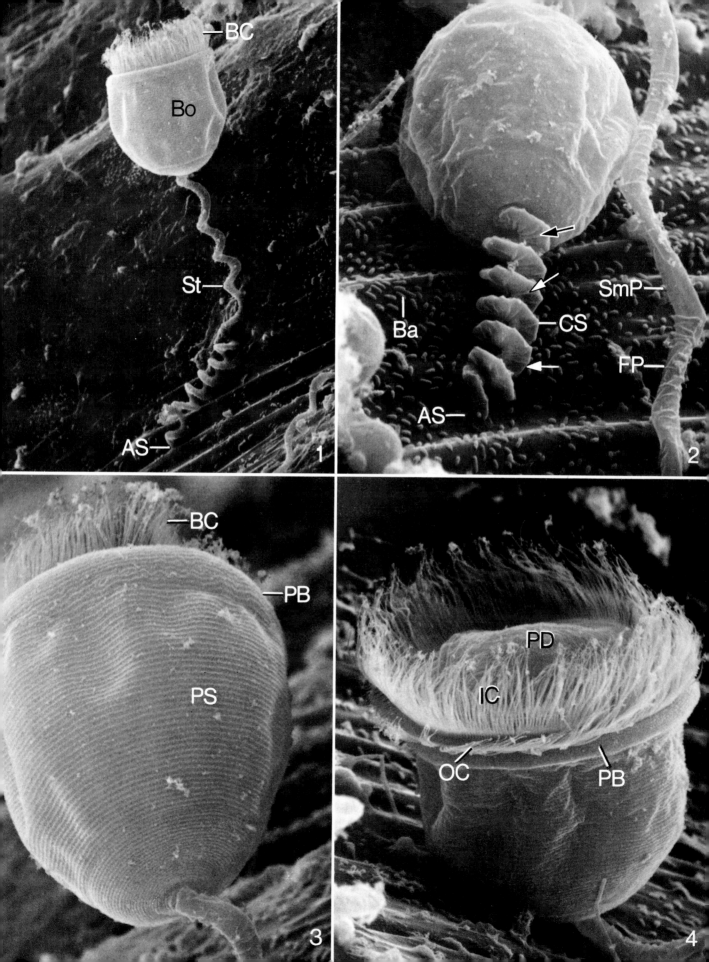

Phylum Protozoa—Subphylum Ciliophora

Class Ciliatea—Subclass Spirotricha—
Order Heterotrichida

Belpharisma undulans

This heterotrich ciliate is shaped much like *Paramecium* and is rose-colored due to rows of pigment granules immediately beneath the pellicle. The size and shape of *Blepharisma undulans* varies with its nutritional state. In general, it is spindle-shaped and the anterior end of the animal tends to be laterally flattened whereas the posterior half tends to be nearly circular in transverse section. This spirotrich ciliate has a complex ciliary pattern associated with a well-developed peristome region. A prominent peristome (Per), or oral groove, extends from the anterior tip of the organism along the ventral (oral) surface for from one-third to one-half the overall length of the animal. The peristome becomes widened posteriorly and is also somewhat curved (Fig. 1).

On the right side of the peristome in a posterior position is an undulating membrane (UM), which consists of a single row of cilia functioning together as a membrane (Figs. 1, 4, 5, 6). On the left side of the peristome is a band of membranelles (MB in Figs. 4 and 6). Each membranelle consists of several transverse rows of coordinated cilia. A number of these membranelles (about 25 to 95 depending upon the particular species) comprise an entire membranellar band (or adoral zone of membranelles).

The floor of the peristomial groove at its posterior end is expanded into a broad concave buccal cavity (BC in Figs. 5 and 6). The floor of the buccal cavity is covered with transverse rows of supporting cilia called a ciliary membrane (CM in Figs. 4, 5, 6). The cilia comprising this membrane arise from the right margin of the peristome near the origin of the cilia comprising the undulating membrane (arrow in Fig. 4). The somatic cilia are arranged in longitudinal rows varying from 10 to 16 in number (SC in Figs. 1 and 3). The rows of somatic cilia extend parallel to the right side of the peristome. The cilia arise from kinetosomes, which are positioned in rows in the ecto-plasm and extend from depressions or grooves (arrows in Fig. 3) in the pellicle (Pe).

40

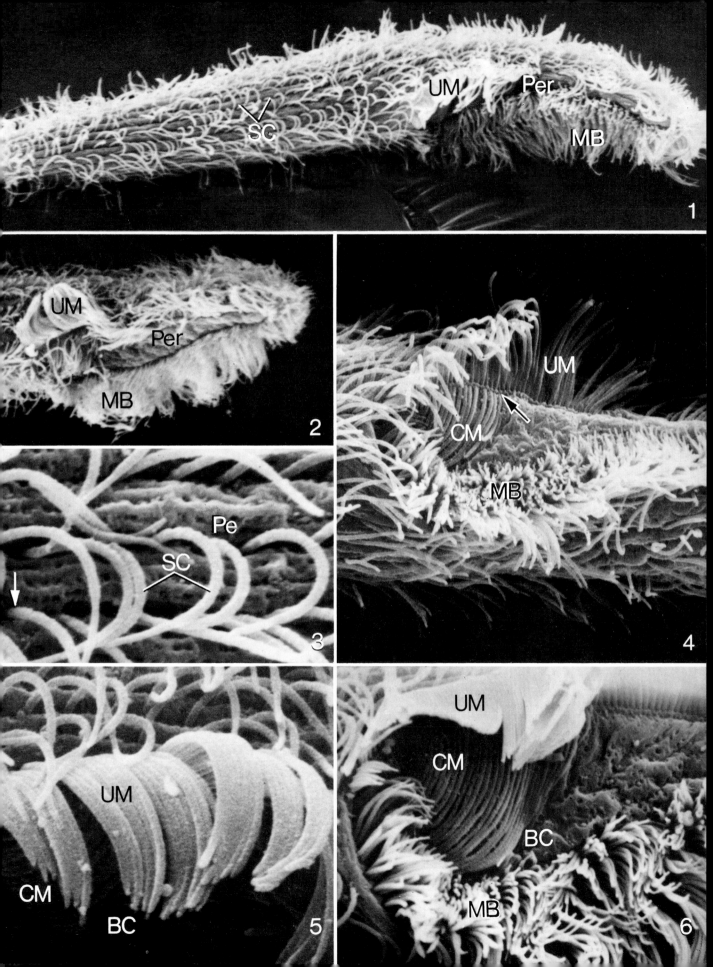

Phylum Protozoa—Subphylum Ciliophora

Class Ciliatea—Subclass Spirotrichia—
Order Heterotrichida

Stentor coeruleus

CVP   Contractile vacuole pore
FF   Frontal field of cilia
MB   Membranellar band of cilia
Pe   Pellicle
SC   Somatic cilia

Fig. 1, × 1200; Fig. 2, × 1445; Fig. 3, × 3330.

Stentor coeruleus is a spirotrich ciliate characterized by an adoral zone of membranelles winding clockwise to a cytostome. *Stentor* belongs to the order Heterotrichida, which is thought to be the simplest group of spirotrichs. This organism has been studied extensively to provide answers to many basic questions in cell biology. The organisms are quite large and generally cone-shaped. When completely extended, they are usually between 1 and 2 mm long. The shape of the animal varies depending upon whether the organism is attached (and its degree of contraction) or in a swimming form. Blue pigment granules are located in rows just under the pellicle (Pe) and they alternate with rows of basal bodies and their attached somatic cilia (SC in Fig. 3). The animal is unusual in that a special ciliated condition appears associated with the adoral zone (Fig. 2). In this region there is present an extensive membranellar band of cilia (MB) which borders a ciliated frontal field (FF), or peristomial disc (Figs. 1, 2). The shape of this membranellar band is variable depending on the degree of contraction of the adoral region (Figs. 1, 2). The buccal cavity consists of an oral pouch associated with the frontal field and membranellar band which extends internally as a short, tubular oral funnel. A contractile vacuole pore (CVP) is typically associated with the surface of the organism close to the adoral zone, and what appears to represent such a structure is illustrated in Fig. 1. The cilia associated with the proximal, or posterior, region of the body have been reported to be longer than those associated with the anterior, or distal, end of the body. Note that the rows of somatic cilia are wider between the contractile vacuole pore and membranellar band in Fig. 1 than elsewhere on the body surface (top, Fig. 1). The cilia are very short on the transverse frontal field. The cilia of the frontal field as well as the membranellar band bordering it beat so as to form a vortex that tends to bring food to the buccal (oral) pouch and oral funnel.

42

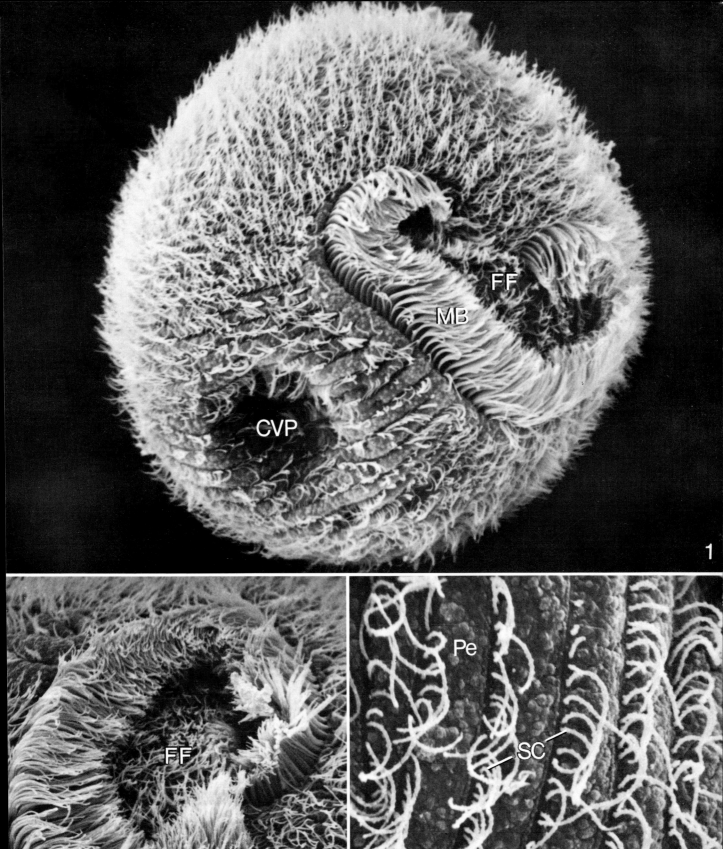

Phylum Protozoa—Subphylum Ciliophora

Class Ciliatea—Subclass Spirotrichia—
Order Hypotrichida

Euplotes eurystomus

BC   Buccal cavity
BrC   Bristle cilium
CC   Caudal cirri
FC   Frontal cirri
KC   Knob-like cilium
MB   Membranellar band
TC   Transverse cirri
VC   Ventral cirri

Arrows: Dorsal cilia in Fig. 1

Fig. 1, ×865; Fig. 2, ×1010; Fig. 3, ×6850.

Euplotes eurystomus belongs to a major group of ciliates, the Hypotrichida, which are considered to represent the peak of evolution in the ciliates and are probably the most highly evolved members of the entire phylum. The body is differentiated into a dorsal and a ventral side, each of which possesses complex buccal and somatic ciliature. On the ventral surface, clusters of cilia called cirri are arranged in asymmetric array. From an anterior to posterior direction there are six frontal cirri (FC), three ventral cirri (VC), five transverse cirri (TC), and four caudal cirri (CC) (Figs. 1 and 2). These complex ciliary modifications permit great variation and mobility of movement.

Also on the ventral surface there is a prominent row of multiple cilia consisting of about 40 membranelles, which are disposed in a curved configuration from the right to the left side of the animal (Figs. 1 and 2). The cilia comprising this membranellar band (MB) beat in a manner so as to direct food (e.g., *Chilomonas paramecium*) along a scoop-shaped buccal cavity (BC) toward a funnel-shaped cytopharynx. Although conjugation can occur in *Euplotes,* the animal typically divides by binary fission. Details in the process of binary fission can be studied in Protozoa with the scanning electron microscope. A late stage in this process in *Euplotes* is illustrated in Fig. 2.

The dorsal surface of *Euplotes* has rows of short cilia or bristles which extend from distinct pits in the cell surface (arrows in Figs. 1). The bristles are spaced along the row at about 3.5 μm (Fig. 1). Two kinetosomes are actually present in the cytoplasmic cortex, one of which is extended into the form of a bristle cilium observable in the scanning electron microscope; the second or posterior kinetosome is continuous externally with a very short, knob-like cilium (arrows in Fig. 1). A portion of an isolated pellicle from the dorsal surface is illustrated in Fig. 3. A bristle cilium (BrC) and a shorter, knob-like cilium (KC) are indicated. (Illustrations by permission of Dr. John J. Ruffolo, Jr.)

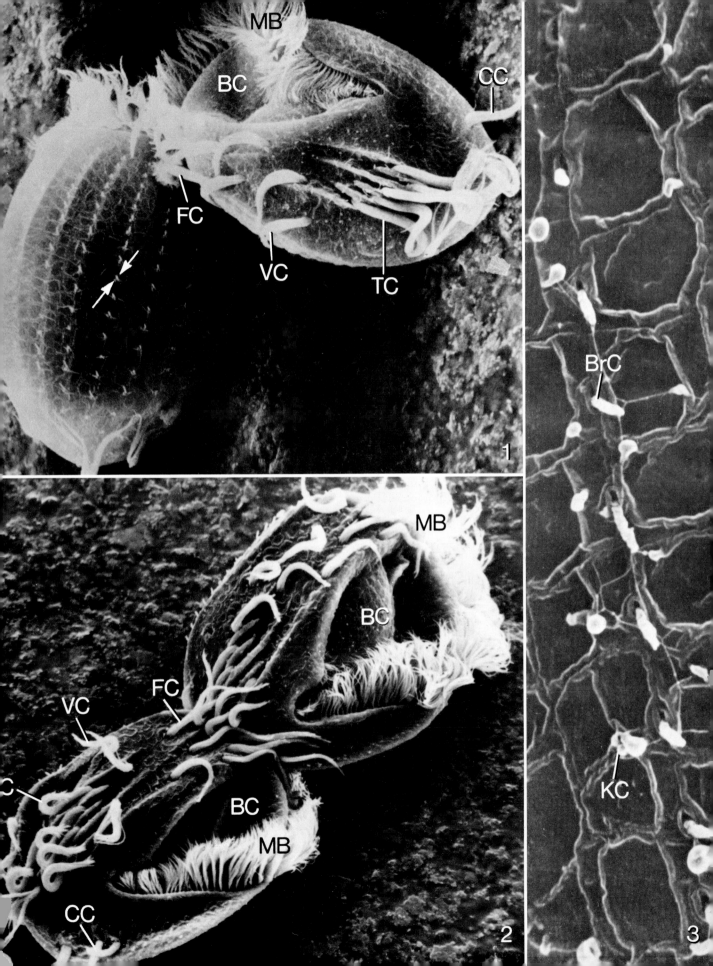

Flagellate Protozoa

Phylum Protozoa—
Subphylum Sarcomastigophora

Superclass Mastigophora—
Class Phytomastigophorea—Order Euglenida

Euglena

AF	Anterior flagellum	Pe	Ridged pellicle
Fl	Flagellum	PF	Posterior flagellum
Mi	Mitochondrion	Re	Reservoir
Mt	Microtubules	SP	Striated pellicle

Arrows: Membranous sacs

Fig. 1 (*Euglena*) ×4900; Fig. 2 (*Euglena*), ×22700; Fig. 3 (TEM, pellicle of *Euglena*), ×50000; Fig. 4 (*Peranema*), ×2475; Fig. 5 (*Peranema*), ×1980.

This organism is considered by many to represent the most primitive of the Protozoa classes. It possesses a single flagellum as a locomotor device (FL in Fig. 1), and internally there are chloroplasts containing pigments utilized in photosynthesis. These organisms are, therefore, phototrophic in their nutrition. *Chlamydomonas* and *Peranema,* in addition to *Euglena,* are also classed as phytoflagellates, and many botanists consider this group of organisms algae. The body of *Euglena* is covered with a pliable and striated pellicle (SP in Fig. 2) which permits some shape changes associated with locomotion. The pellicle at the anterior end of the organism (Fig. 2) is invaginated to form a reservoir (Re) through which the flagellum (Fl) emerges. The fine pellicular striations converge at this anterior point in Fig. 2. The flagellum extends posteriorly during locomotion, and wave-like movements extend from the base of the flagellum to its tip. As this occurs, the body rotates or spirals on its long axis.

A transmission electron micrograph illustrating a small portion of the body surface is illustrated in Fig. 3. The ridged pellicle (Pe) is clearly defined, and it is also apparent that at least two microtubules (Mt) are present in each ridge of the pellicle. A membranous sac (arrow, Fig. 3) is present at the base of each ridge in proximity to the microtubules. It appears that the microtubules run parallel to the pellicle within the ridges along the length of the animal. They may, therefore, play a role in the maintenance of the pellicular striations in this form.

Peranema

This organism also has a faintly striated pellicle, but possesses two flagella (Figs. 4 and 5). One of these locomotor structures extends anteriorly (AF) from the cytostome, whereas the other flagellum (PF) is curved posteriorly and closely adheres to the pellicle. Commonly, only the anterior tip of the anterior flagellum bends during locomotion. *Peranema* often exhibits a gliding or creeping motion when applied to a substratum.

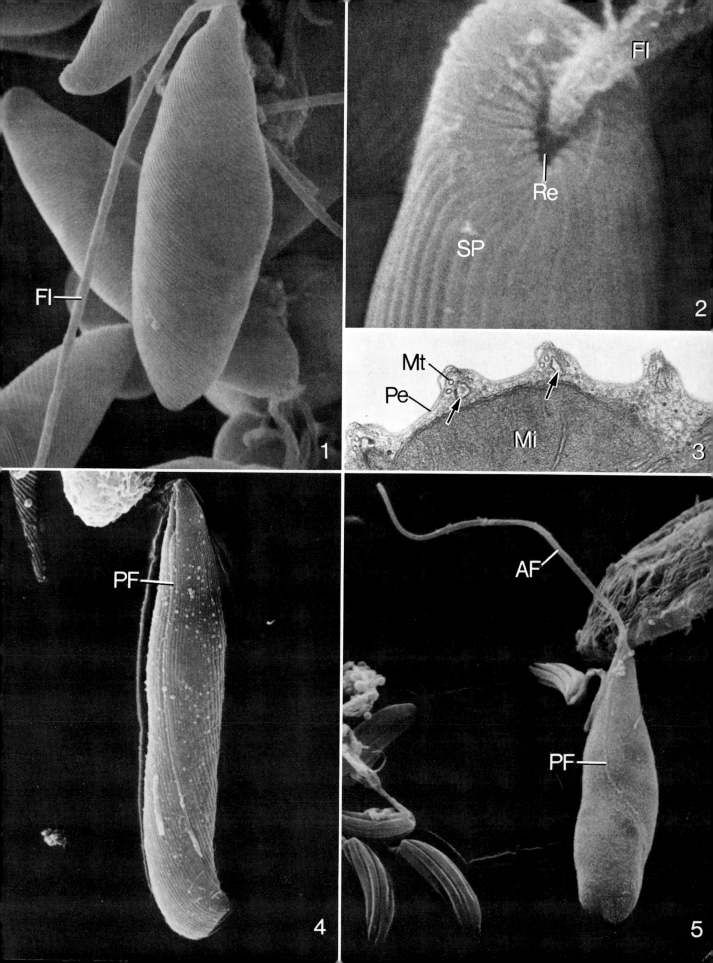

Phylum Protozoa—
Subphylum Sarcomastigophora

Superclass Mastigophora—
Class Zoomastigophorea—Order Hypermastigida

Trichonympha

AC   Apical cap
BF   Body flagella
FP   Disc-shaped structures attached to ends of body flagella
RF   Rostral flagella
Ro   Rostrum

Fig. 1, × 2995; Fig. 2, × 1310; Fig. 3, × 6500.

This complex flagellate commonly inhabits the digestive tract of the termite. These flagellates are relatively large and may range from 50 to 360 µm in length. The anterior-most portion of the body is tapered and is called a rostrum (Ro in Fig. 1). The rostrum is flexible and can bend extensively. It consists of an apical cap (AC) from which many long flagella extend (Figs. 1 and 2). The apical cap is illustrated in side view in Fig. 1 and from the top in Fig. 2. An intermediate region of the body contains most of the flagella. The rounded posterior end is nonflagellated, but this condition is not illustrated in Fig. 1, since many of the long flagella cover the posterior region of the organism. As many as 10000 flagella have been described on a single organism, and they are generally organized into longitudinal rows. The flagella differ in length, but some are more than 25 µm in length. Note the disc-shaped structures, 2 to 3 µm in diameter, that are associated with the distal ends of many body flagella in Figs. 1 and 3. During locomotion, this organism tends to rotate, and the rostral flagella (RF) appear to be more active than the long body flagella (BF).

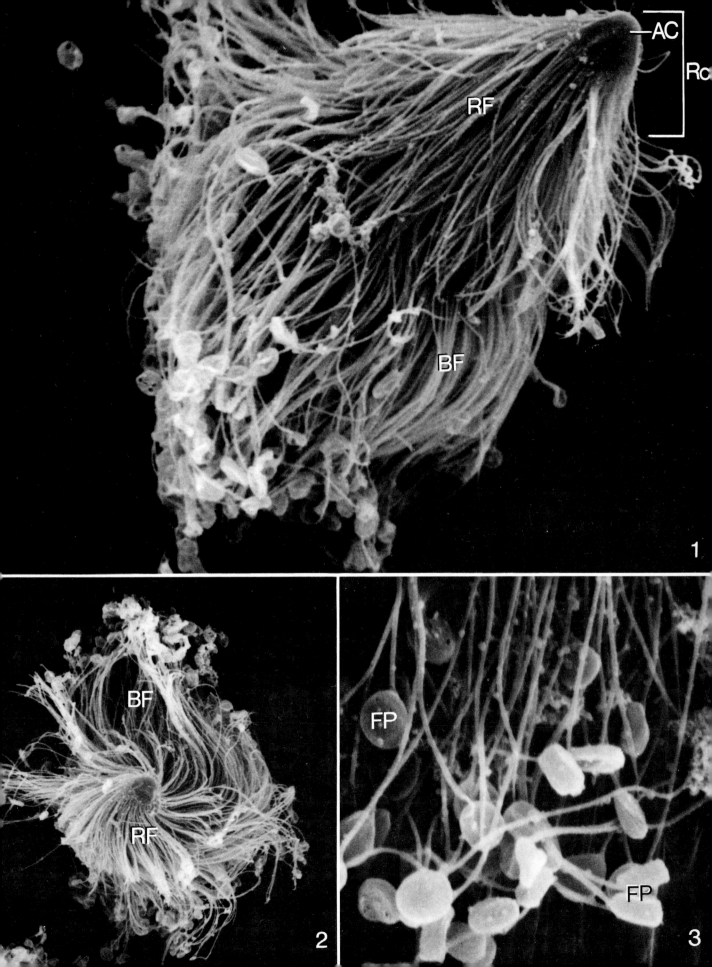

Phylum Protozoa—
Subphylum Sarcomastigophora

Superclass Mastigophora—
Class Zoomastigophorea—Order Diplomonadida

Giardia muris

AD	Adhesive disc	PG	Posterior groove in adhesive disc
AL	Anterior lateral flagella		
CF	Caudal flagella	PL	Posterior lateral flagella
Fl	Flagella	TW	Terminal web (of intestinal epithelial cell)
MF	Marginal fold		
Mi	Microvilli	VF	Ventral flagella
Nu	Nuclei		

Fig. 1 (ventral surface of *Giardia*), $\times 9000$; Fig. 2 (dorsal surface of *Giardia* on intestinal epithelial cell), $\times 11130$; Fig. 3 (TEM cross-section of *Giardia* and apical end of intestinal epithelial cell), $\times 20450$.

The body of *Giardia* is rounded anteriorly, but tapered at its posterior end (Fig. 2). This species parasitizes the intestine of the mouse, and the organism shown in Fig. 2 is attached by its ventral surface to the microvilli (Mi) of an intestinal epithelial cell. The dorsal surface is convex and the ventral surface is flat or concave (Figs. 1, 2, 3). The organisms have two nuclei (Nu in Fig. 3) and four pairs of flagella (Fl), which emerge from the body at different levels along its length. The paired flagella include the anterior lateral flagella (AL), the posterior lateral flagella (PL), the ventral flagella (VF), and the caudal flagella (CF) (Fig. 2). The ventral surface is differentiated into an adhesive disc (AD in Fig. 1), by which the animals are able to secure themselves to the intestinal mucosa. A narrow, deep groove (PG in Fig. 1) is associated with the posterior side of the disc and two ventral flagella (VF in Fig. 1) which have emerged from the body extend along this groove. Extending beyond the margin of the adhesive disc is a marginal fold (MF in Fig. 1), or phalange.

A transverse section through *Giardia* and an underlying intestinal epithelial cell is illustrated in the transmission electron micrograph in Fig. 3. The specialized fibrillar and tubular arrangement of the adhesive disc (AD) and the marginal fold (MF) is illustrated. Two ventral flagella (VF) are contained within the posterior groove (PG) of the adhesive disc, and transverse sections of other flagella (Fl) and nuclei (Nu) are demonstrated. The microvilli (Mi) and terminal web (TW) associated with the apical end of an intestinal epithelial cell are also depicted. The forms of *Giardia* illustrated are termed trophozoites. Encystment of the organism can occur and is the method of transmission of the disease called giardiasis. (Micrographs by permission of Dr. S.L. ERLANDSEN, A. THOMAS, and G. WENDELSCHAFER.)

Reference

ERLANDSEN, S.L., THOMAS, A., and WENDELSCHAFER, G.: "A simple technique for correlating SEM with TEM on biological tissue originally embedded in epoxy resin for TEM", In Scanning Electron Microscopy/1973, O. JOHARI and I. CORWIN, eds. IITRI **6**, 349—356 (1973).

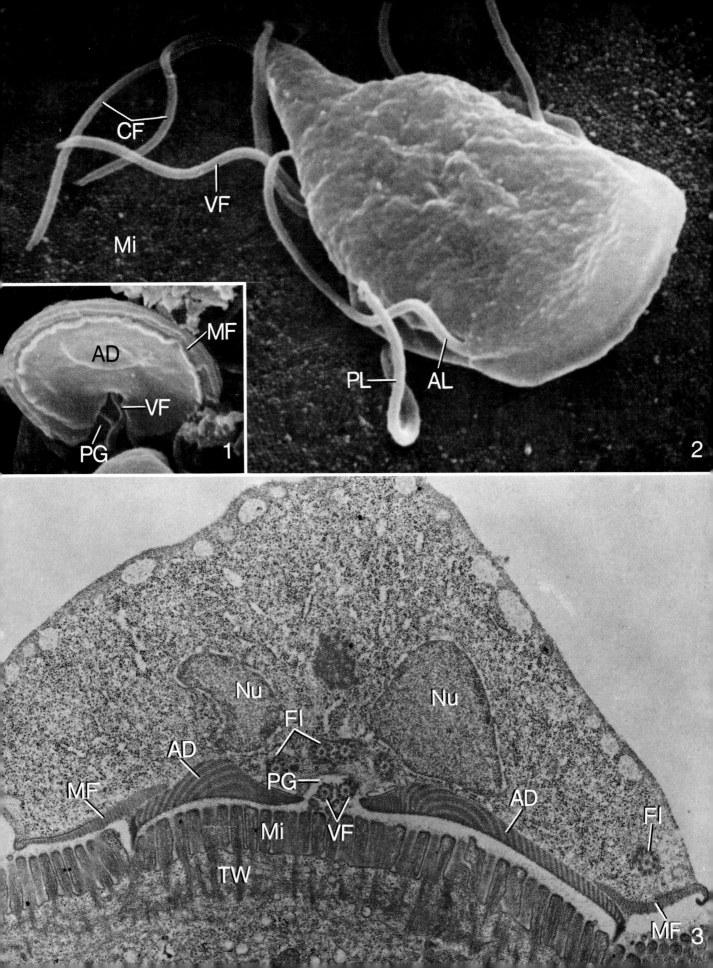

Phylum Protozoa—
Superphylum Sarcomastigophora

Superclass Mastigophora—
Class Zoomastigophorea—Order Diplomonadida

Hexamita muris

AF   Anterior flagella
Ax   Axoneme
FS   Flagellar sheath
Gl   Glycogen
Gr   Groove in body surface containing posterior flagellum
Nu   Nucleus
PF   Posterior flagellum

Fig. 1, ×20700; Fig. 2, ×15000; Fig. 3, ×21500; Fig. 4, ×20500.

This flagellate lives in the intestine of the mouse and is closely related structurally to *Giardia*. The organism measures about 12 to 14 μm in length and 2 to 3 μm in width. The organisms have eight flagella and have been described as being bilaterally symmetrical. However, their symmetry appears to be more rotational, as if the organisms had fused along their length. Three pairs of flagella arise from the anterior end of the organism and emerge laterally (AF in Fig. 2). A fourth pair of flagella, the posterior flagella (PF in Fig. 2), extend posteriorly within the body and are enclosed in a flagellar sheath in the caudal half of the organism (trophozoite) before emerging from the body. The transmission electron micrograph (Fig. 1) shows a transverse section through the anterior end of the trophozoite and includes the paired nuclei (Nu), the posterior flagella (PF), and the flagellar sheath (FS). The anterior flagella (AF) have emerged from the body anterior to this section so are located externally. The transmission electron micrograph in Fig. 3 is in a more posterior direction, so that the paired posterior flagella (PF) have emerged from the body and reside in grooves in the body surface (Gr). A parasagittal section of *Hexamita* is illustrated in Fig. 4. Sections of the anterior flagella (AF) and the posterior flagella (PF) are included in this micrograph. A portion of the axoneme (Ax) and large amounts of glycogen (Gl) are also illustrated. It has been shown that this intestinal flagellate may be phagocytosed and intracellularly digested by Paneth cells present in the crypts of the mouse intestine (ERLANDSEN and CHASE, 1972). (Micrographs kindly provided by Drs. STANLEY ERLANDSEN and DAVID CHASE.)

Reference

ERLANDSEN, S.L., and CHASE, D.G.: "Paneth cell function: Phagocytosis and intracellular digestion of intestinal microorganisms. I. *Hexamita muris.*" J. Ultrastruct. Res. **41**, 296—318 (1972).

52

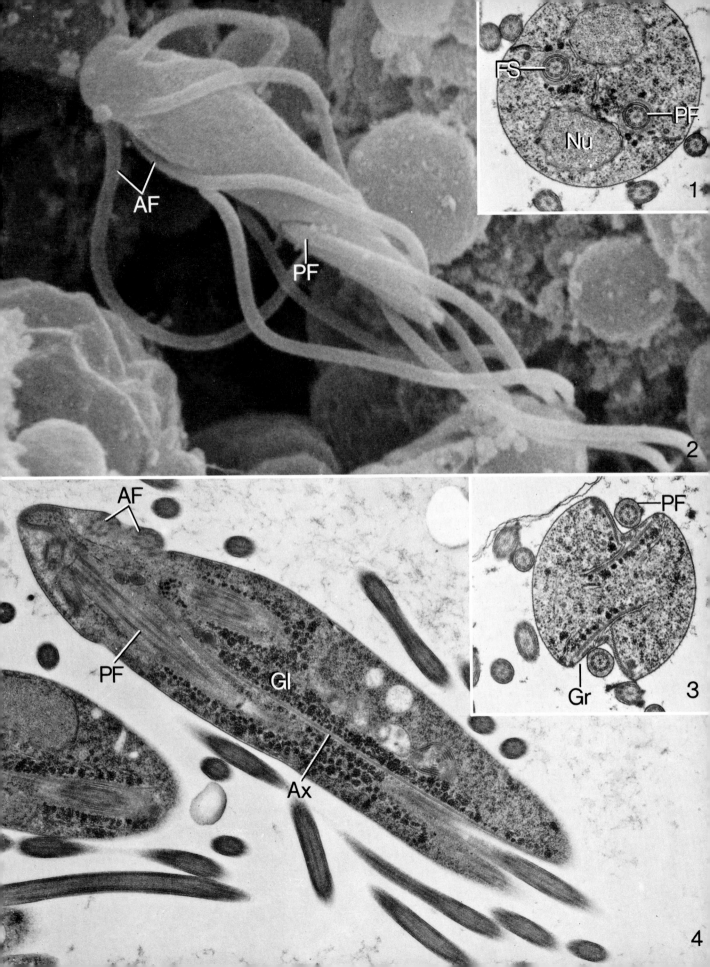

Amoeboid Protozoa

Phylum Protozoa—
Subphylum Sarcomastigophora

Superclass Sarcodina—Class Rhizopodea—
Order Amoebida

Pelomyxa carolinensis

Pelomyxa carolinensis is a large amoeba that constantly changes shape while moving. As a result, variable numbers of pseudopodial projections, called lobopodia (Lo in Figs. 1 and 3), are a characteristic feature of these organisms. The cytoplasm of amoebae consists of an outer, prominent ectoplasm layer that appears hyaline and nongranular. The inner granular cytoplasm (entoplasm) contains the organelles. That por-

tion of the cytoplasm which can be observed to "stream" during locomotion is called the plasmasol while a semisolid gel (plasmagel) surrounds the plasmasol completely with the possible exception of the tips of the pseudopodia. The organism illustrated in Fig. 3 has a folded appearance probably due largely to its response to the fixative. Along portions of the surface, however, there are smoother, rounded eleva-

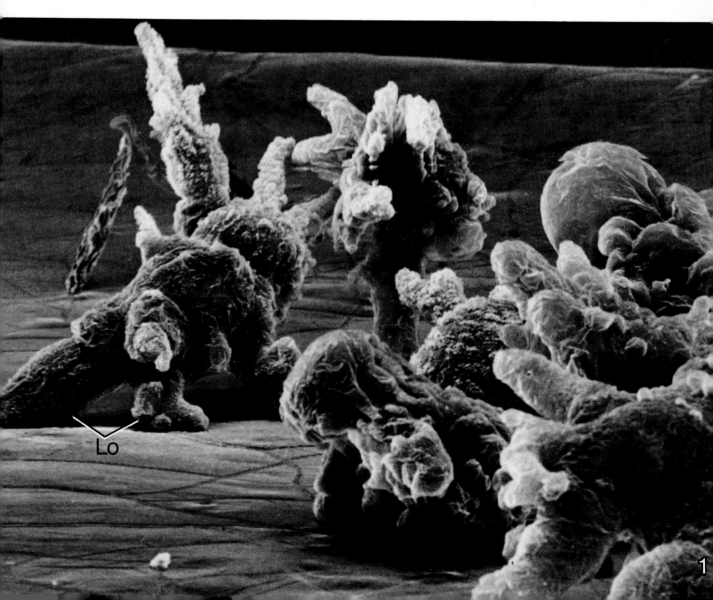

Lo

1

Phylum Protozoa— Subphylum Sarcomastigophora

Superclass Sarcodina—Class Rhizopodea— Order Amoebida

Pelomyxa carolinensis

(continued)

AP   Advancing pseudopod (*Dictyostelium*)
FP   Fingerlike projections
FS   Folded surface (lamellipodia) of migratory *Dictyostelium* amoeba
Lo   Lobopodia (of *Pelomyxa*)
Ps   Pseudopodia

Arrows: Small surface projections of *Pelomyxa*

Fig. 1 (entire *Pelomyxa*), ×210; Fig. 2 (surface of *Pelomyxa*), ×23625; Fig. 3 (entire *Pelomyxa*), ×335; Fig. 4 (*Dictyostelium* amoeba), ×10725; Fig. 5 (*Dictyostelium* amoeba), ×6010.

tions which may represent aggregations of the gelled ectoplasm. At higher magnification the plasma membrane of the amoeba appears covered with minute projections, or hairs (arrows in Fig. 2). Because the surface area of amoebae is constantly changing during locomotion, these organisms appear to be especially adept in the formation and reabsorption of the plasma membrane.

Amoeba Form of *Dictyostelium discoideum*

The amoeboid stage in the life cycle of this myxomycete is illustrated in Fig. 4 and 5. These amoebae are much smaller than *Pelomyxa*. The surface of the rounded cell in Fig. 4 is folded and two types of surface exten-sions are apparent. These include finger-like projections (FP in Fig. 4), which are narrow and variable in length (approximately 2 µm long by 0.2 µm wide), as well as numerous short, wide, and rounded projections which probably represent pseudopodia (Ps in Fig. 4). The amoeba shown in Fig. 5 was attached and in the process of movement when fixed. The advancing pseudopodium (AP) is prominent and smooth, but the remainder of the surface is prominently folded (FS).

56

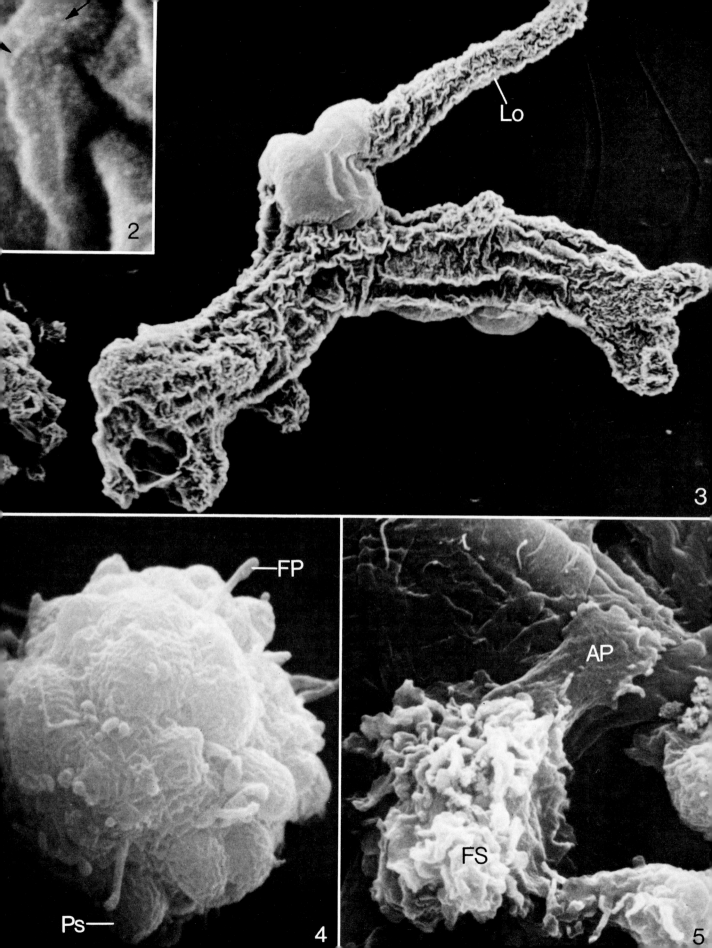

Phylum Protozoa—
Subphylum Sarcomastigophora

Fig. 1, ×200; Fig. 2, ×400.

Superclass Sarcodina—
Class Actinopodia—Subclass Radiolaria

Radiolarian Shells

The Sarcodina include those Protozoans that are amoeboid, and they are thought to be more closely related to flagellate Protozoa than to other classes. The radiolarians represent a group of marine Sarcodina. These amoebae have an axopodial type of pseudopod and are characterized by the fact that they secrete an enclosing shell. Axopodia are slender pseudopods composed of ectoplasm. Numerous parallel fibrils extend the length of the axopodia. The prevalent component of the porous exoskeletal shell is silicon. Because of this, the shells are readily cleaned with acids. The amoebae have a wide range of size; some may be 5 mm in diameter. Even though the amoebae are enclosed in a skeleton, the animals have a floating existence in the ocean due to structural modifications in their peripheral cytoplasm. When the amoebae die, the skeletons become a constituent of the ocean floor, and the shells are especially abundant in the bottom mud in deeper parts of the ocean. Eventually, the shells become incorporated into sedimentary rocks. The shells of radiolaria exist in a vast array of geometries, with numerous variations in the configuration of associated hooks, spines, and thorns (Figs. 1 and 2). The shells are generally well preserved in the fossil record and have been observed in pre-Cambrian rocks. Radiolarian shells constitute a significant portion of siliceous rocks (e.g., chert). Only a few examples of the possible variation in form of radiolarian shells are illustrated.

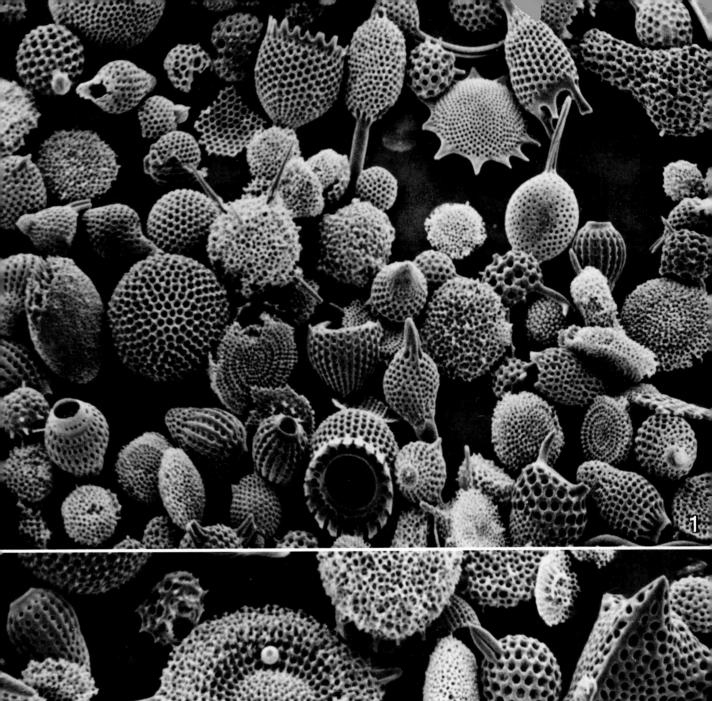

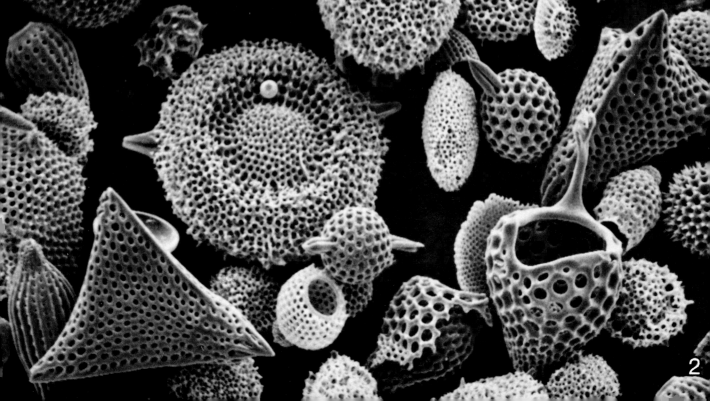

Phylum Protozoa—
Subphylum Sarcomastigophora

Superclass Sarcodina—
Class Rhizopodea—Order Foraminiferida

Foraminiferan Shells

Py   pylome

Arrows: pores in shell

Fig. 1, ×220; Fig. 2, ×96; Fig. 3, ×315; Fig. 4, ×180.

The foraminifera are large, shelled amoebae (rhizo-pods) and almost all are marine. The shells of foraminifera are calcareous and exist in a variety of forms (Figs. 1 through 4). Thus, the great proportion of the earth's limestone and chalk was formed from the shells of these amoebae. Because the shells are widely distributed in sedimentary rocks, the identification of the foraminifera is useful in recognition of rock strata. In addition, foraminiferan shells are used in locating oil deposits, a feature that has prompted considerable amounts of work on fossil foraminifera. The majority of the known families and genera of this group are extinct and preserved in the fossil record. The shells comprise a foraminiferan ooze on the ocean bottom and they are present in the mud of the ocean floor at depths of about 10000 to 15000 feet.

The shells of foraminifera are often composed of a number of chambers of increasing size (Figs. 1 and 2). The initial and smallest chamber is the proloculum. The shell chamber communicates with the exterior by a single aperture termed the pylome (Py). Many of the shells contain small pores through which the amoeboid process are extended in the living condition (arrows in Figs. 1 and 3). A layer of protoplasm covers the surface of the shell and a meshwork of rhizopodia extend from this thin layer of protoplasm. Present day foraminifera have a complicated life history with alternating sexual and asexual generations.

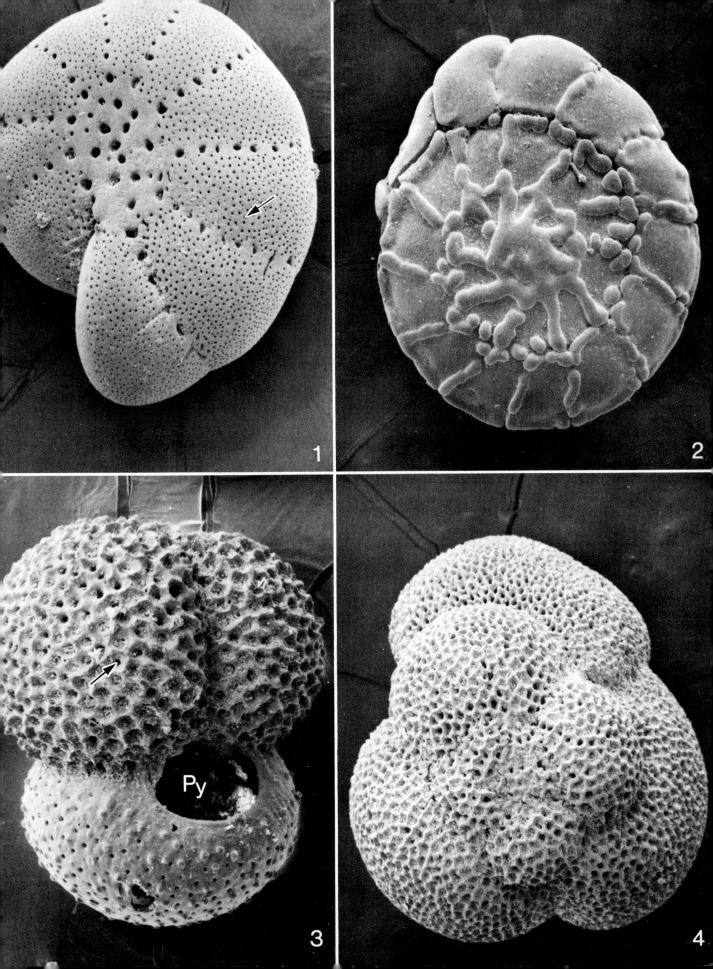

Chapter 3
Cells in Culture

Chapter 3   Cells in Culture

Introduction

Many cell types can now be isolated and maintained in large quantities for long periods as a result of recent improvements in tissue culture techniques. Thus, it is possible to maintain normal and abnormal cells in vitro and under controlled conditions in order to study many questions of major importance in cell biology including cell differentiation, cell-to-cell interactions, cell transformations, and cell chemistry. In the past it has been necessary when desiring details of these cells to examine them with phase contrast and Nomarski optics or to examine large numbers of extremely thin sections with the transmission electron microscope. These techniques have not and cannot provide the degree of information that can be provided by techniques of scanning electron microscopy. In order to study such problems as the surface structure of whole cells and their form under different environmental conditions as well as the processes associated with malignant transformation of cells, the depth of field and the resolution afforded by the SEM is invaluable. To obtain comparable information with the transmission electron microscope, a great amount of time and effort would be required to reconstruct the three-dimensional structure of a single cell under one experimental condition. This means that many serial sections must be studied, because ultra-thin sections studied in the transmission electron microscope provide only a very small representation of the whole cell surface. Further, surface replication techniques are time-consuming and usually do not convey the depth of information comparable to that obtained in the scanning electron micrograph. The ability to view details of whole cell surfaces has been made possible not only by the introduction of the SEM to biology and medicine, but also by improvements in the preservation and preparation of cells for viewing in the SEM.

The micrographs illustrated in subsequent sections will serve to demonstrate the quality and quantity of information obtainable with the SEM when suitable and current preparative procedures are used.

The use of the SEM in the study of transformed or neoplastic cells in culture has grown rapidly in recent times. In part, these studies reflect the increased interest in the nature of the plasma membrane of cells and its specializations. Once suitable methods were developed for preparing cultured cells for SEM observation, the studies became concerned with characterizing the differences in the surface specializations of normal and neoplastic cells, with following the changes in these specializations during the cell cycle in synchronized cultures of transformed cells, and with investigating the effect of sparse cultures on surface specializations and neoplastic cell form. With this information as a background, recent studies have been initiated to study the effects of various drugs and chemicals on surface specializations of cultured cells and with correlating this information with intracellular changes as determined with the TEM.

Cells in culture tend to round up during division and to remain in this configuration throughout the division process. As they progress through the cell cycle (G_1, S, G_2), they progressively flatten over the substratum. Examples of such changes in cell shape are depicted in Fig. 1 which illustrates a population of cultured HeLa cells while in Fig. 2, rounded and flattened KB cells are illustrated. Both types of transformed cells will be used to illustrate the various types of surface specializations associated with such cells. The striking feature of the cells illustrated in Figs. 1 and 2 and the feature that scanning electron microscopy is able to reveal is their normally complex surface. In fact, the SEM studies thus far performed on malignant cells in culture seem to show in most instances an exaggerated surface activity that is reflected in several types of specializations. Before illustrating and commenting on these specializations in detail, the methods by which the cells are prepared for scanning electron microscopy will be briefly described. The KB cells, derived from a human epidermoid carcinoma of the mouth, were grown in round plastic petri dishes in Ham's F12 medium (Grand Island Biological Co., Grand Island, New York) containing 10% fetal calf serum, 0.5% embryo extract, and 2 times the normal amino acid concentration. The HeLa S3 cells (derived from a human cervical carcinoma) were grown in Puck's N16 medium (Grand Island Biological's) containing 10% horse serum and 2% human blood serum. Chick embryo fibroblasts and chondrocytes were grown in F12 medium containing 10% fetal calf serum and antibiotics. Each cell type was grown in plastic petri dishes containing the appropriate tissue culture medium indicated above. Fixation, washing, and dehydration were carried out in this culture dish. During early dehydration stages, small squares were cut from the petri dish with attached cells so that they would conveniently fit in the critical point drying apparatus and could be easily attached to the SEM specimen stubs. The cells were processed in the following manner:

(1) Cells were rinsed 2 times with mammalian Ringer's solution at 37°C to remove tissue culture medium.

(2) Fixation was carried out at room temperature for 0.5 to 2.0 hours in 3% glutaraldehyde using either 0.1 M sodium phosphate buffer or 0.05 M sodium cacodylate buffer (pH 7.2).

(3) The fixed cells were rinsed 2 times with the appropriate buffer for 0.5 hours.

(4) Cells were post-fixed for 30 to 60 minutes with 1% osmium tetroxide buffered to pH 7.2 with either 0.1 M sodium phosphate buffer or 0.05 M sodium cacodylate buffer.

(5) The cells were rinsed with distilled water, 2 changes, 5 to 10 minutes.

(6) The cells were then dehydrated in increasing concentrations of ethanol.

(7) This was followed by 3 changes of absolute alcohol for 30 to 60 minutes.

(8) Dehydrated cells were transferred to liquid carbon dioxide in a critical point drying apparatus and dried.

(9) The petri dish squares were attached to specimen stubs, coated with carbon and gold-palladium, and examined in the SEM.

The specializations illustrated in Figs. 1 and 2 include filopodia (Fi), microvilli (Mi), lamellipodia (La) and zeiotic blebs (ZB). These structures will be described and illustrated in greater detail in subsequent sections.

Fi   Filopodia
La   Lamellipodia
Mi   Microvilli
ZB   Zeiotic blebs

Fig. 1 (HeLa cells), ×2180; Fig. 2, (KB cells), ×4515.

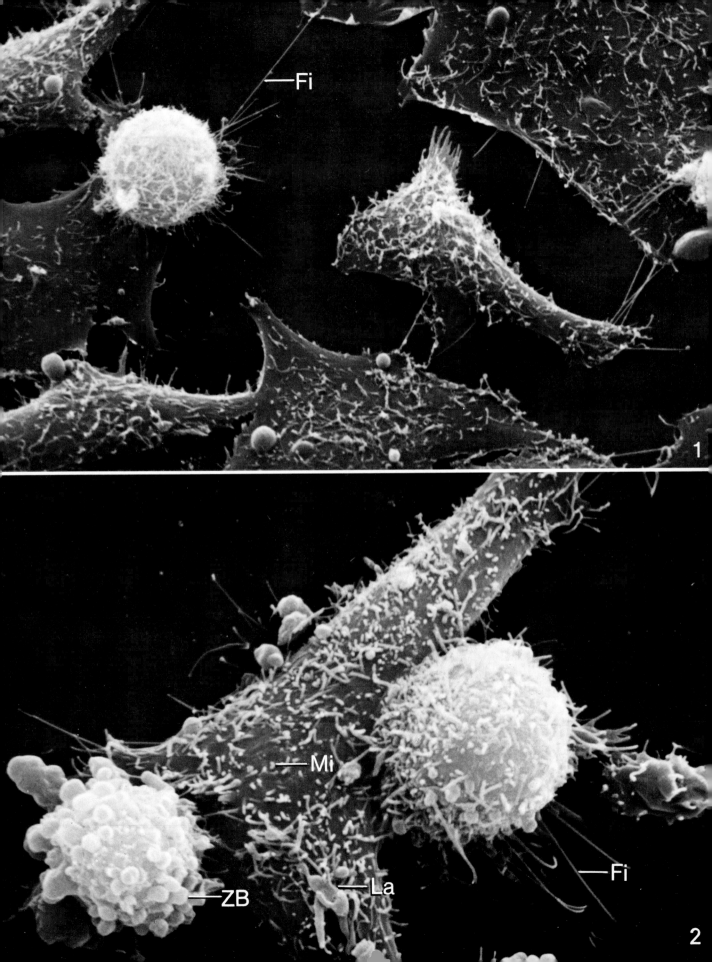

Surface Specializations

Microvilli

Fi   Filopodia
La   Lamellipodium
Mi   Microvilli

Arrows: Microvilli of different lengths (Fig. 3).

Fig. 1 (KB cells), × 4675; Fig. 2 (HeLa cell), × 4065; Fig. 3 (HeLa cell), × 8190.

Many of the transformed or neoplastic cells thus far examined with the SEM possess numerous processes in the form of finger-like projections from the cell surface. These processes are called microvilli (Mi in Figs. 1 through 3). They are present at all stages of the cell cycle in both HeLa and KB cells. The microvilli associated with the surface of the sister KB cells illustrated in Fig. 1 are numerous, widely distributed over the cell surface, and fairly short. In contrast, the microvilli on the surface of the dividing HeLa cell illustrated in Fig. 2 tend to be longer and somewhat more sparsely distributed. These structures have a diameter of 0.1 μm or more, but they vary greatly in their length. The microvilli (arrows) associated with the surface of the flattened HeLa cell in Fig. 3 vary in length from short nubbins to longer extensions. Such a variation suggests that in the living cells, these microvilli may be constantly extending from and withdrawing into the cell. It is clear that from sections of such cells, it would not be possible to guess the number, length or distribution of these microvilli, and it is this type of information that is easily generated in scanning electron microscope images.

The reason for the large number of microvilli that are associated with the surface of transformed cells, compared to a generally more sparse distribution on normal cells, is not too clear. It is possible that these specializations of the plasma membrane may be associated with enhanced transport of nutrients such as glucose and amino acids into the cell, but definite proof for this is not yet available (WILLOCH, 1967). It is also of interest that normal cells in culture which are free of microvilli have been reported to acquire them after transformation by an oncogenic virus. For example, chick fibroblasts that become transformed by Rous sarcoma virus have many more microvilli than the nontransformed cells and they also transport glucose at a higher rate than the normal cells (HATANAKA and HANAFUSA, 1970; BOYDE et al., 1972).

The possibility that microvilli may represent regions of the plasma membrane possessing specific receptor sites of, for example, wheat germ agglutinin has been discussed (BURGER, 1971; PORTER et al., 1973), but sufficient information is not yet available to prove such a function. Additional discussion regarding the possible functional significance of microvilli on transformed cells is available in PORTER et al. (1973).

Other surface specializations illustrated in these SEM micrographs include filopodia (Fi in Figs. 1 and 2) and a lamellipodium (La in Fig. 1).

References

BOYDE, A., WEISS, R.A., and VESELY, P.: "Scanning electron microscopy of cells in culture." Exp. Cell Res. **71**, 313 (1972).

BURGER, M.M.: "The significance of surface structure changes for growth control under crowded conditions." Ciba Foundation Symposium on Growth Control in Cell Cultures. G.E. WOLSTENHOLME and J. KNIGHT, eds. London: Churchill, Ltd. (1971).

HATANAKA, M., and HANAFUSA, H.: "Analysis of a functional change in membrane in the process of cell transformation by Rous sarcoma virus; alteration in the characteristics of sugar transport." Virology. **41**, 647 (1970).

PORTER, K.R., PRESCOTT, D., and FRYE, J.: "Changes in surface morphology of Chinese hamster ovary cells during the cell cycle." J. Cell Biol. **57**, 815 (1973).

WILLOCH, M.: "Changes in HeLa cell ultrastructure under conditions of reduced glucose supply." Acta Pathol. Microbiol. Scand. **71**, 35 (1967).

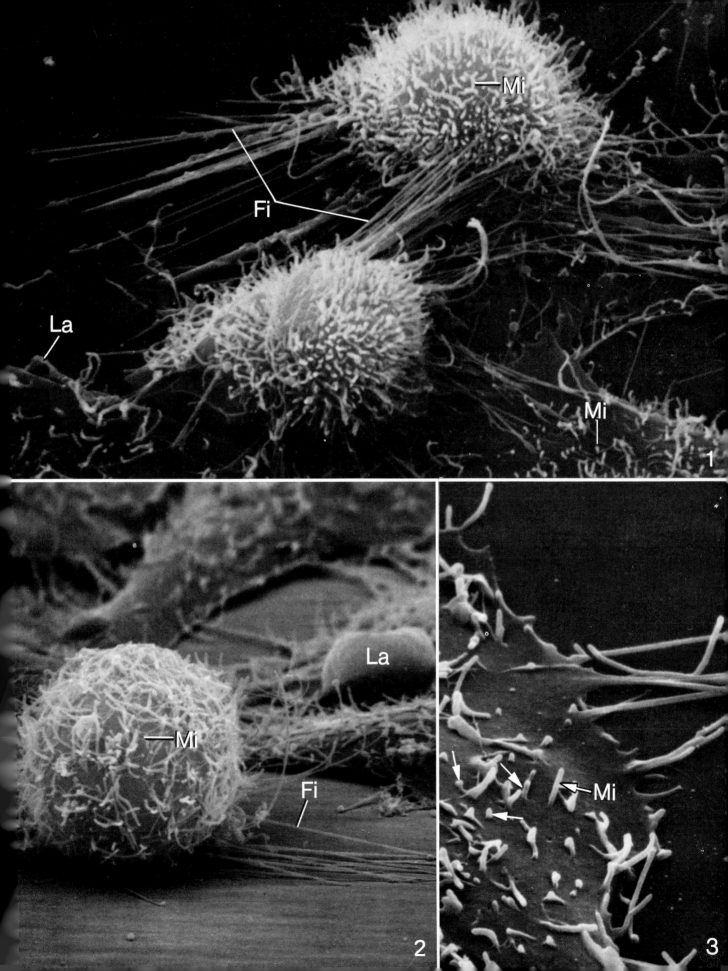

Filopodia

Fi   Filopodia
SB   Secondary branches of filopodia
TF   Thick filopodia
UF   Unattached filopodia

Arrows: Surface excrescences at base of filopodia

Fig. 1 (KB cell), ×5700; Fig. 2 (HeLa cell), ×2430; Fig. 3 (KB cells), ×7600.

A second type of cell extension is much longer and frequently more slender than microvilli. These tentacular processes are called filopodia and are more numerous at the cell margin than elsewhere (Fi in Figs. 1 through 3). The processes are generally present throughout the cell cycle but tend to become more numerous late in G2 phase as the cell rounds up in preparation for division. Such processes undoubtedly aid in anchoring the cell to the substrate during division, and it has also been suggested that the filopodia may provide direction to cell movements. The filopodia extending from the rounded HeLa cell (Fig. 2) are uniform in diameter, but those associated with some KB cells are variable in their diameter (Fig. 1). Some of the filopodia may extend from the cell margin for up to 18 μm in length and many of them are 50 to 100 nm in diameter. However, some of the filopodia may be 4 to 6 times this diameter (TF in Figs. 1 and 3). The significance of the size variation in filopodia is not yet known. Some of the filopodia may have bulbous expansions at their tips or along their length (Fi in Fig. 1). Occasionally a filopodium is observed that is not attached, but is elevated from the surface of the substrate (UF in Fig. 3). Filopodia not only attach to the substrate, but they also attach to adjacent cells.

Filopodia tend to achieve their maximum length on dividing cells. Another feature that can be observed in dividing cells is the extensive terminal arborizations that are sometimes associated with the distal ends of the filopodia. For example, the dividing HeLa cell illustrated in Fig. 2 has a number of long, straight filopodia extending from the lateral cell margin, and these processes then branch into a number of thinner, hair-like process which do not appear to be as rigid. These extremely thin secondary branches (SB in Fig. 2) may be as long as or longer than the diameter of the cell body. In some cases the filopodia appear to emerge from larger excrescences of the cell body (arrows in Fig. 3). It is not yet known, however, if this arrangement is of general occurrence. Filopodia or "microspikes" are known in some instances to contain oriented assemblies of microtubules which are oriented parallel to the long axis of these structures (TAYLOR, 1966; KESSEL and EICHLER, 1966). Drugs such as Colchicine and Colcemid that disassemble microtubules have a pronounced effect on cell form when applied to cells in culture (*cf.* GOLDMAN, 1971; VASILIEV, *et al.*, 1970). Note the distribution of microvilli and zeiotic blebs on the cells in Figs. 1 and 3.

References

GOLDMAN, R.D.: "The role of three cytoplasmic fibers in BHK-21 cell motility. I. Microtubules and the effects of colchicine." J. Cell Biol. **51**, 752 (1971).

KESSEL, R.G., and EICHLER, V.B.: "Microtubules in the microspikes and cortical cytoplasm of grasshopper embryonic cells." J. de Microscop. **5**, 781 (1966).

TAYLOR, A.C.: "Microtubules in the microspikes and cortical cytoplasm of isolated cells." J. Cell Biol. **28**, 155 (1966).

VASILIEV, J.M., GELFAND, I.M., DOMNINA, V., IVANOVA, O.Y., KOMM, S.G., and OLSHEVSKAJA, L.V.: "Effect of Colcemid on the locomotory behaviour of fibroblasts." J. Embryol. Exp. Morph. **24**, 625 (1970).

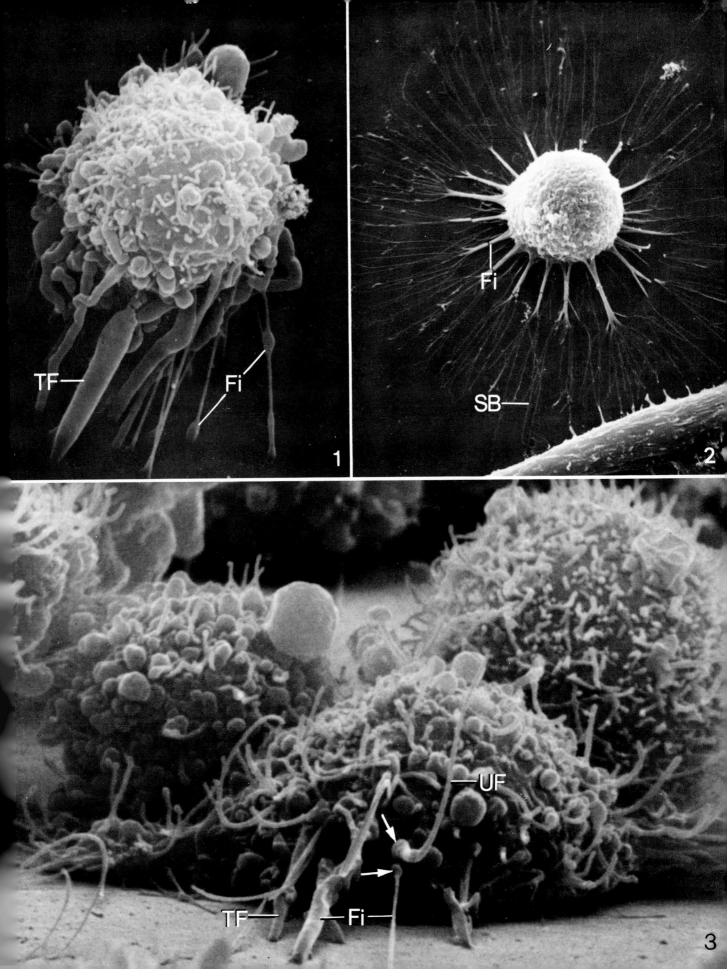

Lamellipodia (Ruffles) and
Zeiotic Blebs (Bulbous Excrescences)

A third type of surface specialization observed on cells in culture consists of thin flaps or folds of the cell surface and are called lamellipodia (La in Figs. 1 and 2) or ruffles. They have been observed with the light microscope, and in the living cell are observed to be in constant motion. They have been described to be involved in cell drinking or pinocytosis (GEY, 1955). The lamellipodia tend to form at the cell margin which is not in contact with other cells and then migrate to the center of the cell where they are resorbed into the cytoplasm (TAYLOR et al., 1971; EDIDIN and WEISS, 1972). Ruffles engulf small amounts of culture medium during pinocytosis. In dense cultures of Chinese hamster ovary cells, increased ruffling occurs during S and G_2 phases of the cell cycle (PORTER et al., 1973), but ruffles are not generally observed on cells in mitosis. However, in low density cultures, lamellipodia may be observed on rounded cells (Fig. 1). In low density cultures, transformed cells can proceed through the cell cycle without expressing the surface changes indicated and it has been concluded that under conditions of low cell density there is insufficient cell contact to induce these changes (cf. PORTER et al., 1973, RUBIN and EVERHART, 1973 for discussion). The significance of the increased ruffling generally noted on cells in dense cultures during the latter part of the cell cycle may indicate that the cells require increased amounts of metabolites for growth during this period (cf., PORTER et al., 1973). Lamellipodia are approximately 0.1 μm thick, but their length and vertical dimension may vary from 8 to 10 μm.

A fourth type of surface modification has variously been called zeiotic blebs (PRICE, 1967), bulbous excrescences (PORTER et al., 1972) or exocytes. These specializations consist of bulbous or spherical structures which vary from a fraction of a micrometer to several micrometers in diameter (ZB in Fig. 3). The bubbling of the cell surface during mitosis has been described many times from observations on the living cell. In normal density cultures of transformed cells for which information is available the bulbous excrescences tend to dominate the cell surface during mitosis and early G_1 phases of the cell cycle (Fig. 3) and they then tend to disappear during the latter half of the cell cycle (PORTER et al., 1973). This disappearance may be due to increased cell contact, for in low density cultures the blebs tend to persist much longer, in some cases throughout the cell cycle. Little information is presently available to suggest how these structures form or their function. Fox et al. (1971) has reported that sites on the cell surface that bind certain agglutinins are transiently exposed during mitoses and this finding has generated increased interest in these structures. The exaggerated presence of blebs on transformed cells appears to be a unique feature requiring further study.

Different transformed cell types in culture thus appear to possess abundant surface specializations as has been illustrated. While the number and kind of surface specializations are generally greater on cancer cells than on normal ones, the number and kind of surface specializations may also vary from one transformed cell type to another. Cells in culture are grown under conditions that can be varied experimentally, and the effects of substances such as Colchicine or Colcemid and dibutryl cyclic AMP on cells in culture have been described (cf., PORTER et al., 1973 for discussion). Such studies should shortly clarify the functional significance of the surface specializations associated with transformed cells in culture.

References

GEY, G.O.: "Some aspects of the constitution and behavior of normal and malignant cells maintained in continuous culture." The Harvey Lectures Ser. L., pp. 154. New York: Academic Press, (1955).

FOX, T., SHEPPARD, J., and BURGER, M.: "Cyclic membrane changes in animal cells: Transformed cells permanently display a surface architecture detected in normal cells only during mitosis." Proc. Natl. Acad. Sci. (U.S.A.) **68**, 244 (1971).

EDIDIN, M., and WEISS, A.: "Antigen cap formation in cultured fibroblasts: A reflection of membrane fluidity and of cell motility." Proc. Natl. Acad. Sci. (U.S.A.) **69**, 2456 (1972).

PORTER, K. R., KELLEY, D., and ANDREWS, P. M.: "The preparation of cultured cells and soft tissues for scanning electron microscopy." In Proceedings of the Fifth Annual Stereoscan Scanning Electron Microscope Colloquium. Kent Cambridge Scientific, Inc., Morton Grove, Illinois (1972).

PORTER, K. R., PRESCOTT, D., and FRYE, J.: "Changes in surface morphology of Chinese hamster ovary cells during the cell cycle." J. Cell Biol. **57**, 815 (1973).

PRICE, Z. H.: "The micromorphology of zeiotic blebs in cultured human epithelial (HEp) cells." Exptl. Cell Res. **48**, 82 (1967).

RUBIN, R. W., and EVERHART, L. P.: "The effect of cell-to-cell contact on the surface morphology of Chinese hamster ovary cells." J. Cell Biol. **57**, 837 (1973).

TAYLOR, R. B., DUFFUS, P. H., ROFF, M. C., and DEPETRIS, S.: "Redistribution and pinocytosis of lymphocyte surface immunoglobulin molecules induced by anti-immunoglobin antibody." Nature New Biol. **233**, 225 (1971).

La   Lamellipodia
ZB   Zeiotic blebs

Fig. 1 (KB cell), ×8310; Fig. 2 (HeLa cell), ×3125; Fig. 3 (KB cell), ×4410.

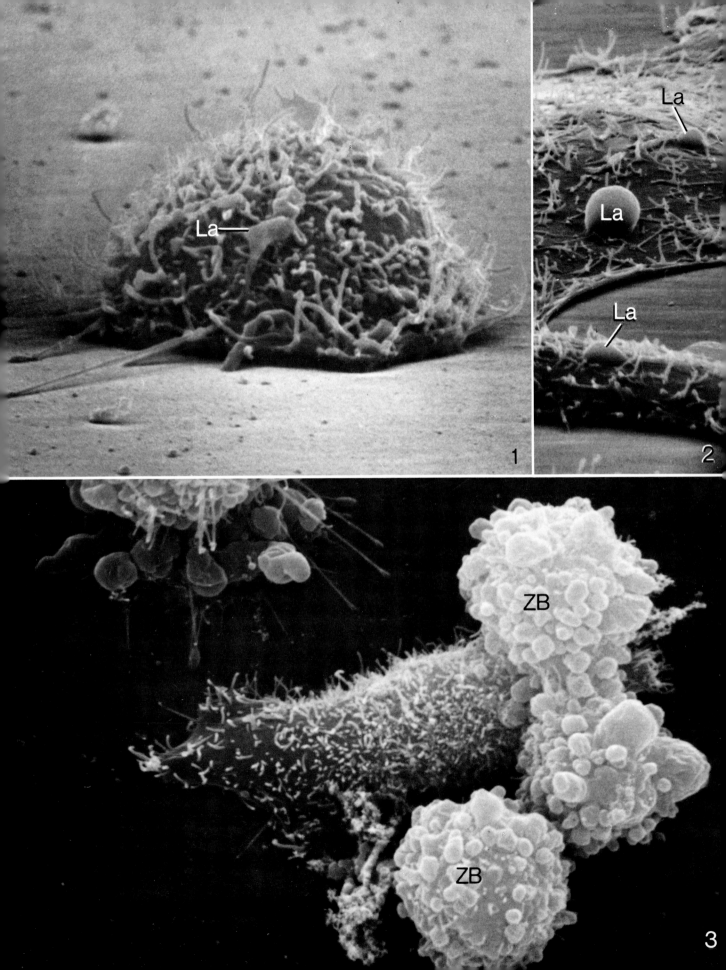

Variations
in Cell Surface Specializations

Fi   Filopodia
Mi   Microvilli
UF   Unattached filopodia
ZB   Zeiotic blebs

Fig. 1, × 6590; Fig. 2, × 6060.

It has been previously pointed out that the surface specializations associated with cultured transformed cells may vary with cell density (cf. RUBIN and EVERHARDT, 1973). Further, some cultured cells may vary in the degree to which they adhere to a substrate. For example, KB cells in culture can exhibit a variation in which certain cell lines do not attach well either to plastic or glass culture dishes (Figs. 1, 2). Under such conditions, many of the cells remain in a rounded configuration with relatively few cells becoming firmly anchored to the substratum. One of the rounded KB cells illustrated in Fig. 1 does not possess filopodia at its margin, while the other cell does possess such structures and many of them appear to be attached to the substratum (Fi in Fig. 1). The more flattened KB cells illustrated in Fig. 2 do possess filopodia, but none are anchored to the substrate (UF in Fig. 2). Rather, the filopodia extend vertically from the cell surface. The inability of these particular cells to attach to the substratum appears to involve an inability of the tips of the filopodia to establish an effective attachment. A characteristic feature of these cells is the large number of zeiotic blebs (ZB in Figs. 1 and 2) associated with their surfaces. The microvilli (Mi in Figs. 1, 2) associated with the surfaces of these cells do not appear as numerous nor as large as those associated with other types of KB cells such as previously illustrated.

The number and kind of surface specializations are not uniform from one time to another during the cell cycle as recently demonstrated by PORTER et al. (1973) for transformed cultured Chinese hamster ovary (CHO) cells. Microvilli were observed at all stages of the cell cycle (G_1, S, G_2, D), but increased in number during G_2. Filopodia were also present throughout the cell cycle, but were especially abundant at the end of G_2 and on cells in mitosis. The filopodia of dividing cells were noted to differ from those on cells in other stages of the cell cycle (1) in being more numerous, (2) in extending for 2 to 3 times the diameter of the cell body, and (3) by their terminal arborization or branching. The cells begin to show ruffles at their margin not in contact with other cells in late G_1 phase, and there is an increased ruffling at the cell margin during S and G_2 phases. Zeiotic blebs were noted on middle and late G_1 cells, but were not prominent on S or G_2 phase cells. In contrast to transformed CHO cells, normal Chinese hamster lung cells in culture have a simple surface fine structure. Blebs, microvilli, and ruffles were found to be generally absent, and only the cellular microextensions (filopodia) were evident (PORTER et al., 1973). It has been determined (RUBIN and EVERHART, 1973) that extensive cellular contacts are required for the cells to undergo those morphological changes just described for the cell cycle. Thus, when cells are plated in very low density so that no intercellular contact occurs, the CHO cells remain in a G_1 configuration (rounded and highly blebbed) as they proceed through the cell cycle (through G_1, S, and G_2).

All cell surface specializations appear to be transient; that is, they are probably continually forming and disappearing. Most cancer cells appear to always have abundant surface specializations at all stages in their life cycle, even in confluent cultures where they would normally be responding to contact (PORTER and FONTE, 1973). Typical of cancer cells also is their failure to be inhibited in growth and movement when in confluence with other cells.

References

PORTER, K.R., and FONTE, V.G.: "Observations on the topology of normal and cancer cells." In Scanning Electron Microscopy/1973, O. JOHARI and I. CORVIN, eds. IITRI 6, 683 (1973).
PORTER, K.R., PRESCOTT, D., and FRYE, J.: "Changes in surface morphology of Chinese hamster ovary cells during the cell cycle." J. Cell Biol. 57, 815 (1973).
RUBIN, R.W., and EVERHART, L.P.: "The effect of cell-to-cell contact on the surface morphology of Chinese hamster ovary cells." J. Cell Biol. 57, 837 (1973).

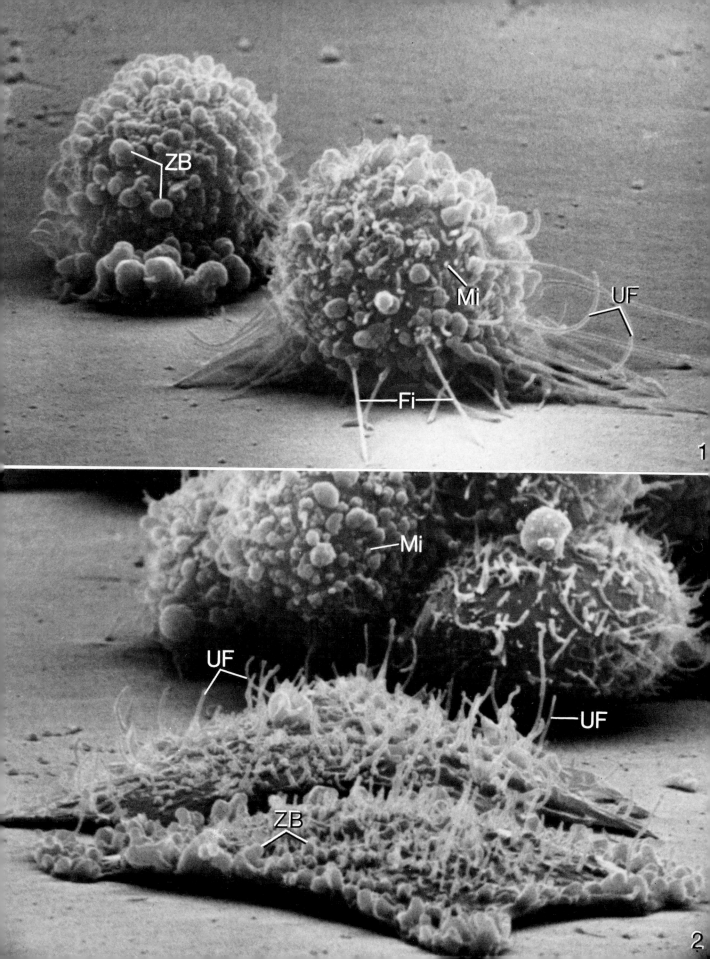

1

2

Cells in Mitosis

An   Anaphase stage of mitosis
IN   Interphase nucleus
Me   Metaphase stage of mitosis
Te   Telophase stage of mitosis

Fig. 1 (fixed and stained HeLa cells), ×450; Fig. 2 (HeLa cell, telophase), ×2000; Fig. 3 (HeLa cell, telophase), ×2000; Fig. 4 (HeLa cell in prophase or early metaphase), ×3750; Fig. 5 (HeLa cell, anaphase), ×3750; Fig. 6 (HeLa cell, early telophase), ×3750; Fig. 7 (KB cell, telophase), ×4500; Fig. 8 (KB cell, telophase), ×3190; Fig. 9 (HeLa cell, late telophase), ×3300.

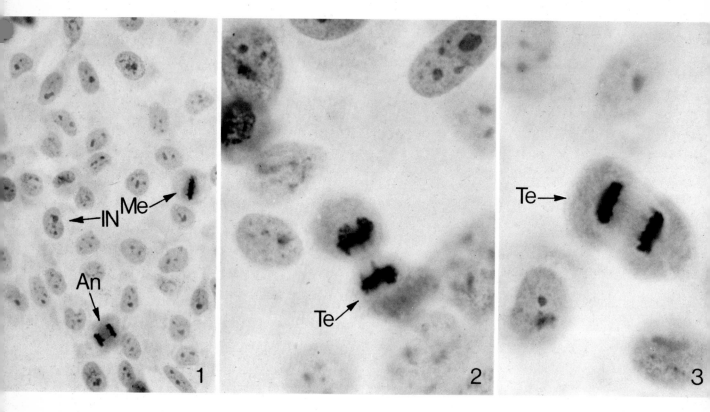

The light microscope appearance of HeLa cells in culture after fixation and staining with Delafield's hematoxylin is illustrated in the photomicrographs of Figs. 1 through 3. Many interphase nuclei (IN) are present in Fig. 1 as well as a metaphase (Me) and anaphase (An) stage of mitosis. Telophase (Te) stages of mitosis are illustrated in Figs. 2 and 3. It is now possible to recognize individual cells in culture not only in relation to stages of their life cycle, but to visualize different phases of mitosis in cultured cells examined by scanning electron microscopy. Several mitotic phases of HeLa and KB cells in culture are illustrated in Figs. 4 through 9. The spherical cell in Fig. 4 posses-sing numerous filopodia and microvilli probably represents a prophase or early metaphase stage of mitosis. The microvilli on this cell appear to be rather evenly distributed over the surface. The more elongated cell in Fig. 5 probably represents an anaphase stage. The microvilli appear to be especially numerous at the poles, but are sparsely distributed about the equator of the cell. Progressive stages of cytokinesis during telophase are illustrated in Figs. 6 through 9. As the cleavage furrow progressively constricts the two cells, a narrow intercellular connection remains between the two daughter cells and becomes longer and narrower as the cells move apart (Fig. 9).

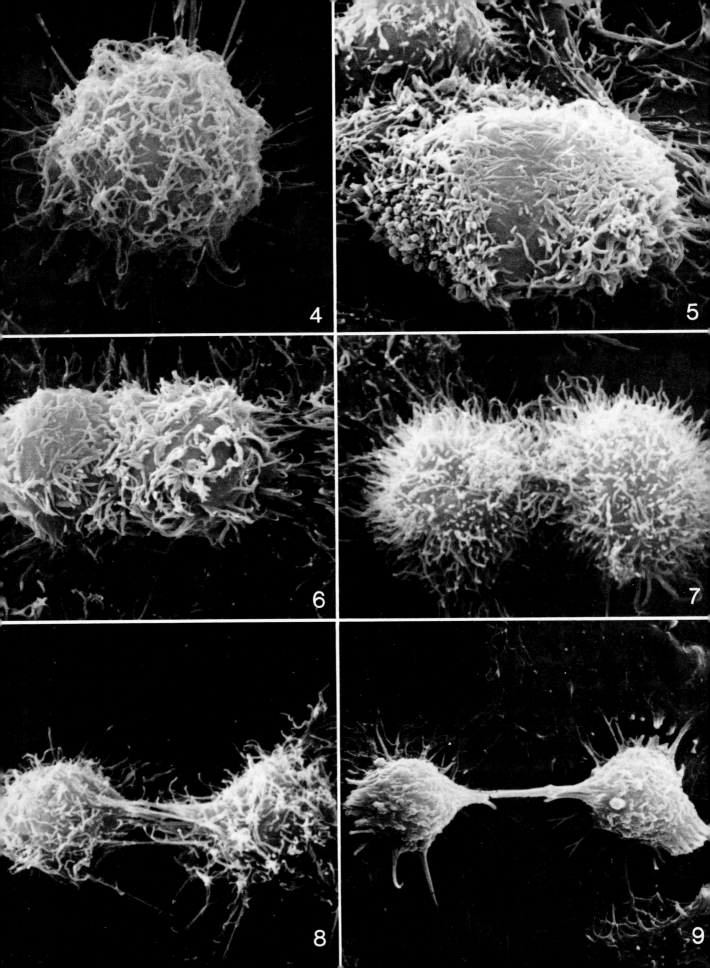

Chick Embryo Chondroblasts

Fi   Filopodia
La   Lamellipodia
Mi   Microvilli
ZB   Zeiotic bleb

Fig. 1, ×8250; Fig. 2, ×6300; Fig. 3, ×3120; Fig. 4, ×3450.

Normal cells in culture, such as chondroblasts that produce cartilage, do not possess as many surface modifications as do transformed cells. A rounded (mitotic) chondroblast cell is illustrated in Fig. 1. This cell has a few slender filopodia (Fi) that are not nearly so numerous as they are on abnormal or transformed cells in a comparable stage. Further, the microvilli (Mi in Fig. 1) that extend from the cell surface are short and sparse. They are, however, more numerous on rounded cells than on flattened cells (Figs. 3 and 4). A portion of the chondroblast surface in Fig. 1 is covered by cartilage matrix that these cells secrete. A chondroblast cell in the process of flattening is illustrated in Fig. 2 and it has several filopodia (Fi), a few short microvilli (Mi) and a single lamellipodium (La) associated with its surface. Normal flattened cells in culture have only a few filopodia (Fi in Figs. 3 and 4) and only occasional microvilli (Mi in Fig. 4), but again they are not nearly so numerous as they are on transformed cells in a comparable stage of flattening. Lamellipodia (La in Figs. 2 and 4) are sometimes observed on cells both in an early and late stage of flattening. Zeiotic blebs are rare (ZB in Fig. 4). In one region of the cell illustrated in Fig. 3 (arrow), a portion of the cell surface appears to be absent. Clearly, the normal cell in culture is much less complex with respect to its surface specializations compared with abnormal or transformed cells.

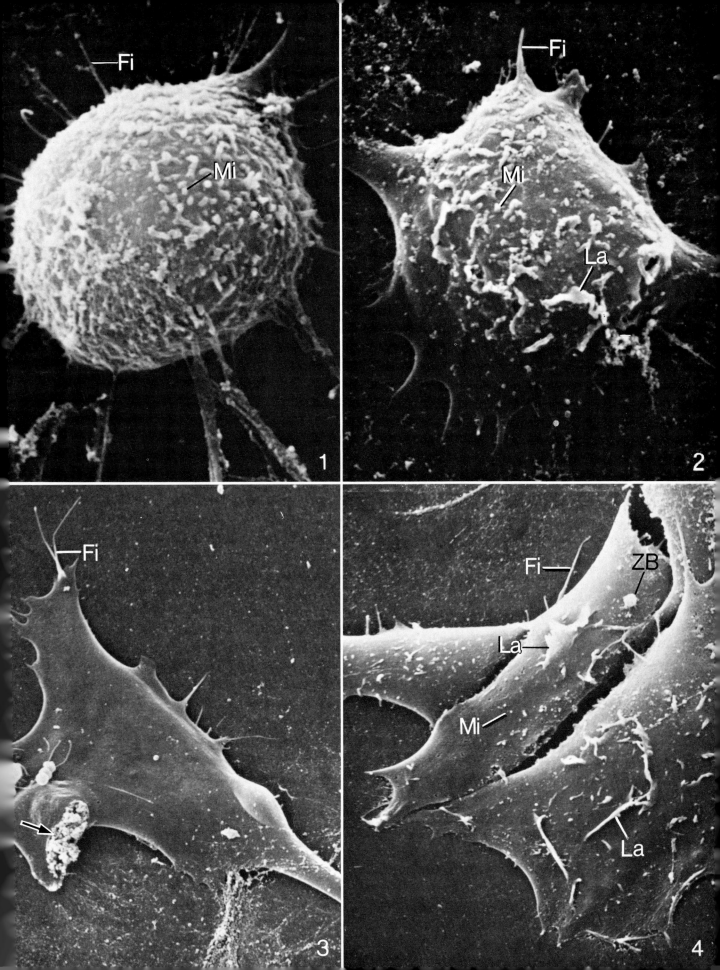

Phagocytosis by Normal and Abnormal Tissue Culture Cells

The immediate milieu of a cell no doubt plays an important role in cell growth, development, and differentiation, both in vivo and in vitro. Since the advent of tissue culture techniques that allow the establishment of a defined and controlled environment for cells, many important basic properties of cell growth, development, and differentiation have been elucidated in vitro. Unfortunately, many in vitro observations cannot be directly applied to the cell in vivo simply because the in vivo milieu of the cell is in a state of constant change. These changes are under the dynamic control of the multicellular organism and are necessary for the survival of the organism as a whole. It should be pointed out that the converse also occurs; namely, the behavior or properties of a cell in vivo can be lost or modified when the cell is transferred to in vitro conditions, and often the changes that then occur are unpredictable.

Though many parameters such as nutrition, growth-stimulating and growth-inhibiting factors, cell-to-cell contact, temperature, and pH are expected to alter the cell's behavior and/or cytomorphology in vitro, the stimulation of phagocytic-like activity in cells that are not normally phagocytes or macrophages is an apt illustration of an unexpected response to a change in the cell's milieu. In addition to the unexpected phagocytic-like activity, the effects of cell-to-cell contact on cytomorphology can be observed as a response to cell density.

To illustrate the varied and sometimes unexpected responses of cells in vitro to even slight changes in their environment, two cell lines were chosen, D550 and C_6. The Detroit 550 (D550), derived from normal human skin, has a diploid chromosome number, a finite life span in vitro, and a fibroblast-like morphology with rather simple surface specializations (all characteristics of a normal cell in vitro). The glial cell line, clone 6 (C_6), derived from a rat glial tumor has an infinite life span in vitro, a varied cytomorphology,

and an ability to reform "tumors" both in vivo and in vitro (all characteristics of an abnormal or transformed cell in vitro). D550 and C_6 cells were grown in Falcon plastic flasks in Ham's F10 medium (Grand Island Biological Co., Grand Island, N.Y.); supplemented with 15% horse and 2.5% fetal calf sera, 100 units/ml penicillin, 100 μg/ml streptomycin; and buffered with 0.015M HEPES, 0.015M $NaHCO_3$, and 5% CO_2 in balanced air. The cultures were fixed, dehydrated, critical point dried, and coated for SEM as described by De Bault (1973). The procedures are somewhat similar to those outlined previously.

One method of evoking phagocytic-like activity in D550 and C_6 cells is simply to add particles to the medium while all other culture conditions remain unchanged. By using polystyrene spheres (Sp, Figs. 1—7) (Styrene-divinyl-benzene copolymer particles, Dow Chemical) coated with an oil-soluble dye (Fluoro-7Ga, Dow Chemical), it is possible to observe the spheres with both the fluorescence microscope and the SEM. The diameters of the spheres are 4 to 18 μm (mean = 12 μm).

In a dense, confluent monolayer the D550 cells exhibit contact inhibition and are considered to be in a stationary phase of growth. Despite the reduced cell proliferation, the cells are still very active and can be stimulated to engage in phagocytic-like activity (Figs. 1 and 2). After only a 24-hour exposure, the cells can be seen in various stages of engulfing the spheres. Note the leading edge of a thin membranous fold of a cell (arrow, Fig. 2) which seems to be coursing over the surface of the sphere (Sp). The surface of the cell itself is rather smooth, with only a few short stubby microvilli (Mi) and small blebs (Bl) (Fig. 2). These simple surface features are characteristic of this and other normal cell lines and are seen in D550 cultures with and without polystyrene spheres present in the medium.

In a very sparse, rapidly growing culture the D550

cells have a more complex surface but not as complex as the surface of abnormal or transformed cells such as the HeLa, KB, CHO, or C_6 cells. The most characteristic feature of the cell surface in these sparse cultures of D550 cells is the increased number of small blebs (Bl in Figs. 4 and 5), which are present in cultures both with and without spheres. The rapidly growing D550 cells exhibit a faster rate of phagocytic-like activity than stationary cultures and they also have the ability to attach spheres to the surface without engulfing them (Fig. 5).

The C_6 cells, rat glioma, exhibited more and varied surface features than did the D550 cells. The C_6 cells are, in general, similar to other transformed cells. However, in sparse cultures a characteristic of this cell line is to grow as flat or fusiform, overlapping cells with very few surface projections (Fig. 6). In sparse C_6 cultures challenged with spheres the most striking responses were the rate of engulfment (approximately 4 times the rate of D550 cells), the large size of spheres engulfed, and the surface features of the cell at the site of engulfment (Fig. 7). Note the irregular surface of the cell (IS) and smooth pseudopodia (SS) in Fig. 8.

It should be emphasized that the phagocytic-like activity of the D550 and C_6 cells is an unexpected response for skin fibroblast and glial cell types. Furthermore, the response to the introduction of spheres into their milieu is modified by cell-to-cell contact. For details and general in vitro properties (morphology, cytochemistry, and TEM) of true phagocytes or macrophages from the chick and mouse, see SUTTON and WEISS (1966) and COHN and BENSON (1965), respectively. For SEM details of mouse macrophages see ALBRECHT et al. (1972). (All micrographs and text material kindly provided by Dr. L. E. DE BAULT.)

Bl Blebs
IS Irregular surface
Mi Microvilli
Sp Spheres
SS Smooth pseudopodia

Arrow in Fig. 2: Leading edge of a very thin lamellipodium

Fig. 1 (dense culture of D550 cells 8 hours after addition of spheres), ×4650; Fig. 2 (enlargement of Fig. 1), ×15400; Fig. 3 (sparse culture of D550 cells 8 hours after addition of spheres—one sphere internal, one sphere external), ×1850; Fig. 4 (enlargement of Fig. 3, angle change), ×3300; Fig. 5 (enlargement of Fig. 4), ×19700; Fig. 6 (sparse culture of C_6 cells 4 hours after addition of spheres), ×1943; Fig. 7 (C_6, sphere half engulfed), ×1275; Fig. 8 (enlargement of Fig. 7), ×9263.

References

ALBRECHT, R. M., HINSDILL, R. D., SANDOK, P. L., MACKENZIE, A. P., SACHS, I. B.: "A comparative study of the surface morphology of stimulated and unstimulated macrophages prepared without chemical fixation for scanning EM." Exptl. Cell Res. **70**, 230—232 (1972).

COHN, Z. A., BENSON, B.: "The differentiation of mononuclear phagocytes." J. Exptl. Med. **121**, 153—189 (1965).

DE BAULT, L. E.: "A critical point drying technique for scanning electron microscopy of tissue culture cells grown on plastic substratum", In Scanning Electron Microscopy/1973, O. JOHARI and I. CORWIN, eds. IITRI **6**, 317—324 (1973).

SUTTON, J. S., WEISS, L.: "Transformation of monocytes in tissue culture into macrophage, epitheloid cells, and multinucleated giant cells." J. Cell Biol. **28**, 303—332 (1966).

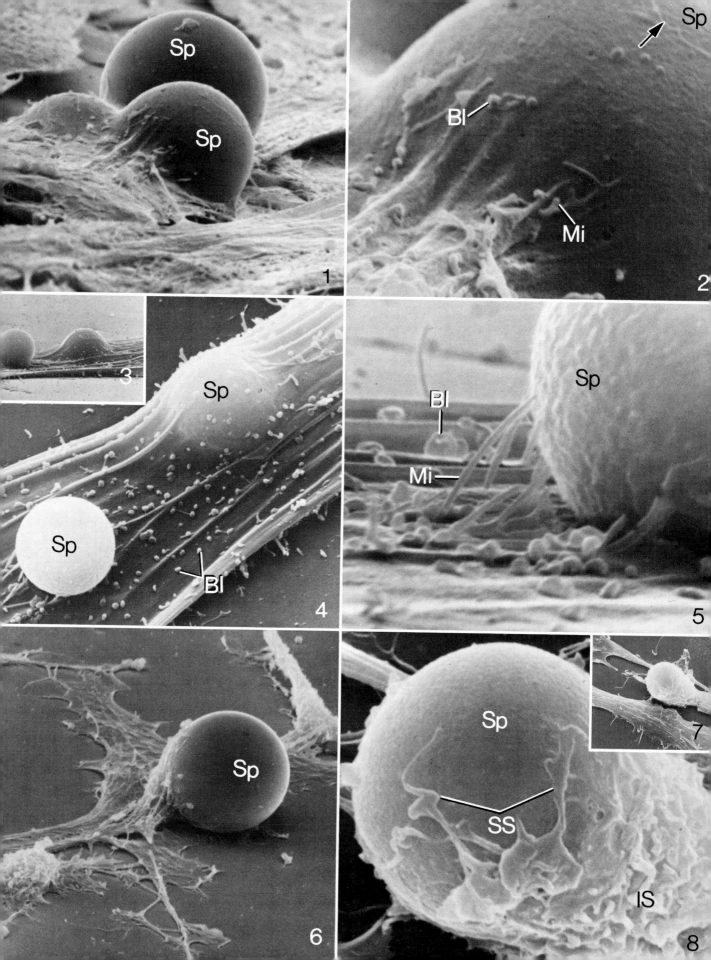

Induced Morphogenesis
of Glial Cells

DC	Dividing cell	Sp	Spherical cell
Fl	Flat cell	Nd	Nodules
Fu	Fusiform cell	PB	Partial bilayer

Fig. 1 (log phase culture of C_6 cells), $\times 1200$; Fig. 2 (advanced stationary phase culture of C_6 cells, 3 months old), $\times 725$; Fig. 3 (C_6 nodule, 4 hours after plating in new culture vessel), $\times 425$; Fig. 4 (C_6 nodule, 18 hours after plating), $\times 250$; Fig. 5 (C_6 nodule, 30 hours after plating), $\times 80$; Fig. 6 (same as Fig. 4, side view), $\times 2675$; Fig. 7 (same as Fig. 5, side view), $\times 1180$.

Another example of how a change in the cell's milieu can alter both the cell's behavior and cytomorphology is demonstrated by the response of C_6 nodules to a cell-free substratum. In log phase cultures (i.e., rapidly growing) of C_6 the cells are flat (Fl in Fig. 1), fusiform (Fu), or spherical (Sp). These forms are in general related to specific stages of the cell cycle. The changes in surface specializations as the cell progresses through the cell cycle are similar to those described in another cell type by PORTER et al. (1973). As the cultures become dense and crowded into bilayers, the cells become spherical and develop a surface of intricate and heterogeneous microextensions (DE BAULT, 1973). If the cultures are maintained for extended periods of time (1 to 3 weeks) without serial transfer, nodules (Nd in Fig. 2), or "microtumors", form. After the nodules have reached a critical size, the cultures become stationary and remain static up to 3 months. The cells in the nodules are spherical and exhibit a heterogeneity of microextentions, including blebs and ruffles and often a few microvilli.

If the nodules are removed from a stationary culture and plated in a new culture vessel, a dramatic sequence of events occurs (Figs. 3 through 5). In the new environment the nodules respond by first attaching themselves to the substratum (4 to 8 hours, Fig. 3). After 12 to 18 hours in the new environment the cells closest to the substratum begin to migrate from the nodule (Fig. 4). The cells in the nodule remain spherical and have complex surface features, but those cells that migrate out of the nodule and onto the substratum revert to flat or fusiform cells (Figs. 4, 6). After 30 hours in the new environment the migration is quite extensive (Figs. 5 and 7), and by 72 hours the nodule has disappeared. Cell proliferation resumes in the monolayer and the culture is again in log phase. (All micrographs and text material kindly provided by Dr. L.E. DE BAULT.)

References

DE BAULT, L.E.: "A critical point drying technique for scanning electron microscopy of tissue culture cells grown on plastic substratum", In Scanning Electron Microscopy/1973, O. JOHARI and I. CORWIN, eds. IITRI **6**, 317—324 (1973).

PORTER, K.R., PRESCOTT, D., and FRYE, J.: "Changes in surface morphology of Chinese hamster ovary cells during the cell cycle." J. Cell Biol. **57**, 815—836 (1973).

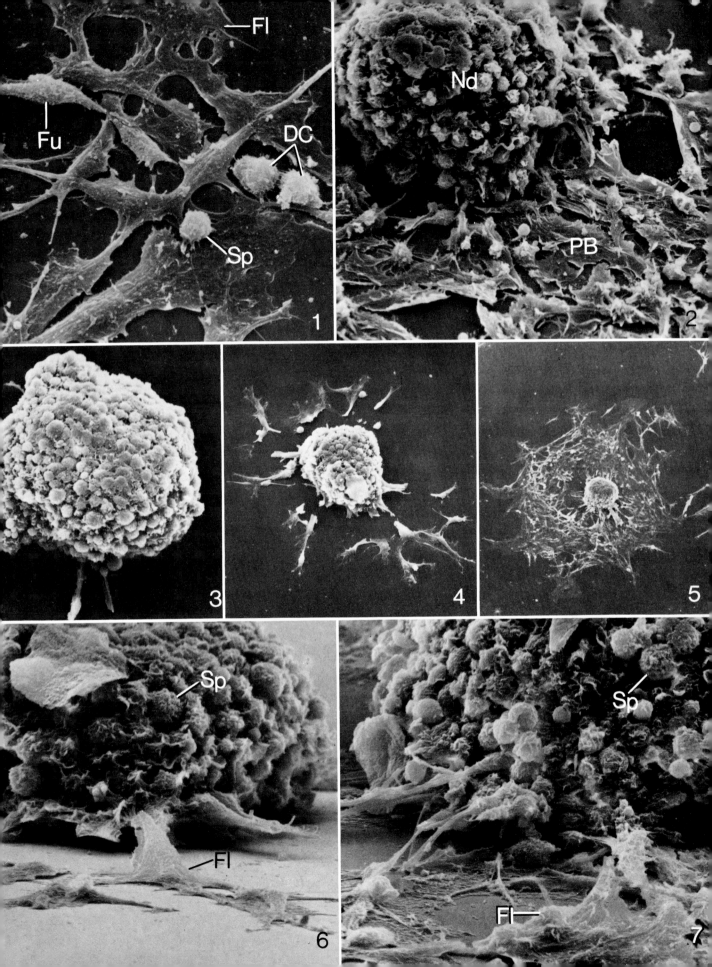

Chapter 4
Prokaryotes

Bacteria

Division Schizomycota

Order Eubacteriales

Cell Form and Structure

The bacteria are probably the simplest cellular organisms. Individual bacteria are visible only under a microscope and measure less than 8 µm in length and as small as 0.5 µm wide. However, the weight of the total number of bacteria in the world exceeds that of all other organisms. It has been calculated that there may be 1 000 000 to 50 000 000 bacteria per gram of top soil.

Bacteria are mainly heterotrophs that require a supply of existing organic nutrients from dead plants or animal remains. These bacteria are commonly in water, soil, sewage, and foods, and are responsible for much of the natural recycling process. Parasitic bacteria obtain their nutrients from living organisms and live at the expense of their host. Many bacteria also form a symbiotic relationship with their host in which both partners benefit. These bacteria are abundant in the digestive tract and on the mucous membranes and skin of animals. Fig. 1 shows rod-shaped bacteria on the surface of *Hydra*. Note the numerous strands (St) connecting the bacteria to the animal surface. Fig. 2 shows spherical bacteria on a rat tongue, and many of these appear to be in the process of cell division (CD). Similar types of bacteria occur on or inside many kinds of plants. A typical example is *Rhizobium*, which exhibits an unusual symbiotic relationship with the roots of leguminous plants such as alfalfa, clover, pea, soybean, and lupines.

The autotrophic bacteria, which include photosynthetic and chemosynthetic species, inhabit soil and water. The photosynthetic bacteria contain photosynthetic pigments which absorb 400 to 900 nm wavelength light. Chemosynthetic bacteria obtain energy through a variety of oxidation-reduction reactions and usually require only carbon dioxide as a carbon source. Such bacteria are abundant in habitats with a supply of the oxidizable substrates such as hydrogen sulfide, methane, ammonium, and nitrate ions. The autotrophic bacteria usually contain chromatophores in parallel membranes or in simple spherical bodies.

The bacteria show relatively little variation in their morphological and cytological features. The commonly encountered forms are rods (bacilli) (Fig. 1), spheres (cocci) (Fig. 2), and curved (spirilla) (see next plate). With shadow-casting and transmission electron microscopic observation the surface features of a *Escherichia coli* are revealed in Fig. 3. The surface of these bacteria possess several long whip-like filaments called flagella (Fl) and short filamentous appendages called fimbriae (Fi). The flagella permit cell movements and are variously placed upon the cell, while the fimbriae radiate from all of the cell surface and are more numerous and shorter than flagella. The function of fimbriae is uncertain, but possibly they help bacterial cells adhere to their host surface, as seen in Fig. 1.

In a longitudinal section of a bacillus (Fig. 4) the cell is seen to be surrounded by a rigid cell wall that contains polysaccharides and mucopeptide. The wall is responsible for maintenance of cell shape. The cytoplasmic matrix is relatively homogeneous and is enclosed by a plasma membrane (PM). Ribosomes (Ri) and glycogen (Gl) are present, but mitochondria of the type found in eukaryotic cells are not known in bacteria. Mitochondrial function possibly is taken over by other cytoplasmic membranes. Under most growth conditions *E. coli* contains several non-membrane-bound fibrillar nuclear regions (Nu). Each region apparently contains identical copies of a circular, double-stranded DNA about 1 000 to 5 000 µm long when biochemically isolated. The strands are considered analogous to the chromosomes of higher organisms although no histones are associated with the DNA strands.

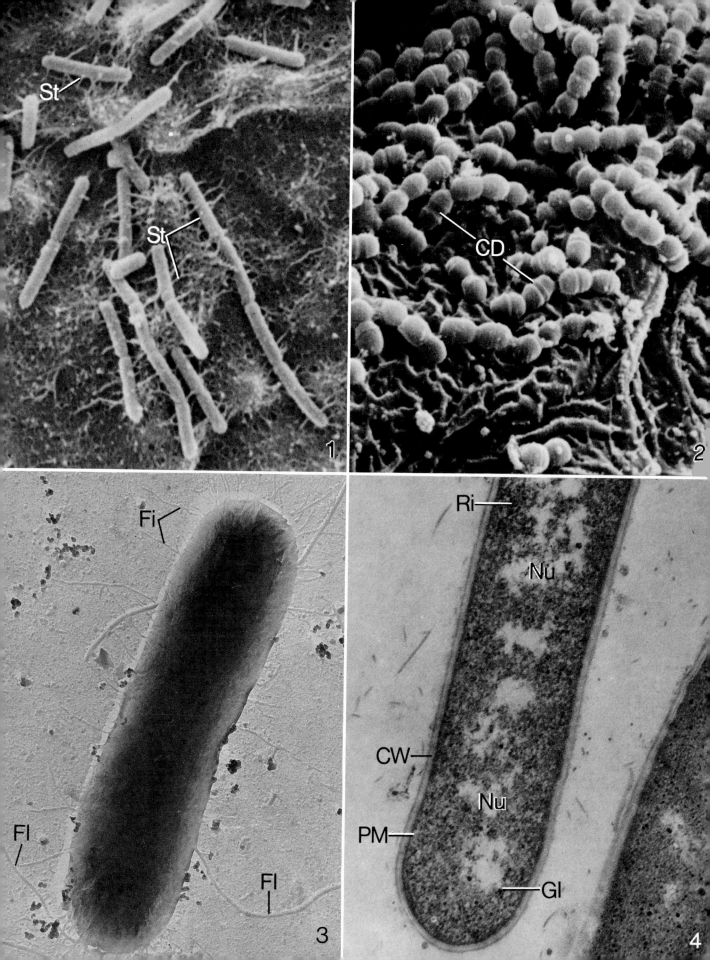

Division Schizomycota

Order Pseudomonadales

Spirillum

(Spiral Microorganism in Rat Intestine)

AF   Axial filaments
CW   Cell wall
FLP Flagella-like processes
Mi   Microvilli
PC   Protoplasmic cylinder
SM   Sheath membrane

Fig. 1, × 30000; Fig. 2, × 35000.

A wide variety of microorganisms reside either temporarily or permanently in the intestinal tract of mammals. In addition to bacteria, examples illustrated in this atlas include the diplomonad flagellates, *Hexamita muris* and *Giardia muris*. In the small intestine of the rat, spiral microorganisms reside in the crypts of Lieberkuhn (Fig. 1). The bacterium is about 7.5 µm long and approximately 0.6 µm in diameter. It consists of a coiled protoplasmic cylinder (PC in Figs. 1 and 2) surrounded by axial filaments (AF) and a sheath membrane (SM) (Fig. 2). Thus, the axial filaments are interposed between the sheath membrane and cell wall (CW in Fig. 2). The axial filaments are arranged at intervals of approximately 130 nm about the periphery of the protoplasmic cylinder so that in transmission electron micrographs (Fig. 2) the organisms have a serrated appearance. The axial filaments take the form of hollow tubules. The polar ends of the spirillum have a number of flagella-like processes (FLP), derived in part from the sheath membrane, which are illustrated in the SEM (Fig. 1) and TEM (Fig. 2) sections. The microvilli (Mi) of an intestinal epithelial cell are included in the TEM section illustrated in Fig. 2. It has been shown that the spiral microorganisms can be phagocytosed and digested intracellularly by Paneth cells at the base of the crypt of Lieberkuhn (ERLANDSEN and CHASE, 1972). Therefore, Paneth cells play a role in the regulation of intestinal flora. (Micrographs kindly provided by Dr. STANLEY ERLANDSEN and Dr. D. CHASE.)

Reference

ERLANDSEN, S.L., and CHASE, D.G.: "Paneth cell function: Phagocytosis and intracellular digestion of intestinal microorganisms. II. Spiral microorganism." J. Ultrastruct. Res. **41**, 319—333 (1972).

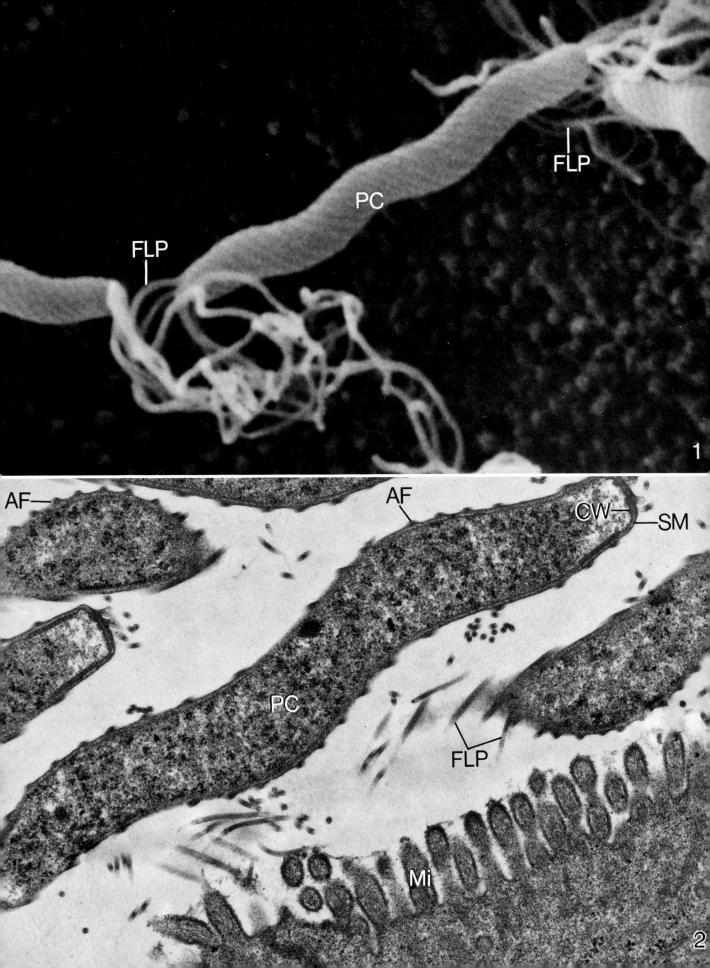

FLP

PC

FLP

1

AF

AF

CW

SM

PC

FLP

Mi

2

Division Schizomycota

Order Eubacteriales

A Probe
for Infectious Disease Research

EC   Elongated cell
IB   Intracellular bridges
RC   Ruptured cell
Se   Septum

Fig. 1 (*Clostridium perfringens*), ×13000; Fig. 2 (*Staphylococcus aureus*), ×20400; Fig. 3 (group B *Streptococcus*), ×14700; Fig. 4 (*Escherichia coli*, 2 hours culture at 37°C in trypticase soy broth), ×36000; Fig. 5 (*Escherichia coli*, 2 hours, exposure to human amniotic fluid at 37°C), ×7250; Fig. 6 (*Escherichia coli*, 4 hours, exposure to human amniotic fluid at 37°C), ×14000.

Man is constantly faced with potentially pathogenic microorganisms in his environment. Bacteria are found in the food he eats, in the air he breathes, upon things he touches, and they colonize his body surface. Because the fetus is an entity that is rapidly undergoing growth and differentiation, it is particularly susceptible to death or defect as a result of encounters with microorganisms. It is important that an understanding of host resistance factors within the fetal environment be gained. Demonstrating that human amniotic fluid is hostile to bacteria is a logical first step because amniotic fluid is easily obtained and because it represents the closest substance to the fetus which may contribute to fetal protection from bacterial invasion. The scanning electron microscope has been found an important avenue for studying deleterious effects of human amniotic fluid upon bacterial cells.

Although very little use of scanning electron microscopy for visualizing bacterial cells appears in the literature, SEM does represent an important tool for some applications. The preparation of specimens requires minimum technical expertise and little time. The procedures for preparing the specimens are also gentle, so that when the specimens are visualized, a three-dimensional view of surface structure is seen in which morphology of the cells is intact and spatial arrangement of the cells is preserved as well. Fig. 1 demonstrates the long bacillary shape of *Clostridium perfringens,* the organism often responsible for gas gangrene. The contrasting morphology of the spherical *Staphylococcus aureus,* in Fig. 2, shows the grape-like clustering of this organism which distinguishes it from other pyogenic cocci. Preservation of spatial arrangement is apparent in Fig. 3, which shows chaining of a group B *Streptococcus*. Intercellular bridges (IB) occurring between cells are evident.

Fig. 4 shows gram-negative *Escherichia coli* to have a rough or fluffy appearing surface, as opposed to gram-positive *Clostridium* in Fig. 1, which appears to have a smooth contour. The gram-positive cell wall is characterized by a thick and homogeneous layer of peptidoglycan, and the wall of the gram-negative cell is composed of multiple layers of peptidoglycan, protein, lipopolysaccharide, and lipoprotein. The differences in cell wall chemistry between gram-positive and gram-negative bacteria are associated with differences in sensitivity to antibacterial agents. For example, gram-positive bacteria are more sensitive to penicillin, as penicillin interferes with incorporation of muramic acid into mucopeptide during cell growth and wall formation. Another detail illustrated in Fig. 4 is septum (Se) formation.

Surface alteration of *E. coli* resulting from exposure to human amniotic fluid is illustrated in Figs. 5 and 6. After two hours' exposure to amniotic fluid (Fig. 5) some cells are elongated. Careful examination of elongated cells (EC) reveals partial septa formed at various points along the length of the elongated cell suggesting that amniotic fluid prevents completion of reproduction although growth (increase in cell mass) continues. After four hours' exposure to amniotic fluid (Fig. 6), elongation is still evident, but more striking is the finding that cells are being ruptured (RC) and losing their characteristic shape.

Although it is not possible to ascertain from this study which factor in amniotic fluid is causing the destruction of bacteria, it is clear that amniotic fluid represents a hostile medium to *E. coli*. (Micrographs kindly provided by Drs. B. LARSEN and R. P. GALASK. Figures used by permission of C. V. Mosby Co., St. Louis, Mo.)

References

GALASK, R. P., LARSEN, B., and SYNDER, I. S.: "Amniotic fluid induced surface ultramicrocytology of *Escherichia coli*." Amer. J. Obstet. Gynecol. In press.
LARSEN, B., SCHLIEVERT, P., and GALASK, R. P.: "The spectrum of antibacterial activity of human amniotic fluid determined by scanning electron microscopy." Am. J. Obstet. Gynecol. **118**, 921—926 (1974).

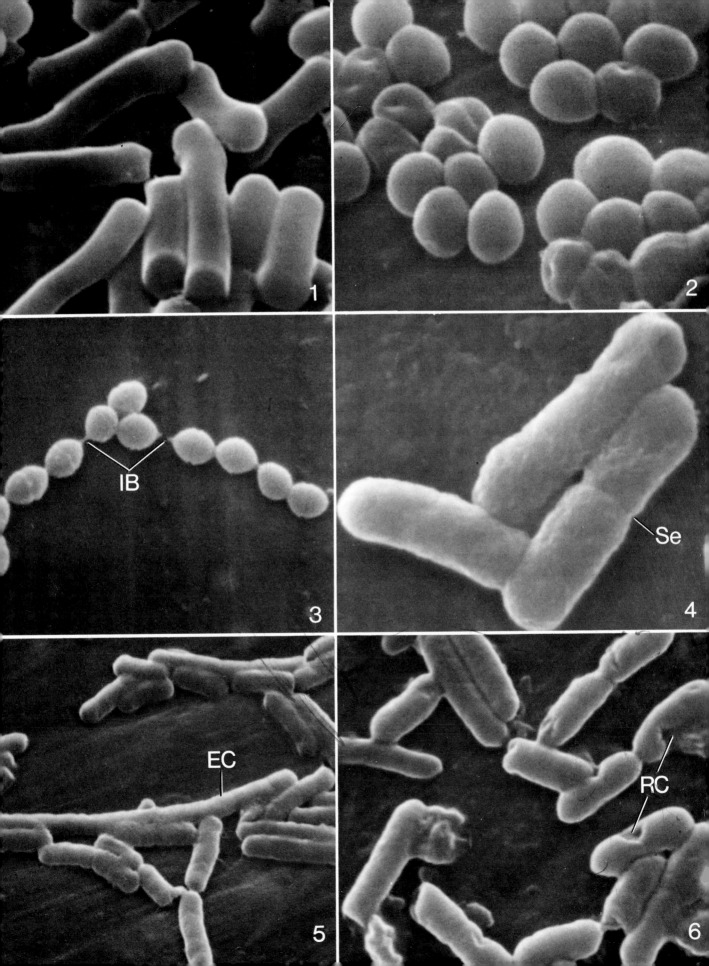

Blue-green Algae

Division Cyanophycophyta

Order Aormogonales

Anabaena sp.

Ak   Akinete
Gr   Granules
He   Heterocyst
Nu   Nucleoplasm
PB   Polyhedral bodies
PC   Protoplasmic connection
PT   Photosynthetic thylakoid
Ve   Vesicles

Fig. 1 (vegetative cells and akinete), ×6200; Fig. 2 (heterocyst), ×5600; Fig. 3 (internal structure of a vegetative cell), ×4000; Fig. 4 (internal structure of a heterocyst), ×4000.

Blue-green algae are probably the most primitive chlorophyll-containing plants existing at the present time. There are about 1 500 known species. They contain special pigments, the phycobillins, which give them their characteristic color. Blue-green algae are present in freshwater pools and ponds. They are also among the first organisms to colonize bare areas of rock and soil, as they can adapt to extremely inhospitable environments. Certain blue-green algae, such as *Anabaena,* are able to use elemental nitrogen from the atmosphere to build their proteins, thereby contributing to the nitrogenous content of soil. This process is important in maintaining soil fertility for instance, in rice paddies.

While some blue-green algae occur as single cells, most form colonies which are usually filamentous (Fig. 1). The cells are cylindrical, giving the filament an appearance of a string of beads. The cell wall is composed of at least two layers. The inner layer is similar to that present in some bacteria (Fig. 2). The outer layer is a mucilaginous sheath (Fig. 1), which is often intensely colored. The cells have no organized nucleus (prokaryotic), no chloroplasts, and no mitochondria. The nucleoplasm (Nu) is dispersed in the cytoplasm and so are the photosynthetic thylakoids (PT), as revealed by the transmission electron micrograph (Fig. 3). Other structures include polyhedral bodies (PB), granules (Gr), vesicles (Ve), and ribosomes.

The chief means of reproduction among the blue-green algae is a simple division process in which the thallus is subdivided into two or more units. In some filamentous genera such as *Anabaena,* a spore-like akinete (Ak) (Fig. 1) develops directly from the differentiation of a vegetative cell. The akinete is considerably larger than the ordinary cell, has a thickened external wall, and contains food reserves. Akinetes can be extremely resistant to desiccation and high temperatures. On germination, the akinete develops into a new filament within the old cell wall. Another specialized structure in *Anabaena* is the heterocyst (He) (Figs. 2 and 4). It is reported that the heterocysts have a higher rate of respiration than the vegetative cells and may be involved in nitrogen fixation. They also function to break up the filament (Fig. 2). The contents of heterocysts are uniform in density and appear clear when viewed with the light microscope. Under the electron microscope, however, it is seen to contain loose photosynthetic lamellae (Fig. 4). A delicate protoplasmic connection (PC) extends through the pore of the heterocyst to the adjacent vegetative cell.

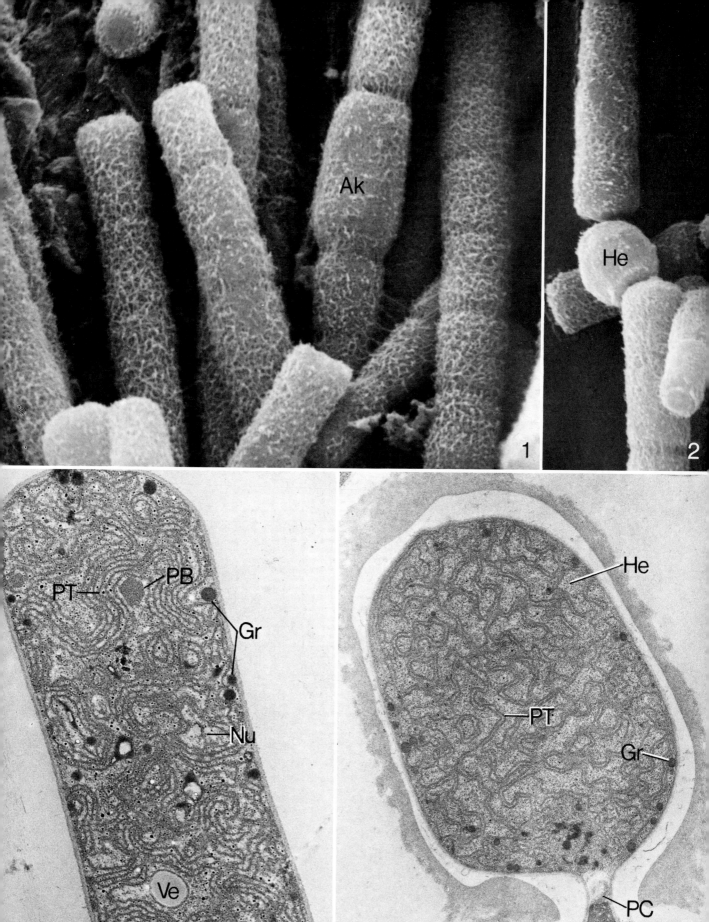

Chapter 5
Fungi and Algae

Slime Molds

Division Myxomycota

Class Acrasiomycetes—Order Acrasiales

Dictyostelium discoideum

Ag   Aggregation of amoebae
Fp   Filopodial-like projections
Ps   Pseudoplasmodium (slug)
Pse   Pseudopodium

Fig. 1, ×7900; Fig. 2, ×3600; Fig. 3, ×195; Fig. 4, ×370.

Most slime molds live in cool, shaded, moist places, as in decaying wood, dead leaves, or other damp organic matter. Taxonomists disagree as to whether the organization of slime molds is more plant-like or animal-like; they possess characteristics of both kingdoms. Zoologists consider them Protozoa because they begin life as individual amoeboid cells. Botanists classify them as plants (myxomycetes) because they reproduce by spores (a plant-like characteristic) and because these spores are reported to have cellulose walls.

The life cycle of *Dictyostelium* consists of a number of major recognizable transitions, including germination of spores into amoebae, aggregation of amoebae into a multicellular pseudoplasmodium, migration of the multicellular aggregate, and the differentiation of this multicellular aggregate into a mature fruiting body (Figs. 1 through 8). This system is especially useful to those scientists concerned with various problems associated with cell differentiation. Further, in this life cycle the events of cell division and cell differentiation are separate phenomenon.

Stages illustrated in the life cycle were grown in plastic petri dishes with a substrate consisting of either agar or filter paper. The organisms were fixed for SEM using (1) osmium tetroxide fumes (Figs. 3 through 5) (2) 1.5% glutaraldehyde in 0.05 M sodium phosphate buffer (Fig. 2), or (3) 2% osmium tetroxide in 0.1 M sodium phosphate buffer (Fig. 1). The myxamoebae that hatch from slime-mold spores are small (approximately 10 μm in diameter) and irregular, but are provided with blunt and active pseudopodia. The free myxamoebae in many cases have a gently scalloped surface (Fig. 1). Some cells have numerous, long, slender filopodial-like projections (FP) from their surface, whereas others have several, shorter but much broader pseudopods (Pse) extending from their surfaces (Fig. 1). Some amoebae have both types of surface projections. When the myxamoebae have food (bacteria), they grow and divide by binary fission. Dur-

ing this time each cell maintains an independent existence.

When the food supply is consumed or withdrawn, the amoebae become attracted to each other and begin to aggregate into numerous "colonies," each consisting of several thousands or several hundreds of thousands of cells (Fig. 2). This aggregation of amoebae results, in part, from the cell surfaces becoming "sticky". As the migratory amoebae become incorporated into the periphery of an aggregate (Ag in Fig. 2), they become flattened and epithelioid in appearance. The cells are closely organized in the colony, but the cell margins have many small finger-like projections (microvilli) which contact adjacent cells in the colony.

The aggregation stage is closely followed by a so-called slug stage (Fig. 3), characterized by a condition in which the aggregation of amoebae (called a pseudoplasmodium) becomes a motile mass of cells. A true plasmodium is a multinucleate mass of cells not divided by cell membranes, as is illustrated later in *Fuligo septica*. The pseudoplasmodium (Ps in Fig. 3) of amoebae continuously secretes a mucopolysaccharide slime covering that remains behind as the pseudoplasmodium migrates. This slug is sensitive to light, temperature, and humidity gradients.

A decrease in humidity of moisture can result in the cessation of slug movement and a vertical column begins to extend from the pseudoplasmodium (Figs. 4 and 5). An initial phase of this process is called the "Mexican hat" stage (Fig. 4).

98

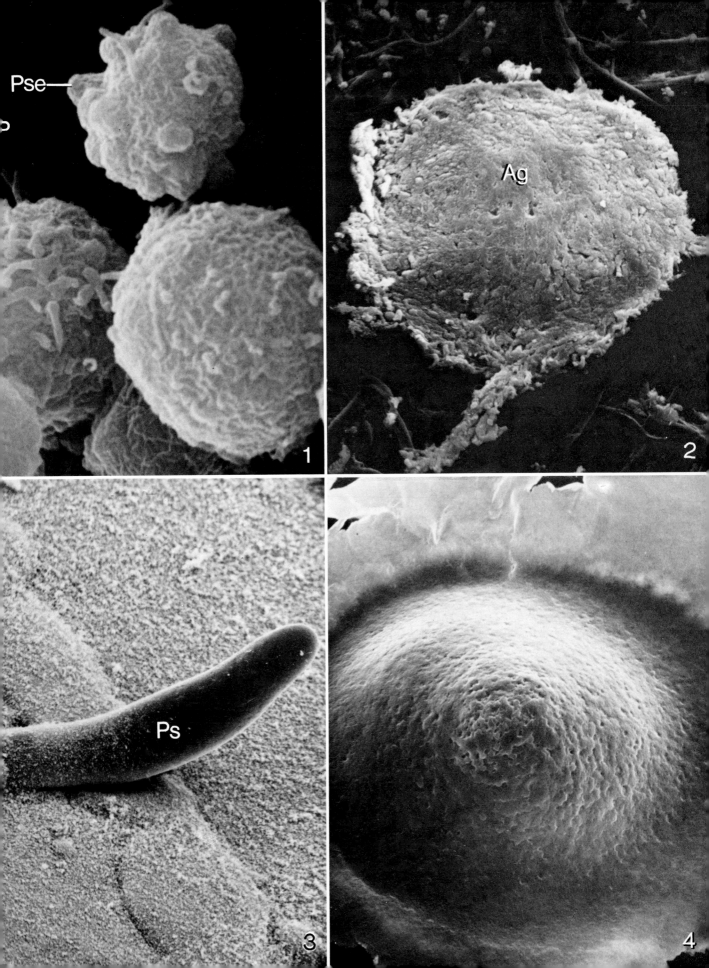

Division Myxomycota

Class Acrasiomycetes—Order Acrasiales

Dictyostelium discoideum
(continued)

B   Base (or basal disc)
Sp   Sporangium
St   Stalk

Arrows: Spores on sporangium (in Fig. 8)

Fig. 5, ×490; Fig. 6, ×430; Fig. 7, ×180; Fig. 8, ×415.

The column that arises from the mass undergoes a series of morphological changes (Figs. 5 through 7), wich eventually result in a mature fruiting body consisting of a basal disc, or base (B in Fig. 7), a long slender stalk (St in Figs. 7 and 8), and a round tip, or sporangium (Sp in Fig. 8), which contains numerous spores (arrows in Fig. 8). The transformation from slug to fruiting body can be influenced by environmental factors such as ionic strength of the medium, pH, and light intensity. The entire developmental cycle can be made to be highly synchronous and reproducible, and under appropriate conditions the slug stage may be omitted.

The cells that were originally at the front of the moving slug become stalk cells while those cells originally at the back end of the slug will become spores of the fruiting body. Thus, formerly identical amoeboid cells comprising a migratory pseudoplasmodium become differentiated into two distinct cell types. The cells at the apex of the extending column (Fig. 5) form stalk cells which synthesize a cellulose-like product. These cells turn inward and push down toward the base through the remainder of the cell mass. The cylinder increases in height as additional cells migrate up the stalk. The cells that will form spores are carried upward on the stalk and become smaller, dry, and rounded. At this time they secrete a covering of slime and a hard protective outer coat of cellulose. In Fig. 8 the cellulose-like covering has been removed, revealing the numerous spores (arrows) associated with the rounded tip or sporangium of the fruiting body. Each of these spores, when shed under appropriate environmental conditions, will hatch or develop into an individual amoeba.

References

ASHWORTH, J.M.: "Cell development in the cellular slime mold *Dictyostelium discoideum.* " Symp. Soc. Exp. Biol. **25**, 27—49 (1971).
GERISCH, G.: "Cell aggregation and differentiation in *Dictyostelium.*" Current Topics Develop. Biol. **3**, 159—197 (1968).

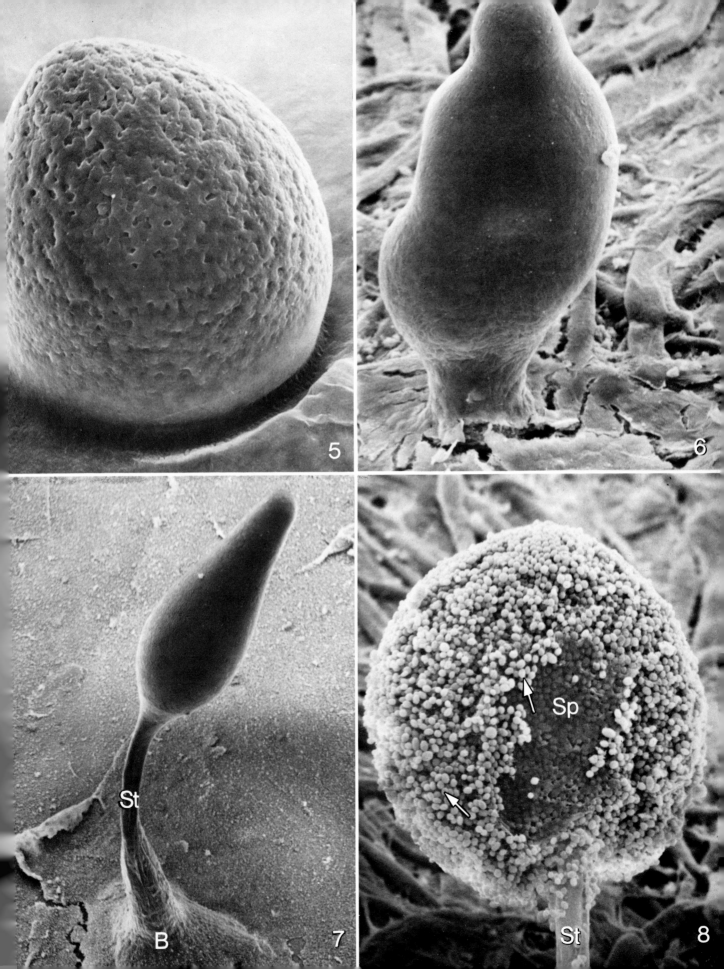

5

6

7

B   St

8

Sp

St

Division Myxomycota

Class Myxomycetes—Order Physarales

Fuligo septica (True Slime Molds)

All true slime molds have a naked, acellular, assimilative body called a plasmodium. It is the main vegetative structure in the life cycle of Myxomycetes. The plasmodium is multinucleate and lacks rigid walls. Some species can form conspicuous protoplasmic sheets many centimeters in extent. Plasmodia can be colorless, white, yellow, or brown, and less frequently violet or black.

Plasmodia are motile by means of slow amoeboid movement. The movements are oriented responses which are generally positive toward increasing moisture and nutrients and negative toward light. As plasmodia move, they ingest mold spores, bacteria (Ba), and other small particles in their path (Fig. 2). The undigested debris is egested at the trailing end. This debris together with secreted slimy material forms a conspicuous trail.

In form, the plasmodium may be amoebae-like, slug-like, or a vein-like network. In *Fuligo* and related genera the plasmodium is a fan-shaped network during migration (Fig. 1). The strands at the posterior end of such a form are relatively large and less branched, while anterior strands are finer and more numerous. Through microscopic examination two zones can be distinguished in the plasmodium. The peripheral layer is transparent and the inner strand is granular. The granular cytoplasm undergoes rhythmic streaming, which is characteristic of all myxomycete plasmodia.

A plasmodium sometimes fragments into two sections. The two sections may again form a single plasmodium if they meet during creeping. Fragmentation seems to be a common method of plasmodial reproduction in nature. Under certain unfavorable conditions such as drying or low temperatures the plasmodium may be converted into a resistant structure called a sclerotium. It is a hardened mass of irregular form consisting of many cell-like compartments. Sclerotia retain their viability for several years and reform plasmodia when favorable conditions return. With other changes in environmental conditions such as temperature, light, and pH, the slime molds will form fungus-like fruiting bodies. Within the fruiting body numerous spores form and are eventually released.

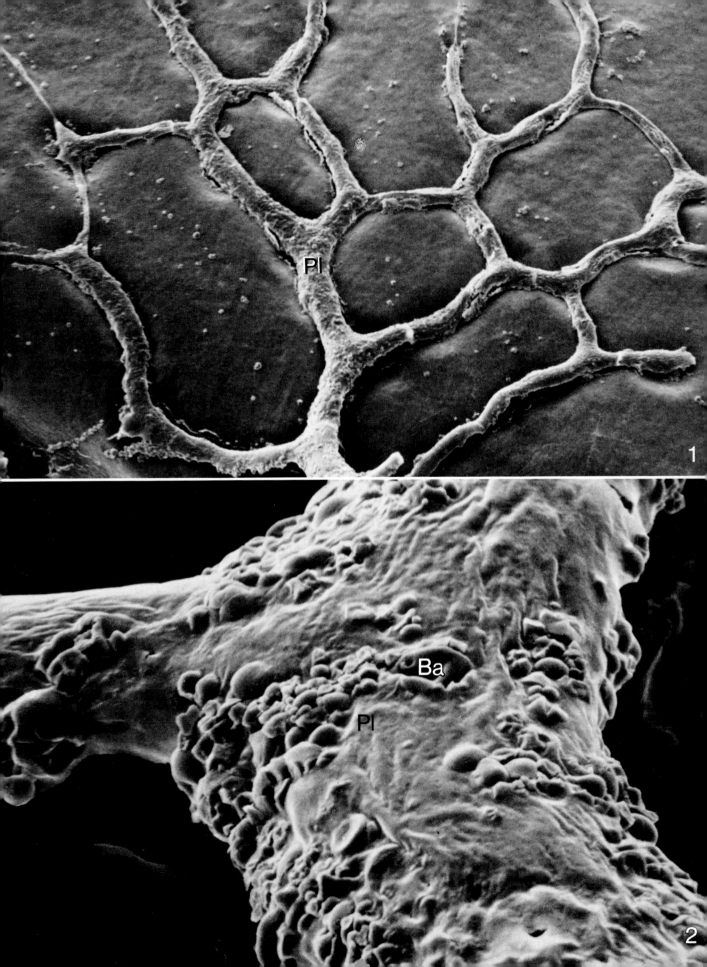

Fungi

Division Eumycota

Class Hemiascomycetes—Order Endomycetales

Sacharomyces cerevisiae (Yeast)

Bu Bud

Fig. 1, ×12375; Fig. 2, ×15000; Fig. 3, ×15000; Fig. 4, ×13125.

Yeast is a single-celled fungus and one of the most important domesticated organisms. It can break down glucose into carbon dioxide and alcohol. The release of carbon dioxide from fermentating yeast cells causes uniform aeration of bread dough, and the transformation of glucose to alcohol turns the juice of fruit and grains into wine and beer. In addition to its use in baking and brewing, yeast is also used to produce vitamin D, ephedrine, enzymes, and many other useful substances.

Asexual reproduction of yeast occurs by budding. When yeast cakes or powdered yeast are added to a sugar solution, the unicellular plants can soon be seen producing smaller buds, which grow (Figs. 1 through 4) and eventually break off. Internally, the nucleus undergoes division during bud formation, and one daughter nucleus migrates into the developing bud. In Fig. 1 it appears that in the budding process a localized softening of the wall occurs, followed by a protoplasmic blow-out. The naked protoplasm is quickly covered by cell wall material. Studies have shown that the wall softening results from enzymatic reduction of disulfide bonds (—S—S) to form sulfhydral groups (—SH) in the wall (NICKERSON, 1963).

S. cerevisiae is heterothallic. Two mating strains are needed to form a zygote, but budding of this diploid zygote occurs and may continue indefinitely. Under certain conditions a single diploid cell can function as an ascus in which meiosis occurs and from which four ascospores develop. The spores are eventually released by breakdown of the ascus wall. When the ascospores germinate, buds are again produced.

Yeast was discovered by PASTEUR in 1860 when he was hired to find a way to make good wine successfully. Later, he postulated from his experiments that fermentation occurred only when the yeast was present. LIEBIG, an influential chemist, thought otherwise and attempted many experiments in which he killed a suspension of yeast cells and then hoped to show the production of alcohol by the fermentation of sugar. He was unsuccessful. The idea was thus developed that fermentation was caused by substances found only in living yeast, and the word enzyme (meaning "in yeast") was adapted.

Reference

NICKERSON, W. J.: "Molecular bases of form in yeasts. Symposium on Biochemical Bases of Morphogenesis in Fungi. IV." Bact. Rev. **27**, 305—324 (1963).

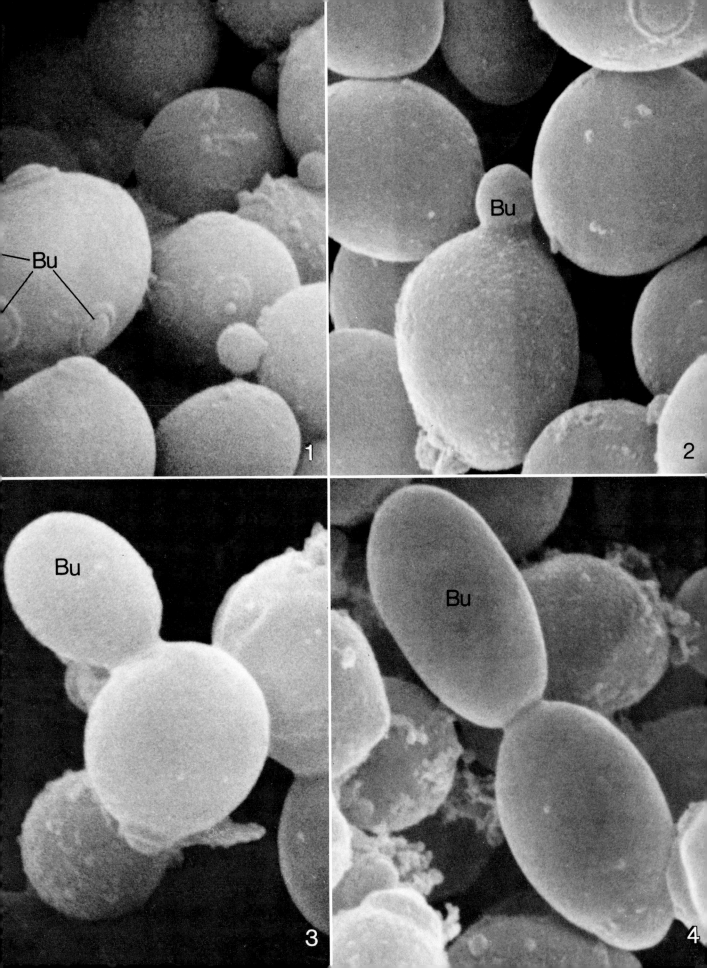

Division Eumycota— Subdivision Ascomycotina

Order Eurotiales

Aspergillus and *Penicillium* (Molds)

Br	Bread	My	Mycelium
CB	Connecting bridge	PS	Primary sterigmata
Co	Conidium	S	Sterigma
Cp	Conidiophore	SS	Secondary sterigmata
CT	Conidium-producing tube	Ve	Vesicle

Fig. 1 *(Aspergillus niger)*, × 175; Fig. 2 *(Aspergillus nidulans)*, × 1650; Fig. 3 *(Aspergillus nidulans)*, × 3200; Fig. 4 *(Penicillium frequentans)*, × 5100.

Many of those fungi called blue and green molds belong to two of the best known genera, *Aspergillus* and *Penicillium*. *Aspergillus* is one of the contaminants of bread (Br in Fig. 1), and it also grows on a wide variety of foods and materials. The fungi are also used in industry. *A. niger* (Fig. 1), a bread mold, is used commercially in citric acid fermentation, and *A. oryzae* has been used for centuries in the Orient for various food and other fermentations. *Aspergillus* is also known to produce disease in many animals including man. For example, *A. fumigatus* causes pulmonary disease while *A. nidulans* (Figs. 2 to 3) has been isolated from diseased fingernails.

Penicillium is abundant in soil and on all kinds of decaying materials. For example, *P. frequentans* (Fig. 4) is believed to play a role in decomposition of vegetable residue in nature. *Penicillium* such as *P. roquefort* and *P. comemberti* also give cheeses the flavor highly prized by gourmets. The spores of *Penicillium* are almost universally present in air and germinate whenever substrata and conditions are suitable for growth. It is a frequent contaminant of cultures which led FLEMING to the discovery of the antibiotic penicillin in 1944. When a culture plate of *Staphylococcus* was contaminated by spores of *Penicillium notatum*, the colonies of the bacteria close to the contaminating mold underwent dissolution. Later it was found that the mold produced a compound that prevented the synthesis of mucopeptide, a polymerized compound responsible for the rigidity of the bacterial cell wall. Since animal cells do not have cell walls, this selective action of the antibiotics was adapted in medicine to kill many bacteria that infect humans.

The mycelium (My in Figs. 1 and 2) of the fungus is septate with perforated cross-walls. The cells are either uninucleate or multinucleate depending on species. The hyphal walls are chitinous. Asexual reproduction begins with the enlargement of certain cells in the mycelium into thick-walled foot cells. Each of these cells produces a single conidiophore (Cp in Figs. 1 to 2) as a vertical branch and ultimately forms a globose vesicle (Ve in Fig. 3) at its apex. The vesicle then gives rise to one or two layers of conidium-producing cells, the sterigmata. Each of the first series of cells, or primary sterigmata (PS in Fig. 3), can bear two or more secondary sterigmata (SS in Fig. 3). The actual spore-producing cells, or sterigma (S in Fig. 4), ordinarily consists of a cylindrical body that narrows at its apex into a conidium-producing tube (CT in Fig. 4). The nucleus in the sterigma divides and one of the daughter nuclei passes into the conidial tube that is then separated from the sterigma by a septum to form a conidium (Co in Figs. 3 and 4). With repeated divisions of the sterigma nucleus, new spores are formed and the older spores are pushed outward to form a spore chain. After the conidium is separated by means of a septum from the sterigma, it normally continues to draw nutrients from the parent cell until it assumes the general size and shape characteristic of the species. In the chain, conidia remain attached to each other by a connecting bridge (CB in Fig. 4) that may be conspicuous or almost invisible. Each spore when shed can germinate to form a mycelium when the environment is suitable.

Most species in genera *Aspergillus* and *Penicillium* reproduce only by means of conidia, i.e., *A. niger* and *P. frequentans*. They are considered to be imperfect and are grouped into a class called Fungi Imperfecti. Because their conidia are very similar to those of the Ascomycetes it is believed that most of them are conidial stages of Ascomycetes whose ascus stage no longer exists or has not been discovered. A number of species such as *A. nidulans* do form an ascus stage which involves the fusion of nuclei, meiosis, and formation of eight ascospores in each ascus.

Division Eumycota—
Subdivision Basidiomycotina

Class Hymenomycetes—Order Agaricales

Agaricus bisporis (Mushroom)

Ba   Basidia
Bs   Basidiospores
Cu   Cup
Gi   Gill

Fig. 1 (gills), ×338; Fig. 2 (basidiospore), ×840; Fig. 3 (cross-section of a gill), ×860; Fig. 4 (surface of a gill), ×3525.

Basidiomycetes consists of 25000 species including mushrooms, toadstools, puff balls, rusts, smuts, and bracket fungi. They are most commonly seen in fields, forests, wooded stream banks, and as parasites on many vital economic plants. They reproduce sexually by an enlarged, club-shaped hypha cell called a basidium. Like an ascus, the basidium is a cell in which nuclear fusion and meiosis occur. However, the spores are formed external to the basidium whereas ascospores develop within the ascus. Each basidiospore when released will develop into a new mycelium when it reaches a proper environment.

The main vegetative phase of growth in *Basidiomycetes* is by dikaryotic hyphae. Dikaryotic hyphae are established by hyphal fusions between different strains. Once formed, the dikaryon usually is long-lived. The mass of these white, branching, thread-like hyphae grow mainly below ground. After a time compact masses of hyphae called buttons appear at intervals on the mycelium, and they grow into fruiting bodies commonly known as mushrooms. The fruiting body consists of a stalk (stipe) and an umbrella cup. On the underside of the umbrella cup (Cu in Fig. 1) many thick perpendicular plates called gills (Gi) extend radially from the stalk to the edge of the cap. The basidia (Ba in Fig. 3) develop on the surface of these gills. Each basidium contains two nuclei which fuse to form a diploid nucleus. This, in turn, divides by meiosis to form two or four haploid basidiospores (Bs in Figs. 2 and 4).

The term "mushroom" usually refers only to edible *Basidiomycetes* that have a fruiting body differentiated into cap, stipe, and gills. Toadstool usually refers to morphologically similar but inedible varieties. However, there is no clear dividing line between the two groups, since a fungus that is poisonous to one person may not affect others. In addition, many edible mushrooms are closely related to poisonous ones.

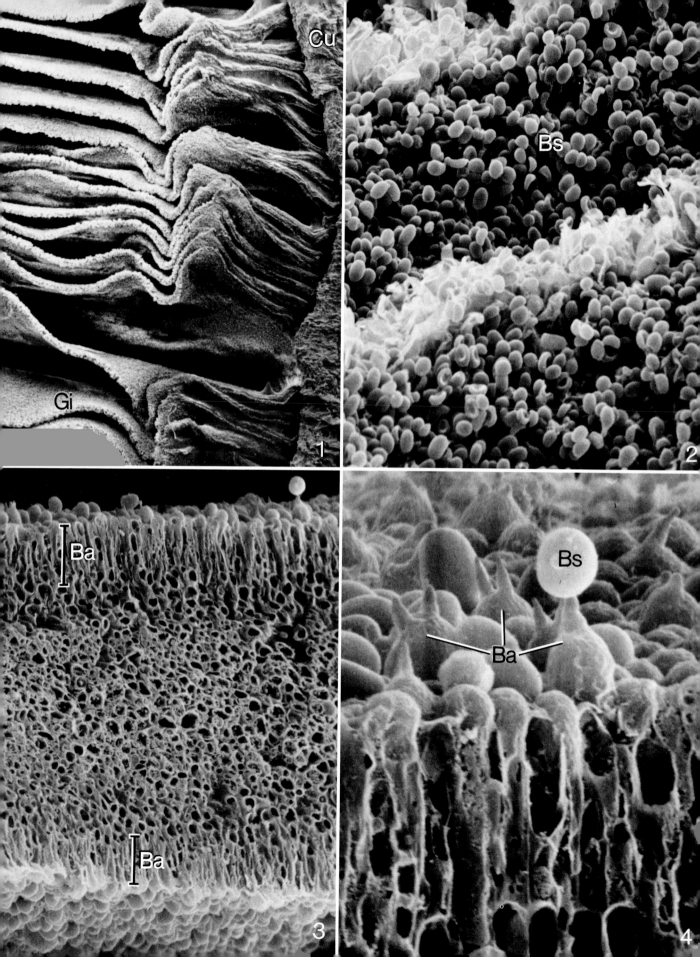

Algae

Division Chlorophycophyta

Class Chlorophyceae—Order Volvocales

Chlamydomonas rhinhardtii
and *Pandorina morum*

Green algae are extremely diversified morphologically and range from unicellular, motile, or nonmotile forms to multicellular, branched filaments. The characteristic green color of this group is due to the presence of chlorophylls a and b. In addition, there are carotenoid pigments, some of which are the same as those in higher plants. Most of the green algae are autotrophic, with starch as their carbohydrate reserves.

Chlamydomonas, a motile unicellular organism, is one of the most common green algae in nature. *Chlamydomonas* has a relatively simple cellular structure. Motility is provided by two anteriorly inserted, whip-lash flagella (Fl) (Fig. 2). Two contractile vacuoles occur in the cytoplasm near the bases of the flagella. A large cup-shaped chloroplast is the most conspicuous intracellular structure. One or more pyrenoids associated with starch synthesis occur in the basal part of the chloroplast. A carotenoid-containing eyespot is also present at the anterior end and is thought to be involved in phototropic behavior.

Although simple in structure, *Chlamydomonas* possesses all the metabolic machinery necessary for photosynthesis, growth, and reproduction. Since only carbon dioxide, water, light and a few inorganic ions are needed for growth, *Chlamydomonas* is a very useful research organism. In recent years it has been used to advantage in physiological, genetic, and biochemical research.

During asexual reproduction the nucleus usually divides twice mitotically, and the cell divides to form four separate cells within the parent cell. The daughter cells are released when the parent cell wall ruptures. Under certain environmental stimuli, cells of *Chlamydomonas* become nonmotile, forming the so-called palmelloid state, and they become embedded in a gelatinous matrix (Fig. 1). Usually the cells in this stage are nonflagellated. Flagella reappear and the cell swims away when favorable conditions return.

In sexual reproduction when cells of opposite mating types are mixed, gamete pairing is initiated by flagellar cohesion and is followed by gamete fusion. Ultimately the zygotes lose their flagella, secrete a thick wall, and become dormant. Meiosis occurs during the early stages of zygospore germination and four haploid, flagellated meiospores are produced.

In the volvocine line of green algae a series of colonial forms has evolved from the *Chlamydomonas*-like type. This is exemplified by *Pandorina* in Fig. 3. In the colonial forms each cell in the colony is typically similar to motile vegetative *Chlamydomonas* (Fig. 4). The number of cells comprising colonial forms is usually constant. *Pandorina* is a spherical or ellipsoidal colony which may consist of 8, 16, or 32 tightly packed cells. All cells of the colony are capable of asexual and sexual reproduction. During asexual reproduction, each cell of the parent colony divides repeatedly to form a daughter colony. Upon breakdown of the parent colony, daughter colonies are liberated and the cells are held together by a matrix. Cells of the daughter colonies then enlarge to the adult size and no further cell division occurs.

Gamete formation in *Pandorina* resembles daughter-colony formation except that the gametes are liberated from the colony. Details of gamete pairing and fusion are similar to those described for *Chlamydomonas*. Germination of zygospores by meiosis results in the formation of four flagellated cells. Each of the cells then gives rise to a new colony by means of asexual reproduction.

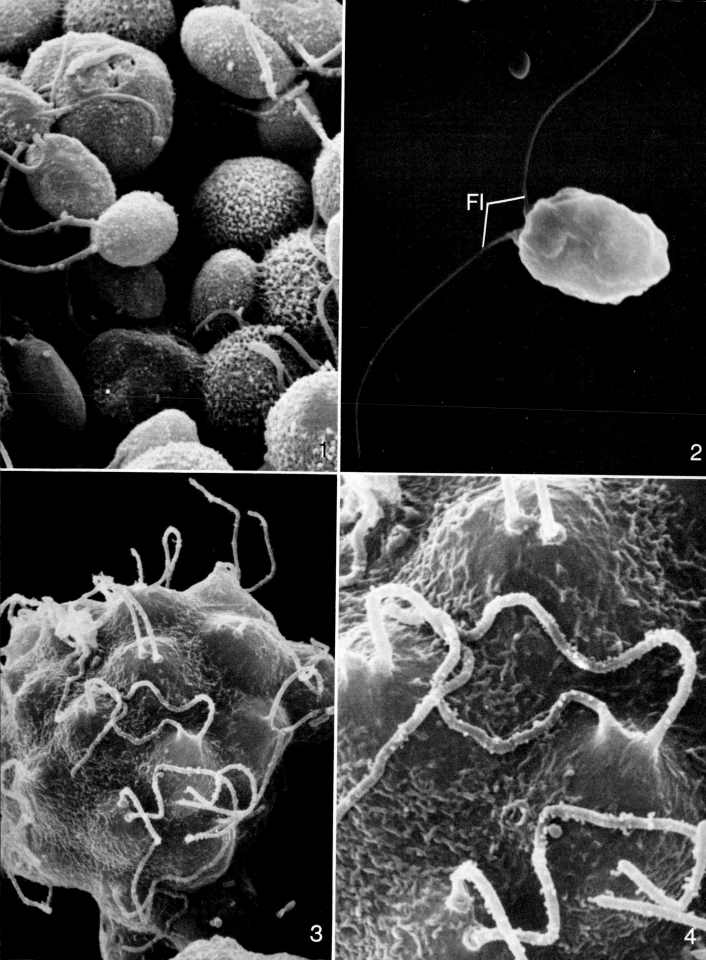

Division Chlorophycophyta

Class Chlorophyceae—Order Zygnematales

Staurastrum, Closterium and *Cosmarium* (Desmids)

Sc   Semicell
Sp   Spine

Fig. 1 *(Staurastrum)*, ×1650; Fig. 2 *(Staurastrum)*, ×6500; Fig. 3 *(Closterium)*, ×310; Fig. 4 *(Cosmarium)*, ×1500.

Commonly found free-floating in ponds and streams, desmids are widely distributed freshwater algae. The cells are uninucleate and composed of two halves with a constriction in the middle of cells. They may be spiny (Sp), such as *Staurastrum* (Fig. 1 and 2); curved, such as *Closterium* (Fig. 3); or ellipsoidal, such as *Cosmarium* (Fig. 4). A single nucleus lies embedded in the middle of the cytoplasm that separates the two chloroplasts into semicells (Sc).

Vegetative reproduction is by cell division. After nuclear division the two semicells separate, and each regenerates itself to produce two new identical daughter cells. Sexual reproduction is by fusion of somewhat amoeboid gametes. The zygote is again a resting cell. No flagellated cells are produced by any member of this order.

112

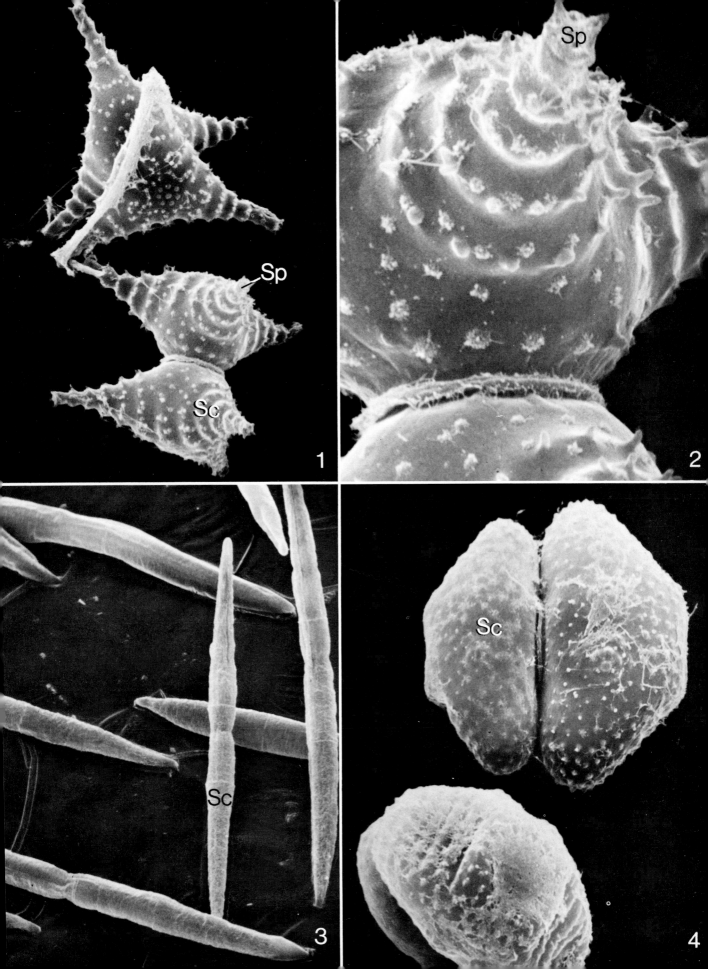

Division Chlorophycophyta

Class Chlorophyceae—Order Oedogoniales

Oedogonium cardiacum

An   Antheridium
Ca   Cap
CC   Cap cell
Oo   Oogonium
VC   Vegetative cell

Fig. 1 (vegetative filaments), ×355; Fig. 2 (cap), ×2300; Fig. 3 (oogonium), ×1750; Fig. 4 (antheridium), ×1875.

Oedogonium is a widespread freshwater alga that grows epiphytically in ponds and pools. The filaments are unbranched and consist of cylindrical vegetative cells (VC in Fig. 1). Each cell is covered with a thick wall, which is differentiated into an inner cellulosic, middle pectic, and outer chitinous layer. The cells are uninucleate and when mature have a central vacuole containing cell sap as well as an elaborate reticulate chloroplast. *Oedogonium* has a unique feature in cell division that leads to the formation of caps on certain cells. When the cell divides, the nucleus moves from a lateral position to the center of the cell. Soon a transverse ring of wall material forms on the inner lateral wall just below the anterior end of the cell. The ring grows in thickness and forms a groove while the nucleus divides, and a floating septum is formed between the daughter nuclei. The middle and outer wall layers external to the groove then rupture, allowing the ring to extend so as to form a new piece of cell wall. Ultimately, the floating septum becomes fixed near the joint of the old and new wall. The cell wall may rupture several times at the same spot so that a striation strip called a cap (Ca) (Fig. 2) is formed. The cell bearing it is called a cap cell (CC).

In sexual reproduction the gametes are produced in specialized gametangia. A single egg is produced in an oogonium (Oo in Fig. 3), while multiflagellate sperm are produced in the narrow antheridial cells (An in Fig. 4). The asexual reproduction of *Oedogonium* is by formation of a zoospore within a cap cell.

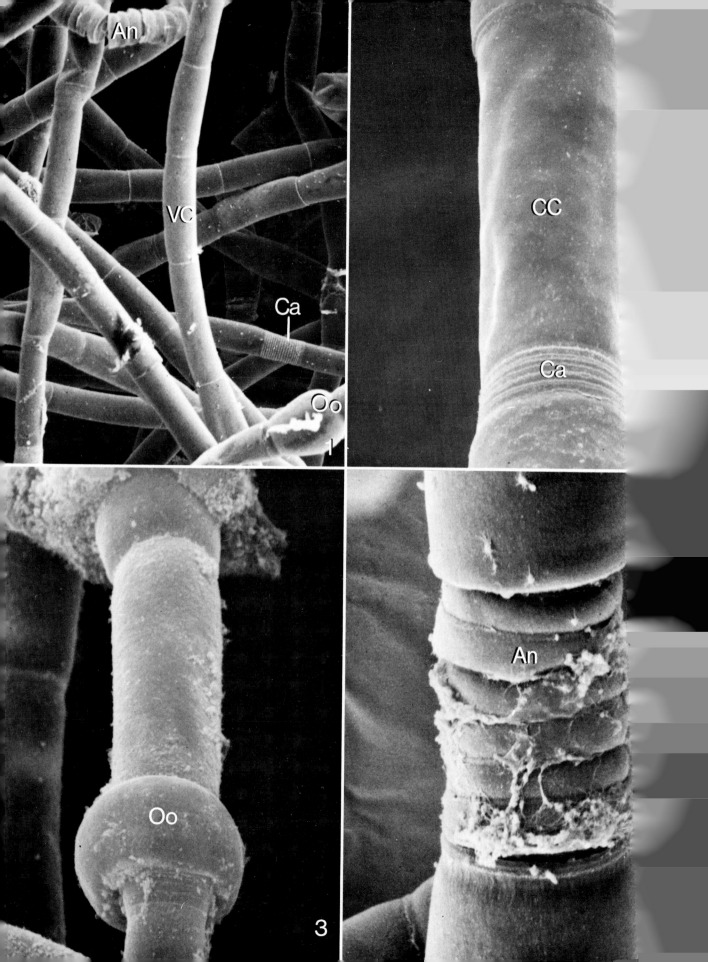

Division Chrysophycophyta

Class Bacillariophyceae

Diatoms

CB   Connecting band
No   Nodule
Ep   Epitheca
Hy   Hypotheca
Po   Pore
Ra   Raphe

Fig. 1 *(Pinnularia)*, ×2310; Fig. 2 (inner view of epitheca), ×1600; Fig. 3 *(Biddulphia)*, ×370; Fig. 4 *(Biddulphia)*, ×340; Fig. 5 *(Cyclotella)*, ×780; Fig. 6 *(Triceratium)*, ×1060.

The diatoms are conspicuous elements of both present-day marine and freshwater environments and are extensively represented in the fossil record. There are about 190 genera with over 5500 species. Diatoms resemble golden algae in the possession of both chlorophyll a and fucoxanthin. In addition, diatoms possess chlorophyll c.

The most striking characteristic of diatoms is the presence of a silicified cell wall which consists of two overlapping portions fitted together like two halves of a box. The half that partially overlaps the other is called the epitheca (Ep); the inner half is the hypotheca (Hy) (Figs. 3 and 4). The joint of the two halves is called connecting band (CB). The siliceous walls are ornamented with fine ridges, lines, nodule (No), and pores (Po). The markings are either radially or bilaterally symmetrical on either side of the long axis of the cell. The cell shape and ornamentation have been used to classify diatoms. Those bilateral symmetric cells appearing more rectangular in valve view are pennate diatoms; for example, *Pinnularia* (Figs. 1, 3, and 4). Those radial symmetric cells appearing round or triangular in valve view are classed as centric diatoms and include *Cyclotella* (Fig. 5) and *Triceratium* (Fig. 6).

Diatoms are capable of a slow gliding movement, apparently produced by the streaming of cytoplasm through the raphe (Ra) on the surface of the cell wall (Figs. 1 and 2). Diatoms store their food as the polysaccharide leucosin and as oil rather than as starch. It is widely believed that petroleum is derived from the oil of diatoms that lived in past geologic ages.

Diatoms reproduce sexually or asexually. The presence of the hard silica-containing wall complicates the process of asexual reproduction. When the diatom cell divides, two cells are formed within the original cell wall, and then two new cell walls form, back to back, between the two cells. Thus, each daughter cell ends up with two cell walls, with one inherited from the parent and a new one that fits inside the old one. As a result, each successive generation becomes a little smaller. This decrease in size can be restored by sexual reproduction, which involves the fusion of two protoplasts and the production of new cells. Sexual reproduction may be isogamous, anisogamous, or oogamous. The vegetative cell is diploid and meiosis occurs during gametogenesis. The zygote increases two to three times the size of the original gamete-producing cell so that is has the maximum cell dimension possible in the species.

When diatoms die, the remains of the silica-containing cell walls accumulate as sediments in the oceans. Later, geologic uplifts may bring these to the surface. The diatomaceous earth can be mined and used in making insulating bricks, as a filtering agent, or as a fine abrasive used in toothpaste.

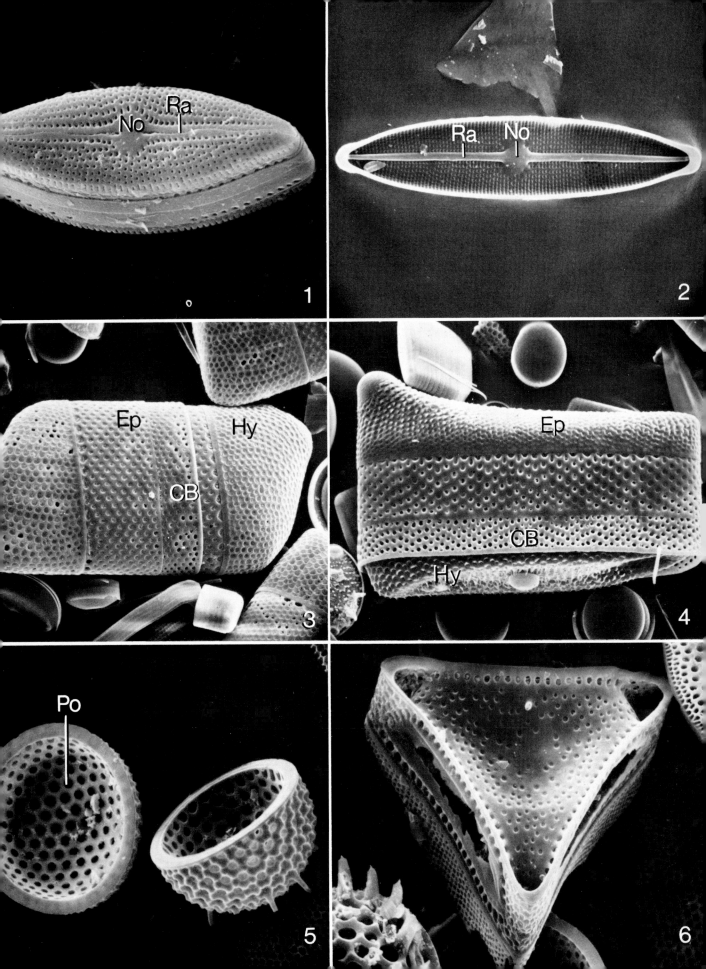

Division Charophycophyta

Class Charophyceae

Chara (Stoneworts)

CC   Crown cell
FG   Female gametangium
In   Internode
MG   Male gametangium
No   Node
SC   Sheath cell

Fig. 1 (shoot apex), ×45; Fig. 2 (reproductive organ), ×119; Fig. 3 (female gametangium), ×112; Fig. 4 (male gametangium), ×245.

The aggregation of unicellular forms into discrete units is quite conspicuous among the various algal divisions. *Chlorophycophyta* are representative of many divergent evolutionary trends beginning with walled and flagellated unicells and culminating with multicellular filaments and even three-dimensional leaf-like plant bodies.

The *Charophyceae* such as *Nitella* and *Chara* are quite complicated in structure. They occur submerged and attached to the bottom of the freshwater ponds. This group is morphologically distinguished by an apical growth and the differentiation into nodal (No) and internodal (In) regions (Fig. 1). Most of the cells are uninucleate, but the large internodal cells are often multinucleate. At the base of the thallus, branched rhizoidal filaments are also differentiated.

Vegetative reproduction is by fragmentation of specialized groups of cells, and the chief method of sexual reproduction is oogamous. In *Chara* the oogonium is borne at a node of the thallus on a short, stalk-like cell. Five sterile vegetative sheath cells (SC in Fig. 3) originate beneath the oogonium and grow upward in a spiral fashion to protect the oogonium. At maturity five crown cells (CC) are formed on the top of the sheath cells (Fig. 3). Each oogonium contains a single egg. The male reproductive organ of *Chara* is the most complex of all algal structures. It is also attached by a short stalk cell to the node and appears as a spherical structure (Fig. 4). Sometimes a male and a female gametangia are attached to the same node, with the female organs located above the male organ (Fig. 2). Early in male gametangium development an outer sterile layer of cells and an inner fertile group of cells are differentiated. Attached to the inner surface of the sterile cells is a small cell that forms one or more antheridial filaments. Each cell of the antheridial filament produces a single coiled sperm.

Fertilization occurs after the sperm passes between the protective sheath cells. The zygote secretes a heavy wall and remains surrounded by sheath cells during a resting stage. Upon germination, the zygote divides meiotically to produce four haploid nuclei. Three of these degenerate and the functional nucleus produces filamentous protonema which eventually gives rise to the typical thallus with apical growth.

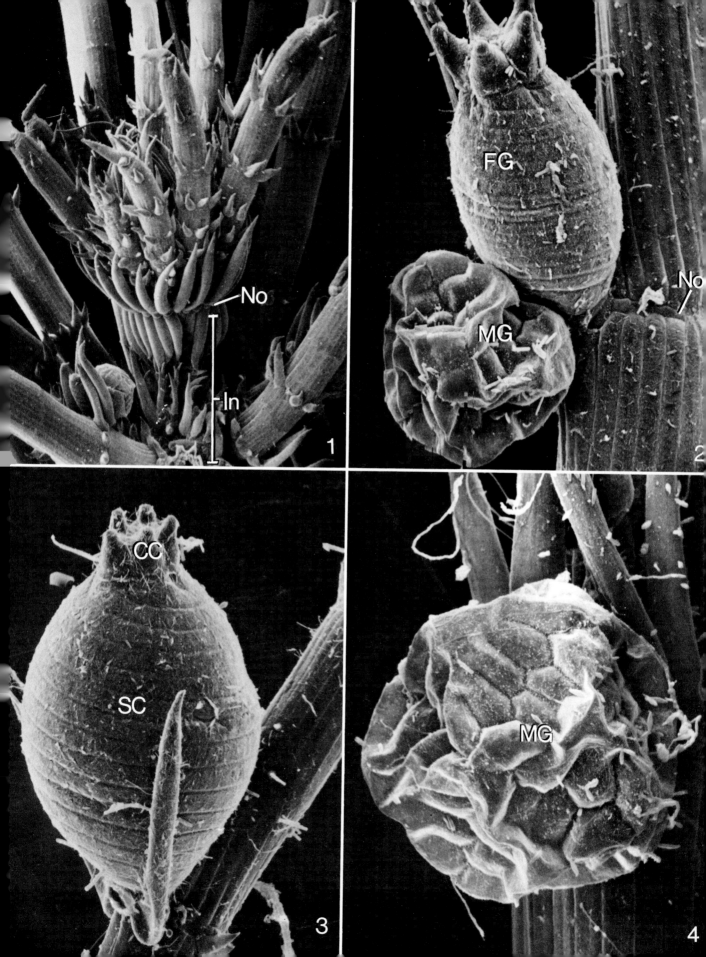

Chapter 6
Multicellular Plants

Liverworts and Mosses

Division Hepatophyta

Class Hepatopsida-Order Marchantiales

Marchantia (Liverwort)

Ep   Epidermis
FT   Food-storage tissue
GC   Gemmae cup
Ge   Gemmae
PF   Photosynthetic filaments
Po   Pores
Rh   Rhizoid

Fig. 1 (gametophyte), ×45; Fig. 2 (section through a gemma cup), ×62; Fig. 3 (cross-section of a thallus), ×210.

The Bryophytes include mosses, liverworts, and hornworts, and probably represent transitional plants through which aquatic algae evolved to become fully terrestrial. In this respect they are evolutionarily comparable to amphibians. The liverwort thallus is a simple and flat, sometimes branched, ribbon-like structure. It lies flat on the ground with one surface exposed to the air. The plant has developed an epidermis (Ep), which is provided with stomata-like pores (Po) to permit gas exchanges (Fig. 1). *Marchantia*, which is considered to be the most evolved of the liverworts, exhibits considerable internal differentiation. Beneath the air pores are air chambers filled with photosynthetic filaments (PF in Fig. 3). A thick layer of food-storage tissue (FT), which lacks chloroplasts, is present in the lower surface of the thallus. Rhizoids (Rh) grow from the cells covering the lower surface and serve to anchor the thallus and are utilized in absorption.

Marchantia reproduces asexually in a specialized way. It produces small cups, known as gemmae cups (GC) on the upper surface of thallus (Fig. 1). Multicellular bodies called gemmae (Ge) form within the gemmae cups (Fig. 2). When mature, the gemmae are dispersed by wind and water, and upon germination they give rise to new gametophytic plants.

Sexual reproduction in *Marchantia* occurs only under wet environmental conditions, a phenomenon that illustrates the transitional nature of these plants. The gametophytes are either male or female. The male thallus develops an antheridiophore consisting of a slender stalk and a disc head that bears antheridia. The female develops an archegoniphore with an eight-lobed head containing archegonia. After fertilization the diploid zygote will develop into a sporophyte. The sporophyte differentiates a foot, which grows into the gametophyte so as to obtain nutrients. When mature, the sporophyte produces haploid spores which will germinate to form the gametophyte thallus.

122

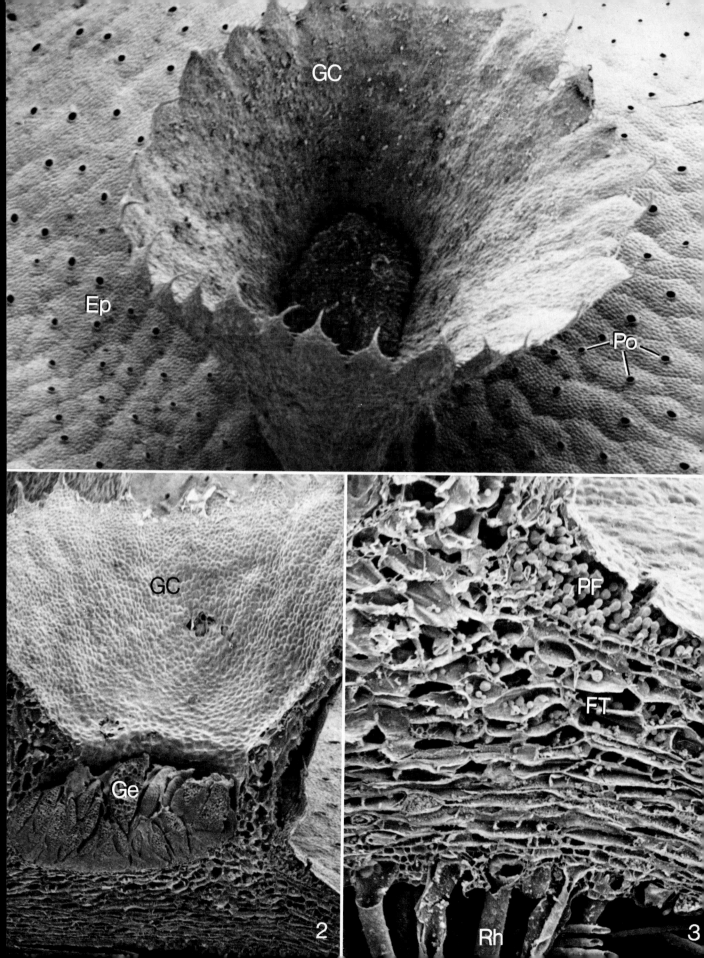

Division Bryophyta

Class Bryopsida—Order Bryales

Leptobryum pyriforme (Moss)

An	Antheridium	Op	Operculum
Ar	Archegonium	Pe	Peristome
CaW	Calyptra wall	Sk	Stalk
Cp	Capsule	St	Stoma
Ft	Foot		

Fig. 1 (gametophore), ×88; Fig. 2 (antheridia), ×100; Fig. 3 (sporophyte), ×43; Fig. 4 (operculum), ×60; Fig. 5 (peristome), ×125; Fig. 6 (archegonia and foot), ×70; Fig. 7 (stomata), ×400.

Mosses are abundant in moist environments such as woods, fields, and shaded stream banks. They are also one of the "pioneer" organisms to colonize rocks and bare places. Once mosses have become established and soil begins to accumulate, other plants will follow. Like all other Bryophyta, mosses are characterized by an alternation of generations, with a long-lived gametophyte alternating with a short-lived sporophyte. The familiar small green leafy moss plant is the sexual or gametophyte, and sporophytic plants develop as a partial parasite on the gametophyte.

The gametophytic moss plant, which consists of a leafy shoot and rhizoids, is commonly called a thallus, while the leafy shoot is called a gametophore (gamete bearer) (Fig. 1). The typical moss leaf consists of a single cell layer except in the midrib section. The stem consists of an epidermal layer, a cortical region, and a central core. There is no vascular tissue in moss plants although the central core consists of elongated thick-walled cells. The leaves are spirally arranged on the stem. The rhizoids are multicellular branched filaments which serve to attach the plant to the substratum.

At maturity the gametophore produces archegonia and antheridia in a cluster at the apex of the stem (Fig. 2). Depending on the species, antheridia (male organs) (An in Fig. 2) and archegonia (female organs) (Ar in Fig. 6) may occur on a single plant or on separate plants. When the antheridium matures, the double flagellated male gametes are discharged and swim to the egg, which is located at the base of an archegonium. Thus, the presence of water is essential for fertilization to occur. The diploid zygote resulting from the fusion of the two gametes is the beginning of the sporophytic generation. The embryo develops rapidly within the enlarged archegonium, which is now called the calyptra, and differentiates into a capsule (Cp in Fig. 4), a stalk (Sk in Fig. 3), and a foot (Ft in Fig. 6). The foot grows into the gametophyte anchoring the sporophyte (Fig. 6) and facilitates movement of nutrients from the haploid plant to the diploid sporophyte. As development proceeds, the stalk elongates and ultimately ruptures the calyptra wall (CaW in Fig. 3). The upper part of the wall is carried aloft like a hat, undergoes a limited amount of growth, and is necessary for capsule development. The capsule eventually differentiates into three regions. A basal portion is specialized for photosynthesis and possesses stomata (St in Fig. 7). The central portion of the capsule is specialized for spore formation. The apical region develops a lid called an operculum (Op in Fig. 4), which will fall off when the capsule is mature. Directly under the operculum is a peristome (Pe in Fig. 5), which is composed of teeth-like units. The number of teeth is species specific. The peristome is hygroscopic and when the weather is damp, the teeth are positioned inward. In dry weather the teeth curve outward and the spores are discharged.

The haploid spore germinates to produce a branched filamentous protonema, which is the beginning of the gametophytic generation. Some of the filaments comprising the protonema develop into rhizoids, whereas others on the upper surface produce buds that eventually grow into leafy shoots.

Lower Vascular Plants

Division Psilophyta

Class Psilopsida—Order Psilotales

Psilotum nudum (Whisk Fern)

Lv   Leaves
Ph   Phloem
S   Stem
Sg   Sporangium
Sp   Spore
St   Stoma
VaT   Vascular tissue
Xy   Xylem

Fig. 1 (aerial system), ×44; Fig. 2 (stem), ×105; Fig. 3 (transverse section of a stem), ×236; Fig. 4 (transverse section of a sporangium), ×60.

Psilotum is the most primitive of all living vascular plants. It closely resembles *Psilophyton,* a fossil that lived 420 million years ago and is considered ancestral to all vascular plants. *Psilotum* occurs world-wide in the tropics and subtropics. It is green and upright, reaching a height of about one foot. The conspicuous sporophyte consists of an aerial system and an underground rhizome system, both of which repeatedly divide into equal branches.

The leaves (Lv) are actually very small scales without vascular tissues (Fig. 1). The trilobed sporangium (Sg) is terminally placed on a short branch and appears to reside axillary in position with reference to the leaves (Fig. 1). Longitudinal section reveals the presence of vascular tissues at the base of the sporangium (Fig. 4).

Groups of stomata (St) may be found on the epidermis of the stems (Fig. 2). The cells comprising the outer cortex contain chloroplasts and this region accounts for most of the photosynthesis. Vascular tissues (VaT) occupy the center core of the stem (Fig. 2). The phloem (Ph) forms a cylinder around the xylem (Xy) strands (Fig. 3).

Meiotic divisions occur in the sporangia and result in haploid spores, which are released by the splitting of sporangia along longitudinal slits. The spores (Sp in Fig. 4) germinate to form very small gametophytes, which are devoid of chlorophyll and grow on trunks of trees or underground. Two types of gametangia are scattered over the surface of the gametophyte, producing eggs and sperms. Since the sperms can only reach the eggs in water, *Psilotum* has not completely escaped its dependence on water for the completion of its life cycle.

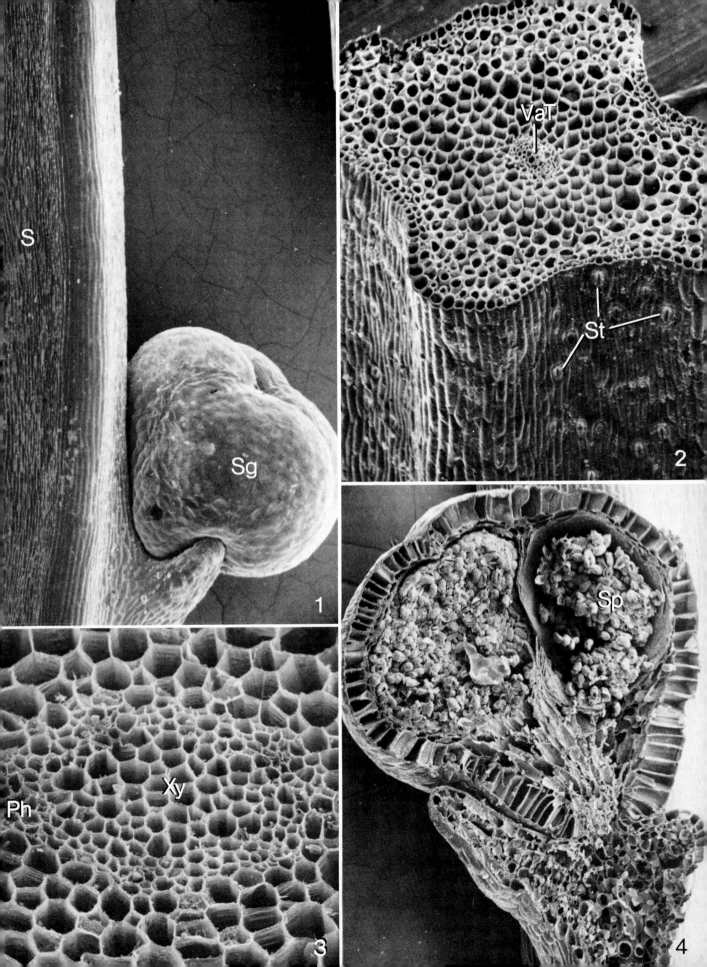

Division Pterophyta

Order Filicales—Family Parkeriaceae

Ceratopteris thalictroides (Fern)

An	Antheridium	Rh	Rhizoids
Ar	Archegonium	Rt	Root
Ga	Gametophyte	Sg	Sporangia
IN	Indusium	Sp	Sporophyte
Lf	Leaf	St	Stomata

Fig. 1 (sorus of Boston fern), × 70; Fig. 2 (spore germination), × 600; Fig. 3 (gametophyte), × 110; Fig. 4 (young sporophyte), × 45.

Ferns are widely distributed in both tropical and temperate regions. They are generally shade-loving plants and small in size, although some tropical ferns are tall and have an erect, woody unbranched stem with a cluster of compound leaves. During the Carboniferous Period there were great forests of fern trees. These ferns had tall, slender trunks composed of stems and an eveloping mass of roots. These plants formed a significant portion of our present coal deposits.

All Pterophyta have a definite alternation of generations, with both the sporophyte and gametophyte being autotrophic plants. The diploid sporophyte (Sp) is the dominate generation of all ferns. It consists of a stem, roots (Rt), and leaves (Lf in Fig. 4), but the leaves are the most conspicuous. The leaves are coiled in the bud, but unroll to form the mature leaves. Internally, the leaf contains mesophyll, which in some species is differentiated into palisade and spongy parenchyma. The epidermis is cutinized and virtually all the epidermal cells contain chloroplasts. The stomata (St) are usually confined to the lower surface (Fig. 1). The root of a fern has a root cap as well as a meristematic elongation and maturation zones similar to the root of a seed plant. The vascular tissue in the stem is organized into one or more strands, each having a core of xylem surrounded by phloem and endodermis. The xylem contains only tracheids.

The sporophyte (Sp) plant may live for several years and produce haploid spores each year. The spores are born in sporangia (Sg in Fig. 1) which usually develop on the lower surface of the leaf. The sporangia may be grouped into sori and protected by an unbrella-like indusium (IN in Fig. 1). The mature spores are dispersed by wind and germinate into gametophytes in an appropriate moist environment (Fig. 2).

The gametophytes (Ga) are small (2 to 3 mm in *Ceratopteris*), green, and heart-shaped structures, with rhizoids (Rh) on their lower surfaces (Fig. 3). Antheridia (An in Fig. 3) and archegonia (Ar in Fig. 4) also develop on the under surface of the gametophyte. Antheridia are formed and mature early, discharging their sperm before the archegonia of the same gametophyte have matured. Archegonia are usually formed in clusters close to the notch. They consist of a short neck and a venter, which encloses the egg cell. Fertilization occurs only under moist conditions, so that the sperm can swim to the neck of the archegonium. The resulting diploid zygote develops rapidly into a sporophyte (Fig. 4). The embryo is dependent upon the gametophyte for nutrients until it develops its own root (Rt) and leaf (Lf in Fig. 4).

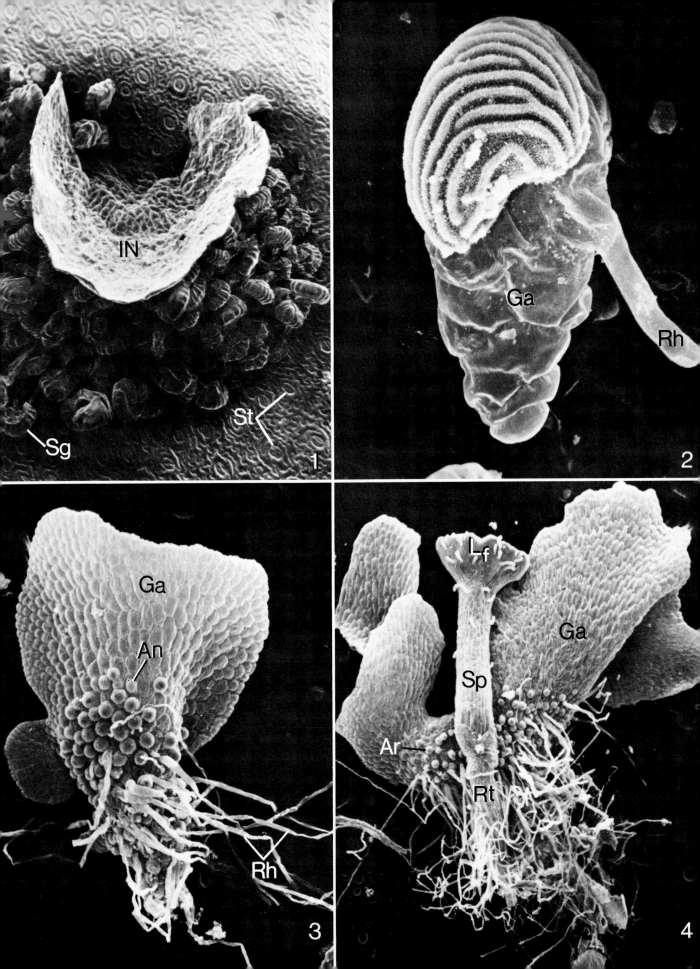

Gymnosperms

Division Cycadophyta

Class Pteridospermopsida—
Order Pteridospermales—Family Medullosaceae

Alethopteris sullivanti
(Cuticle of Fossil Seed Fern)

LV   Lateral vein
MR   Midrib of leaflet
St   Stoma
GC   Guard cell
SC   Subsidiary cell

Fig. 1 (outer surface of epidermis), ×68; Fig. 2 (outer surface of epidermis), ×184; Fig. 3 (single stoma), ×1620; Fig. 4 (inner surface of foliar cuticle), ×184.

The epidermal cells of aerial parts of plants produce a tightly adhering waxy layer or cuticle over their outer surface. This relatively impervious layer of cutin is very important to land plants in that it reduces water loss through the epidermis when the stomata are closed.

Cuticle and cutinized plant parts, such as spores and pollen, are quite resistant to decay and may persist as plant fossils for hundreds of millions of years. Cuticle from fossilized leaves can be freed from enclosing rock or sediment by various chemical treatments and can often yield information on epidermal cell characteristics, stomatal structure and distribution, and the structure and distribution of hairs or trichomes. Such features can provide insight into the taxonomic affinity and paleo-ecological adaptions of a fossil plant.

The accompanying illustrations are views of the cuticle from the lower surface of leaves of a seed fern (pteridosperm) that lived some 250 million years ago during the Carboniferous Period, or coal age. It is possible to reconstruct the venation of these leaves because the epidermal cells that overlaid the veins clearly differed in their appearance from those between the veins (Figs. 1, 2). Those epidermal cells located between the veins bore one or two prominent papillae, whereas those overlaying the veins bore no papillae or were only weakly papillate.

In Figs. 1 and 2 scattered stomata (St) can be seen in the areas between lateral veins (LV). Each stomata (Fig. 3) has a pair of guard cells (GC) surrounded by 5 to 9 subsidiary cells (SC). Generally each subsidiary cell has a single, large papilla that overarches the guard cells.

Figs. 1, 2, and 3 are the outer surface of the cuticle, and in Fig. 4 the inner surface is exposed. The high relief on the inner surface resulted because the cuticle was produced not only on the outer surface of the epidermal cells, but also projected inward along their lateral walls.

The thick cuticle, sunken stomata, and overarching papillae that are seen in this fossil leaf are features characteristically seen in the leaves of living xerophytic plants that are trying to conserve water. (Micrographs and text kindly provided by Dr. JEFFRY T. SCHABILION and MARY ANN FAULWELL.)

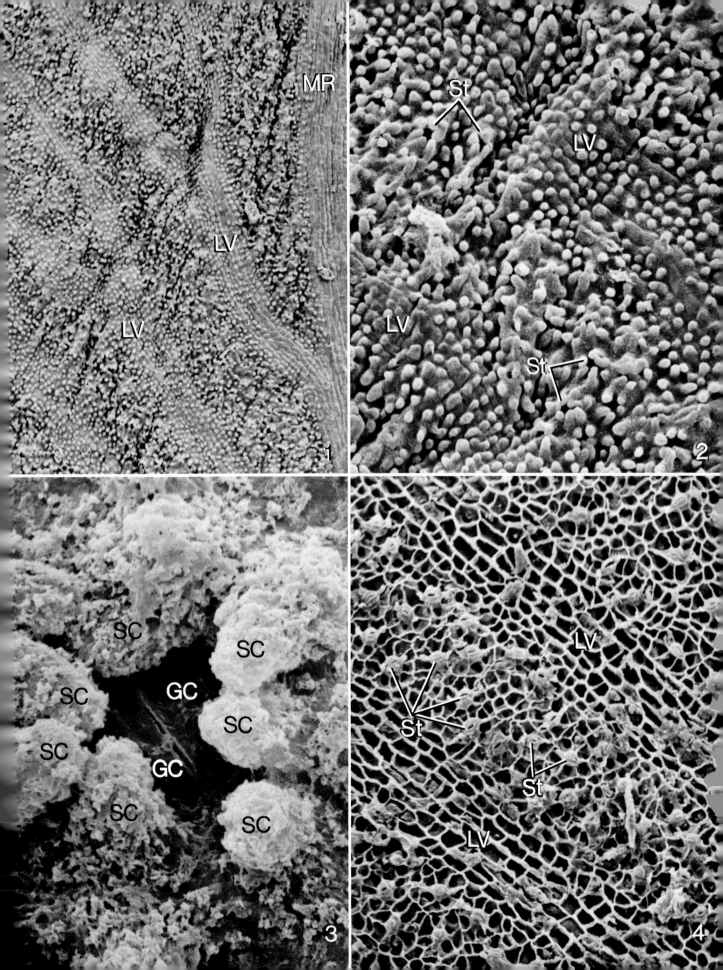

Division Coniferophyta

Class Coniferopsida—Family Pinaceae

Pinus strobus (Pine Needles)

AC	Albuminous cells	RD	Resin duct
Ep	Epidermis	St	Stoma
En	Endodermis	SF	Sclerified fiber
Me	Mesophyll	TT	Transfusion tissue
Ph	Phloem	Xy	Xylem

Fig. 1 (pine needle), ×146; Fig. 2 (mesophyll and resin duct), ×450; Fig. 3 (stoma), ×530.

Most gymnosperms are evergreen. The leaves of the conifers, which comprise the largest number of species among the gymnosperms, have been studied most frequently, especially those of the pine.

Pine needles commonly grow in groups of two or three and are oval or triangular in transverse section (Fig. 1). The needle has a thick-walled epidermis with a heavy cuticle and deeply sunken stomata (St in Fig. 3). The stomata occur in vertical rows on all sides of the needle. A sclerified, fibrous (SF) hypodermis occurs beneath the epidermis (Ep in Fig. 2). The mesophyll (Me) consists of parenchyma cells with folding ridges. There is no differentiation of mesophyll into palisade and spongy parenchyma in pine leaves, as there is in the dicot leaves. Two or three resin ducts (RD) occur in the mesophyll (Figs. 1 and 2).

The vascular tissue occupies a central position in the needle. The xylem (Xy) is on the adaxial side while the phloem (Ph) is on the abaxial side (Fig. 1) The vascular bundle is surrounded by a peculiar tissue known as transfusion tissue (TT), which is composed of nonliving tracheids and living parenchyma cells. Some cells with dense cytoplasm, interpreted as albuminous cells (AC), occur next to the phloem. The vascular bundle with its associated transfusion tissue is surrounded by a thick-walled sheath of cells called the endodermis (En). No intercellular spaces are present in the endodermis or in the tissues enclosed by it.

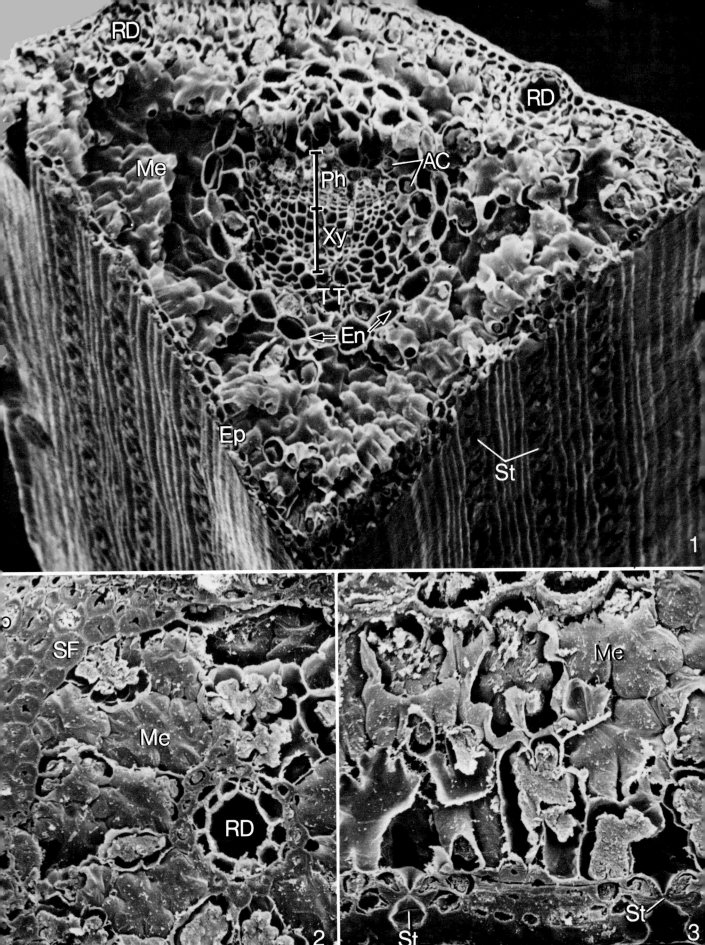

Division Coniferophyta

Class Coniferopsida—Family Pinaceae

Pinus strobus (White Pine Wood)

Ap	Aperture	RP	Ray parenchyma
AR	Annual ring	TaS	Tangential section
Pi	Pit	To	Torus
RaS	Radial section	Tr	Tracheid
RD	Resin duct	TrS	Transverse section

Fig. 1 (wood in three dimension), ×103; Fig. 2 (transverse section), ×250; Fig. 3 (radial section), ×250; Fig. 4 (torus and pit) a, ×2750; b, ×1840; Fig. 5 (tangential section), ×250; Fig. 6 (newspaper), ×250.

Pine wood is a typical gymnosperm conifer wood. It has no vessels, but contains tracheids (Tr), ray parenchyma (RP), and resin ducts (RD). A transverse section (TrS) illustrates the extremely regular structure of wood resembling a wire netting (Figs. 1 and 2). The homogeneous structure makes it easily workable and is commonly called soft wood. For the same reason conifer woods are suitable for paper manufacture. Fig. 6 is a magnified newspaper showing wood fibers. The tracheids formed in the spring are the largest in diameter and have a thin wall. As the season progresses, the tracheids become smaller and their walls thicken. The morphologically distinct woods, produced annually in spring and late summer, contribute to a so-called annual ring (AR), which reveals the age of a tree (Fig. 2).

The wood rays, composed of parenchyma cells, are relatively long-lived and serve to conduct material transversely in the stem. The ends of the rays when viewed in tangential section (TaS) are uniseriate and composed of several cells (RP in Fig. 5). The radial view of wood reveals strands of fiber-tracheids with pits (Pi) on the radial walls (Fig. 3). A torus (To) between two pits serves as a flap to cover the aperture (Ap), thereby regulating the flow of water (Figs. 4a and b). Pits also occur between ray cells and the tracheids (Fig. 3).

Pines produce resin, as do many other gymnosperms. This substance occurs in specialized cavities which are surrounded by a group of parenchyma cells. Together they are called resin ducts (RD in Fig. 2).

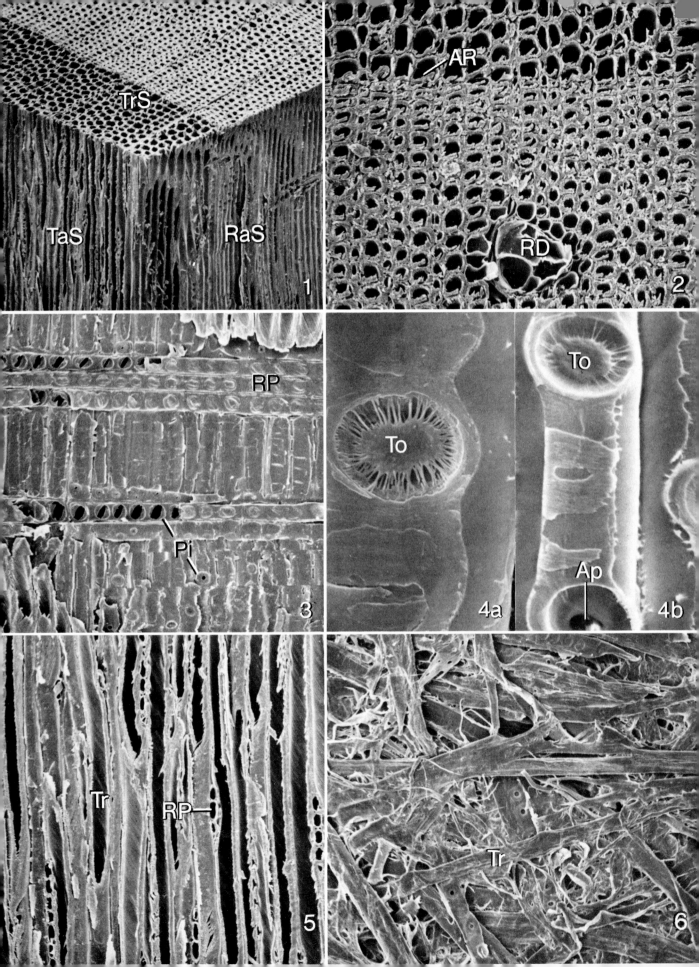

Chapter 7
Organ System of Angiosperms

Roots

Root Hairs and Root Tip

Ep   Epidermis
EZ   Elongation zone
RH   Root hairs
RHZ   Root hair zone
RT   Root tip
SC   Seed coat

Arrow: Root hair from epidermis

Fig. 1 (radish seedling), ×35; Fig. 2 (wheat root), ×120; Fig. 3 (corn root, cross-section), ×710; Fig. 4 (wheat root), ×270.

The root system is one of two fundamental parts of a land plant. When a seed germinates, the root (RT) grows downward into the soil to anchor the plant and hold it upright (Fig. 1). The root also functions to absorb water and minerals from soil.

At the tip of the root is a root cap (RC), which protects the meristematic zone (Fig. 2). The cells of root cap secrete mucus which facilitates extension of the root through soil. Since the root cap tends to be stripped away by mechanical abrasion during root growth, it must continually be regenerated from the growing point.

Immediately behind the growing points is the elongation zone (EZ) (Fig. 2). Here the cells grow rapidly in length but remain undifferentiated. The elongation zone, 3 to 5 mm long, together with the meristematic zone, account for the increase in root length.

Water absorption by roots is facilitated by the differentiation of the root hairs, which are slender tubular extensions from the epidermal cells (Ep) indicated by arrows in longitudinal section (Fig. 3) and in surface view (Fig. 4). Actually, any root epidermal cell is capable of absorption. However, root hairs serve to markedly increase the surface area of epidermal cells and, consequently, most of the water and minerals for the plant are absorbed in these regions. The root hair zone (RHZ in Fig. 2) is also called the maturation zone, since internally the cells become differentiated into conducting tissues to transport water and sugars. An individual root hair has only a temporary existence. Root hairs wither and die as the root elongates and new ones are differentiated (Fig. 4). Only a short segment of root, perhaps 1 to 6 cm long, possesses root hairs.

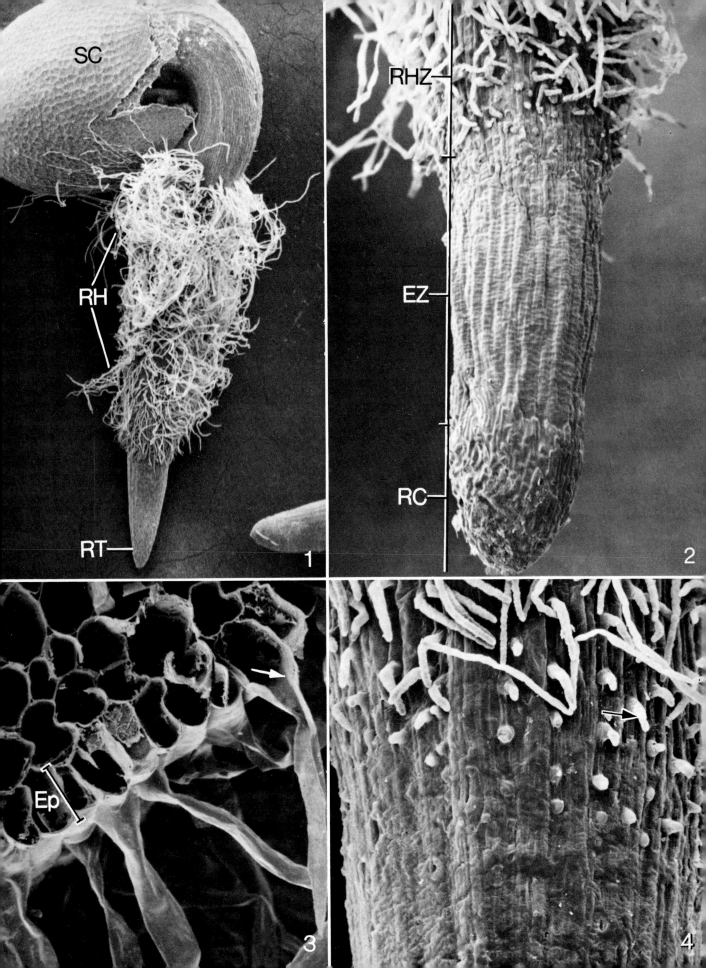

Internal Structure of a Root and Lateral Root Initiation

Co   Cortex
En   Endodermis
Ep   Epidermis
LR   Lateral root
Pe   Pericycle
Ph   Phloem
Pr   Procambium
PX   Primary xylem
VC   Vascular cylinder

Fig. 1, ×120; Fig. 2, ×470.

A transverse section through the differentiating region of a lily root shows an orderly organization of various tissues (Fig. 1). From the exterior toward the center they are epidermis (Ep), cortex (Co), endodermis (En), and vascular cylinder (VC), The vascular tissue, in turn, consists of pericycle (Pe), primary xylem (PX), phloem (Ph), and a procambial (Pr) layer. Primary xylem not only makes up the central core of the vascular tissues, but also extends outward with radiating arms between which the phloem elements are grouped. The procambial layer is sandwiched between phloem and xylem (Fig. 2).

As the shoots of a plant grow and branch, so too must the roots in order to increase the absorbing surface and enforce anchorage. Unlike the branching of a stem that occurs by growth and differentation of the outer cell layers near the shoot tip, lateral roots (LR in Fig. 1) in gymnosperms and angiosperms originate from pericycle cells in the vascular cylinder. During the initial stage of lateral root formation, several contiguous pericyclic cells divide periclinally to form meristematic cells. The resulting cells divide again periclinally and anticlinally to eventually form a root primordium. As the root primordium elongates, it pushes through the cortex and the epidermis to reach the surface (Fig. 1). The endodermis often divides anticlinally to keep pace with the growth of the primordium, but other cortical cells are mechanically compressed, crushed, or partially dissolved by the substances secreted from the cells of the lateral root primordium.

During growth of the young primordium through the cortex, the differentiation of the promeristem and root cap are initiated. When phloem and xylem are differentiated in the lateral root, they become connected with the vascular elements in the parent root. This connection is achieved by the differentiation of pericyclic cells into conducting elements at the proximal end of the primordium.

140

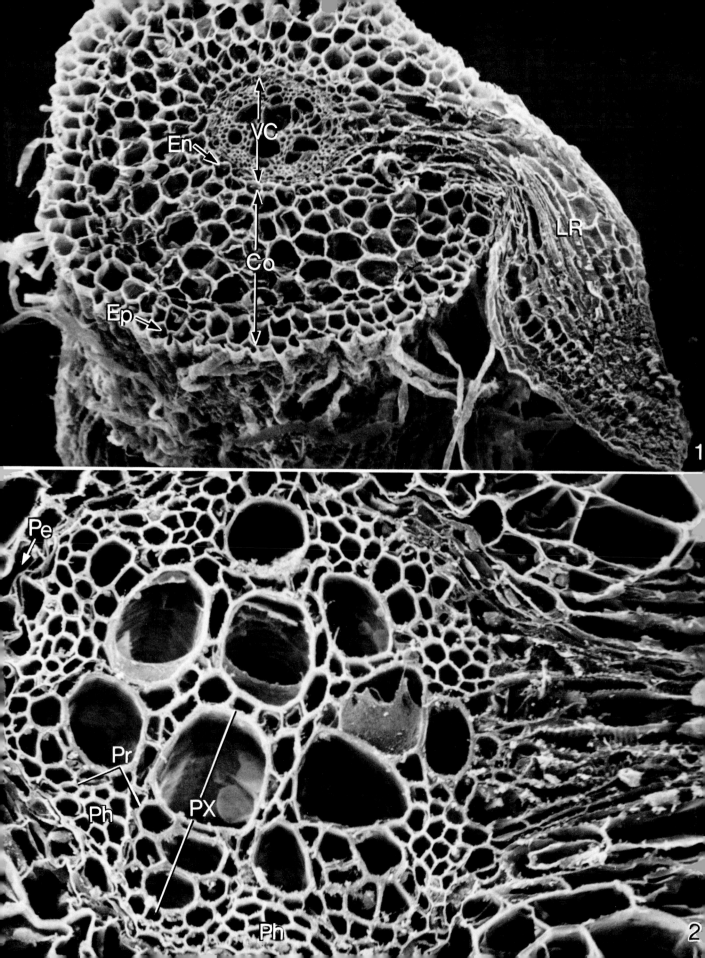

Shoot Apex

Buds of Potato Tubers

AB   Axillary buds
AM   Apical meristem
Co   Cork
LP   Leaf primordia
SG   Starch grain
VT   Vascular tissue

Fig. 1 (external structure), $\times 110$; Fig. 2 (internal structure), $\times 110$.

The shoot apex is a permanent growth region of the plant. It is remarkably self-determining, autonomous, and it is the organizing center for the plant. Buds of potato tubers (storage stem) provide a good example of the shoot apex. Like typical shoot tips, they are not covered by any cap analogous to the root cap. However, they are protected by bud scales or overarching young leaves (Fig. 1).

A longitudinal section of the bud reveals its internal structure (Fig. 2). The central cone of meristematic tissue is called the apical meristem (AM). An embryonic leaf called a leaf primordium (LP) emerges from the flank of the apical meristem as a bump. This leaf primordium will elongate so that it soon towers above the apical meristem. Differentiation of vascular tissues (VT) also occurs with the early appearance of elongated cells in strands. Phloem and xylem will differentiate from these elongated cells.

Small axillary buds (AB in Fig. 2) are also formed in the angle between the leaf primordium and the stem. This bud normally remains dormant until well after growth of the adjacent leaf and internode have been completed. These buds can later grow out to produce lateral branches or specialized shoots such as tubers or flowers. Also shown in Fig. 2 are several layers of corky skin (Co) which serve to protect the tuber. Starch grains (SG) can also be seen scattered in the cells adjacent to the vascular tissue.

When the potato tuber is fully grown, the buds may remain dormant for up to ten weeks even under conditions favorable to their development. Ethylene chlorohydrin and gibberellin are able to break bud dormancy in potato tubers. Recent studies reveal that bud dormancy may involve an interaction between abscisic acid, gibberellins, and possibly ethylene.

Stem

Herbaceous Monocot and Dicot Stem

Stems of plants serve many functions. Some are specialized as storage organs and others carry out photosynthesis. But most of them serve to transport and support, since the stem is the connecting link between the roots and leaves. Plant stems are either herbaceous or woody. The soft, green, thin herbaceous stems are typical of annual plants. Two main groups of flowering plants, the Dicotyledoneae and the Monocotyledoneae, have different arrangements of conducting tissue in the stem.

In the stem of a dicot such as the sunflower, tobacco, or coleus (Fig. 1) the circular arrangement of the xylem and phloem bundles subdivides the stem into three concentric regions: the outer cortex (Co), the vascular bundles (VB), and the central core, or pith (Pi). Each vascular bundle has an outer cluster of phloem cells (Ph in Fig. 1) and an inner cluster of xylem cells (Xy) separated by a layer of meristematic tissue, the cambium (Ca). The pith is composed of colorless parenchyma cells which function as a storage area. Despite the presence of a cambium, secondary tissues are poorly developed and the stem remains nonwoody. The herbaceous dicot stem normally does not live through the winter. Any growth in diameter is restricted to one season only.

The stem of a monocot such as corn (Fig. 2b) or lily has an outer epidermis composed of thick-walled cells. The vascular bundles are scattered throughout the stem instead of being arranged in a ring as in the dicots. The bundles are smaller and more numerous in the outer part of the stem, and each bundle contains xylem and phloem but no cambium (Fig. 2a). The vascular bundle is usually enclosed in a sheath of supporting sclerenchyma cells (SC). A great portion of the stem consists of parenchyma tissue, with some collenchyma and sclerenchyma found near the epidermis for added support. In some monocots, such as wheat and bamboo, the parenchyma cells in the center of the stem disintegrate and leave a central pith cavity.

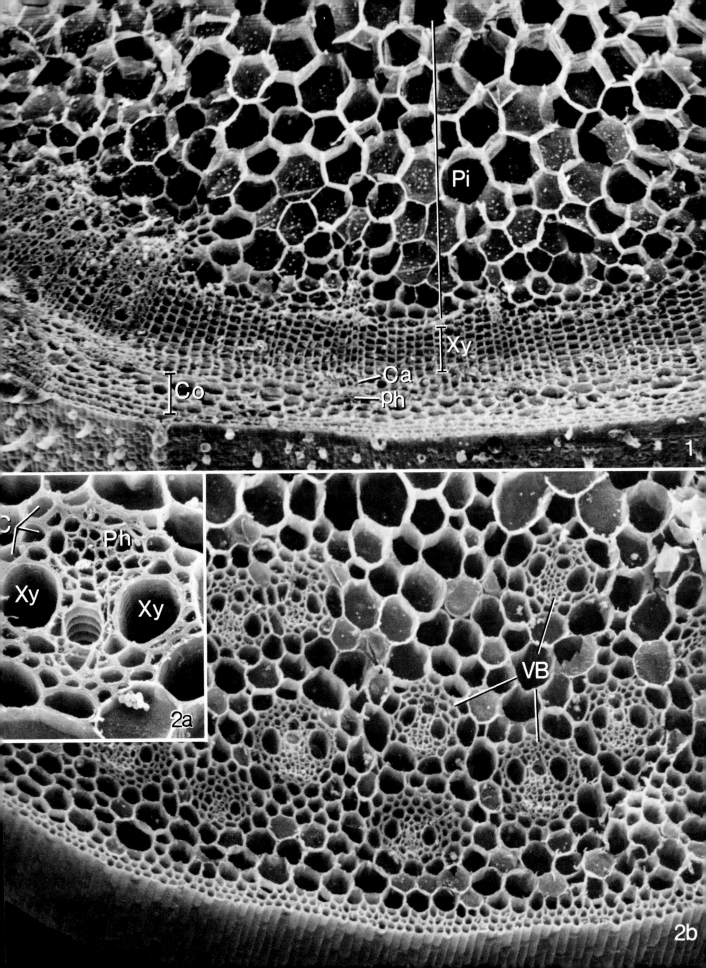

Woody Dicot Stem (Maple and Oak)

The woods of dicotyledons are more varied and highly evolved than those of conifers. Their wood structure shows much greater variation with respect to cell size as well as cell arrangement. Maple wood (Fig. 1) and oak wood (Fig. 2) provide examples of this diversity in structure. The distinguishing cell type of these woods is the vessel. The individual vessel members (VM) are relatively short, but their diameter is variable. The end walls (EW in Fig. 1) of the cell break down to form a system of continuous tubes, and the side walls are perforated with various kinds of openings (Fig. 1). Tyloses (Ty) are seen in vessels of oak wood (Fig. 2). They are actually outgrowths of parenchyma cells through the wall pits into the lumen of vessels. This condition usually occurs in older heartwood.

Parenchyma rays are common to all dicotyledonous wood species. Maple wood has multiseriate rays (MR in Fig. 1), whereas oak wood consists of uniseriate rays (UR) and massive multiseriate rays (MMR) (Fig. 2). Vertical parenchyma (VP) is also very common in dycotyledonous wood and can be observed scattered around the vessel members in oak woods (Fig. 2).

The fibers (Fi) of dicotyledonous wood are the chief supporting and strengthening cells. They are long cells with tapered ends. In heavy woods the fiber walls are very thick and in low-density woods they are thin-walled.

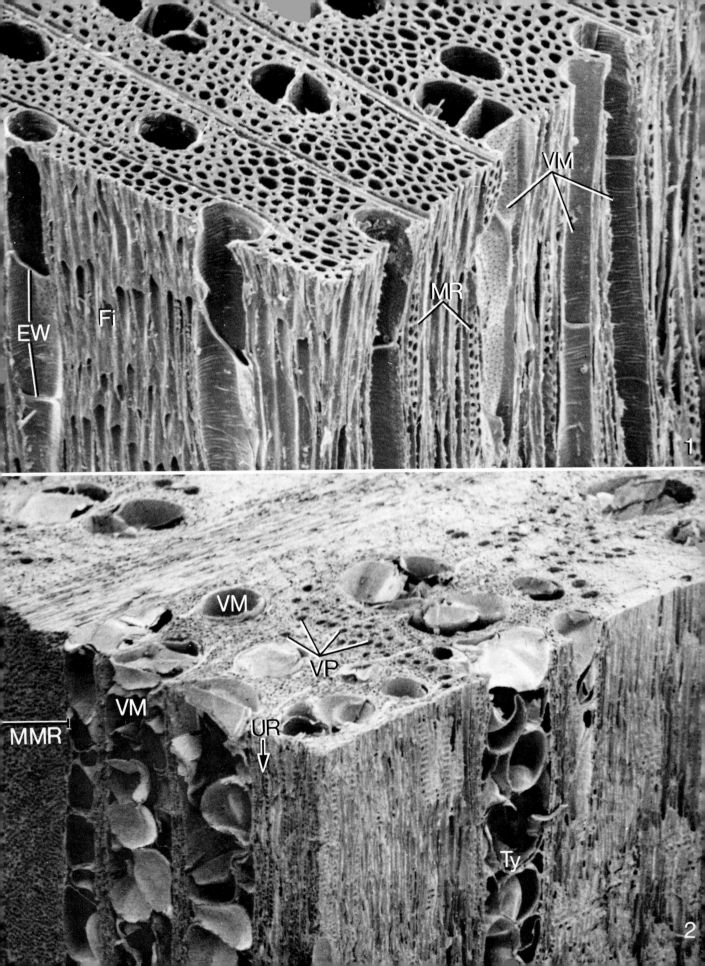

Leaf

A Monocot Grass Leaf (Corn)

BC   Bulliform cells
BS   Bundle sheath
GC   Guard cell
SC   Subsidiary cells
St   Stoma
Tr   Trichomes
VB   Vascular bundle

Fig. 1 (corn leaf surface), ×220; Fig. 2 (a stoma), ×4660; Fig. 3 (cross-section of a mid-rib), ×370; Fig. 4 (a bundle sheath), ×460; Fig. 5 (cross-section of corn leaf), ×470.

The corn leaf has a typical grass leaf structure consisting of a narrow blade and a sheath enclosing the stem. The epidermis contains a variety of cells. The ground tissue consists of narrow and elongated cells. Enlarged bulliform cells (BC in Fig. 5) are present in the upper surface which serve as motor cells when the leaf folds during water stress. Stomata (St) consists of narrow guard cells associated with subsidiary cells (SC) (Fig. 2). Sharp trichomes (Tr) are present on the leaf margin and leaf surface (Fig. 1).

The mesophyll is not differentiated into palisade and spongy parenchyma. However, the vascular bundles (VB) are surrounded by a layer of mesophyll cells called a bundle sheath (BS in Fig. 4). The bundle sheaths are characteristic structures in many grasses of tropical origin including maize, sugar cane, and others. Current study indicates that the grasses of hot climates possess a C-4 pathway of photosynthesis.

Through the careful isolation of bundle sheaths from the surrounding mesophyll cells it was found that each carries out different metabolic pathways. In the mesophyll cell the CO_2 is first fixed into C-4 acids (malic or aspartic). These acids are then transported to bundle sheath cells, which decarboxylate the acids to release CO_2. The CO_2 is then refixed by the regular C-3 pathway.

In regular C-3 plants up to 30 percent of the photosynthetically fixed carbon may be lost through photorespiration, a side reaction from the Calvin cycle. In the tropical grasses, however, no such loss is observed, because the bundle sheath is entirely surrounded by mesophyll cells (Fig. 4), and the CO_2 released by photorespiratory side reactions in the bundle sheath can be completely recaptured by a CO_2-fixing mechanism of the C-4 pathway in the mesophyll cells.

The coordination of C-4 and C-3 pathways in two distinct parts of the leaf serves a metabolic adaptation of the grass to hot weather and low water supply. The prevention of CO_2 loss during photosynthesis means that the plant can continue to photosynthesize at a very low partial pressure of CO_2. Such a capability is of great ecological significance. (Fig. 4 kindly provided by Drs. C. C. BLACK and T. M. CHEN, and used by permission of Quartly Review of Biology.)

Reference

BLACK, C.C., Jr., CAMPBELL, W.H., CHEN, T.M., and DITTRICH, P.: "The monocotyledons: Their evolution and comparative biology. III. Pathways of carbon metabolism related to net carbon dioxide assimilation by monocotyledons." Quart. Rev. Biol. **48**, 299—313 (1973).

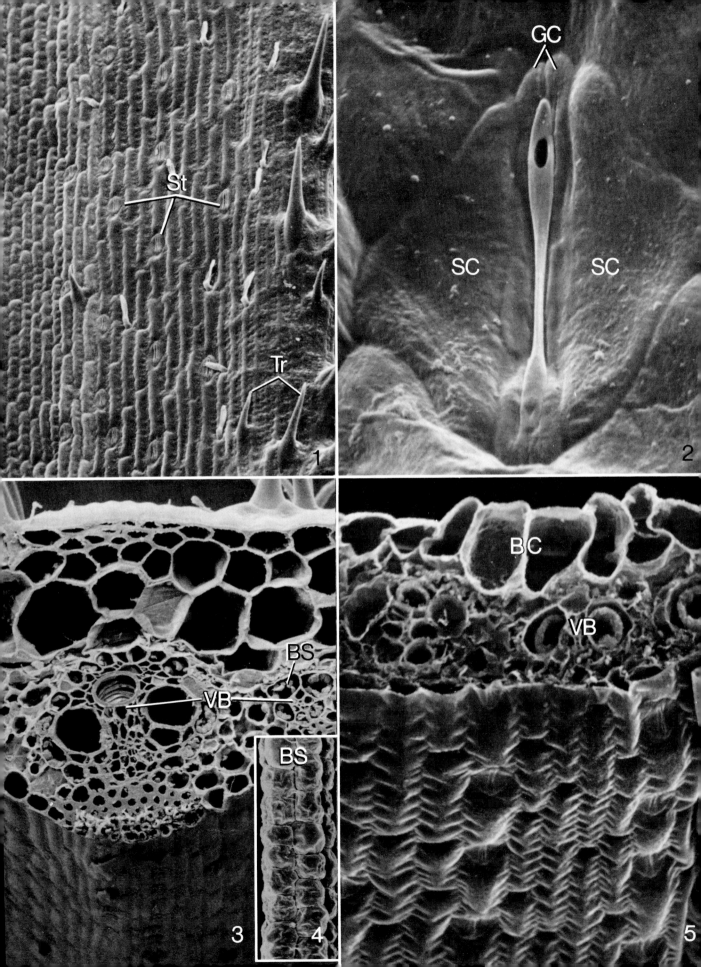

A Dicot Leaf (Tobacco)

AS	Air space	SG	Secretory gland
Ch	Chloroplast	Sp	Spongy layer
Ep	Epidermis	St	Stoma
Ha	Hair	VB	Vascular bundle
Pa	Palisade layer	Xy	Xylem
Ph	Phloem		

Fig. 1 (epidermis), × 326; Fig. 2 (cross-section), × 467; Fig. 3 (tangential section), × 357.

The leaf is a specialized nutritive organ of the shoot system. It carries out photosynthesis, a process requiring a constant supply of water, carbon dioxide, and radiant energy to produce sugar and oxygen. A leaf is covered on both sides with a layer of epidermal cells (Ep in Fig. 2), which are perforated by numerous small stomata (St) for gas exchange and by hairs (Ha) and secretory glands (SG) for protection (Fig. 1).

In most dicotyledonous plants the bulk of the internal leaf tissue consists of two layers: a palisade layer (Pa) of elongated cells packed in the upper half, and the spongy parenchyma tissue (Sp) of irregularly shaped cells in the lower half of the leaf (Fig. 2). The palisade cells are filled with chloroplasts (Ch in Fig. 2) and appear to be specialized for photosynthetic efficiency. The elongated cell form ensures that a direct light path from the upper epidermis down through each photosynthesizing cell is uninterrupted by light-scattering air spaces. Cells of the spongy layer usually possess fewer and smaller chloroplasts. Large air spaces (AS in Figs. 2 and 3) are located between the spongy parenchyma cells and sometimes extend between the palisade cells.

As the external layers of the leaf, including the epidermis and palisade cells, are peeled off, vascular bundles (VB) become clearly visible (Fig. 3). The vascular bundles, consisting of phloem (Ph in Fig. 3) and xylem (Xy in Fig. 2), are finely branched to supply the individual leaf cells with water on one hand and to transport the synthesized sugars on the other.

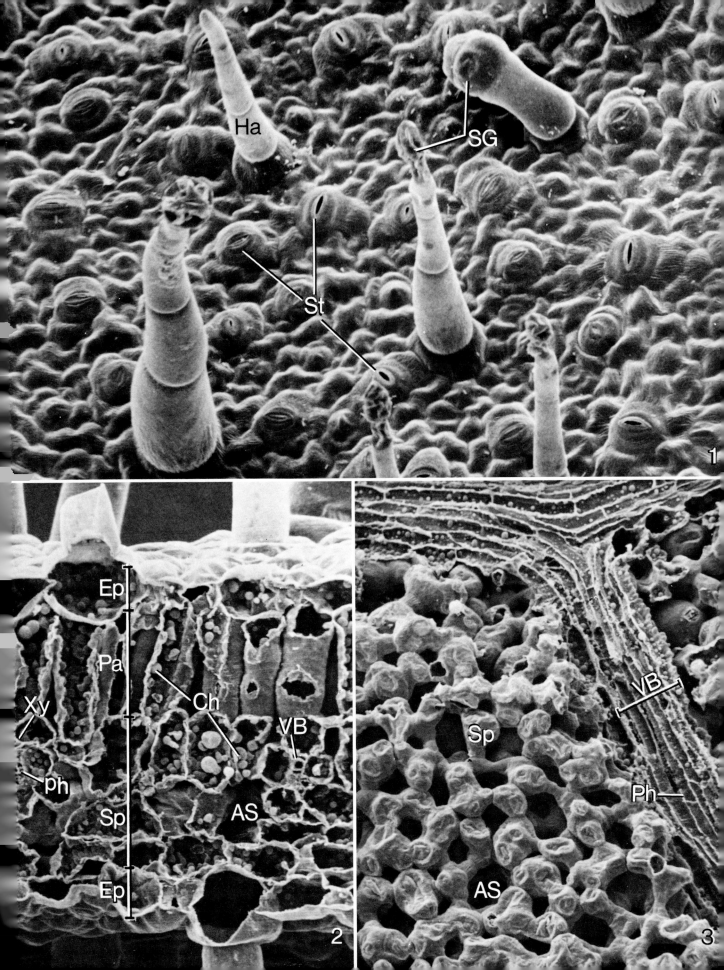

Gas Exchange and Stomata (Lily)

AS   Air space
CI   Chloroplast
EC   Epidermal cell
GC   Guard cell
St   Stoma

Fig. 1 (leaf surface), $\times 455$; Fig. 2 (cross-section of a stoma), $\times 1470$; Fig. 3 (guard cells), $\times 1620$.

The life processes of plants involve respiration and photosynthesis, for which there must be an exchange of gases between the plants and their environment. In lower, simple plants gases are simply exchanged through the cell or plant surface, and no special structures are required. In a tissue of tightly packed cells plants incorporate numerous small spaces which interconnect with each other and with the outside, often by pores. In lower land plants pores are simple holes (see *Marchantia*), but in higher plants the pores have become specialized stomata (St in Fig. 1).

The trend of plant evolution has been from water to land. Therefore, the problems of water availability are always associated with higher land plants. The exchange of gases with the environment for the life processes consequently accelerates the loss of water. To cope with excess water loss, plants have developed a cuticle to cover their body surface. Also the stomata of plants higher than the Bryophytes are regulated by two guard cells (GC in Fig. 2). The variable thickening on the two sides of kidney-shaped guard cell walls allows them to vary their shape according to their turgidity. In the turgid state the pore is fully open and in the wilted state it is closed.

Stomata respond to two stimuli: light and water stress. Under water deficiency, turgidity is reduced or lost and the stomata close. The opening response of stomata to light results from an increased solute concentration within the guard cells. This, in turn, leads to an increase in water uptake and an increase in the turgor pressure of the guard cells. Although guard cells always possess chloroplasts (Cl in Fig. 2) and are capable of photosynthesis, the rate at which their solute concentration increases upon illumination is much too rapid to be explained by photosynthetic production of sugar.

The mechanism by which light increases solute concentration of guard cells has been a matter of dispute. The traditional view is that the change is caused by a conversion of starch to sugar. Some recent experiments, however, support the notion that the increase is the result of active uptake of ions, especially potassium. There is also experimental evidence suggesting that guard cell behavior is hormonally conditioned. Abscisic acid causes stomatal closure whereas cytokinins promote stomatal opening. Guard cells have also been shown to be sensitive to CO_2 concentration. Treatment of a leaf with high concentrations of CO_2 causes stomatal closure, whereas stomatal opening may be induced in the dark by introducing CO_2-free air into the leaf. Thus, it may be assumed that illumination causes stomatal opening as a result of the depletion of CO_2 within the leaf by photosynthesis.

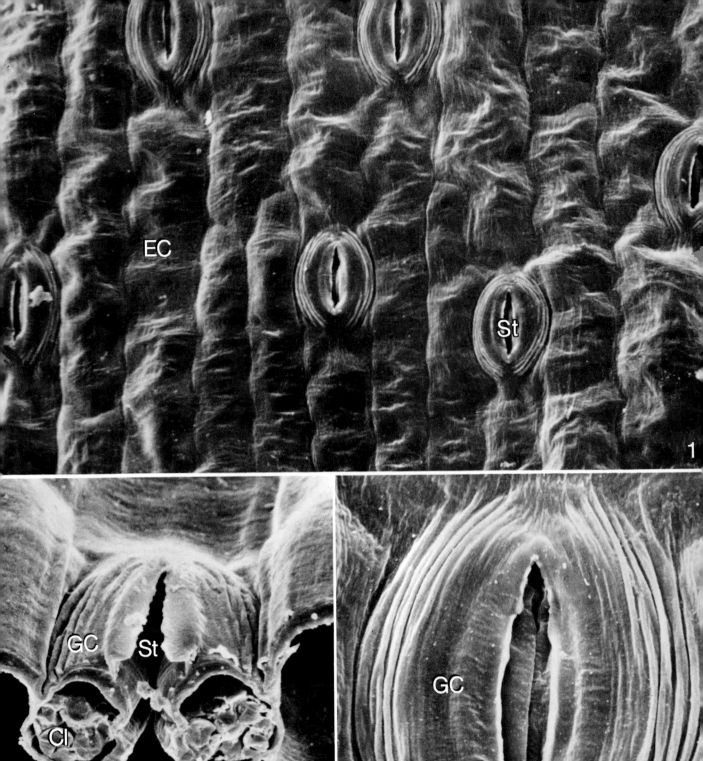

Hair and Glandular Trichomes
on Leaf Surface (Hemp and Snapdragon)

Ex   Exudation
GT   Glandular trichome
HT   Hair trichome

Fig. 1 (hemp leaf trichomes), ×360; Fig. 2 (hemp glandular trichome), ×1340; Fig. 3 (petal trichomes of snapdragon), ×240; Fig. 4 (enlarged tip of a petal trichome), ×1275.

Trichomes are epidermal appendages. They may be unicellular or multicellular; glandular or aglandular; straight, spiral, hooked, or tortuous; simple, peltate, or stellate. They may occur on any part of the plant and vary in form and density from one organ to another. Two or more kinds of trichomes may intermingle on the same surface. This is illustrated on the leaf surface of hemp (*Cannabis*), where glandular (GT) and hair trichomes (HT) are mixed (Figs. 1 and 2). The trichomes may also be dispersed (Fig. 1) or grouped, as is illustrated by the glandular hairs on the petal surface of snapdragon (*Antirrhinum*) (Fig. 3).

Trichomes have been widely used for taxonomic purposes as they are sometimes uniform among groups of plants. Trichomes play a role in plant defense, especially against phytophagous insects. Studies have shown that there is a negative correlation between trichome density and insect feeding on the leaf.

Glandular trichomes secrete terpenes, phenolics, alkaloids, or other exudations (Ex) which are olfactory or gustatory repellents (Fig. 2). These trichomes can be considered to be the chemical defense device of plants against their enemies. The glandular trichomes on flowers, however, produce fragrant chemicals which attract insects to carry out pollination. The sweet scents of flowers are attributable to the presence of essential volatile oils such as esters, alcohols, aldehydes, ketones. The oil secretion of flowers or leaves takes place in the epidermal cells as well as in the glandular hairs projecting from the surface (Figs. 3 and 4). In some cases the odors originate in a lobed disc or a tiny depression.

The secretions from glandular trichomes have been used by humans as drugs, stimuli, and scents for ages. Nicotine in tobacco, a resin in marijuana (dried hemp leaves), and mint are all secretory products of glandular hairs. The exudations, although powerful, are produced in minute quantities. For example, it requires 15 tons of violet flowers to produce a single pound of violet oil.

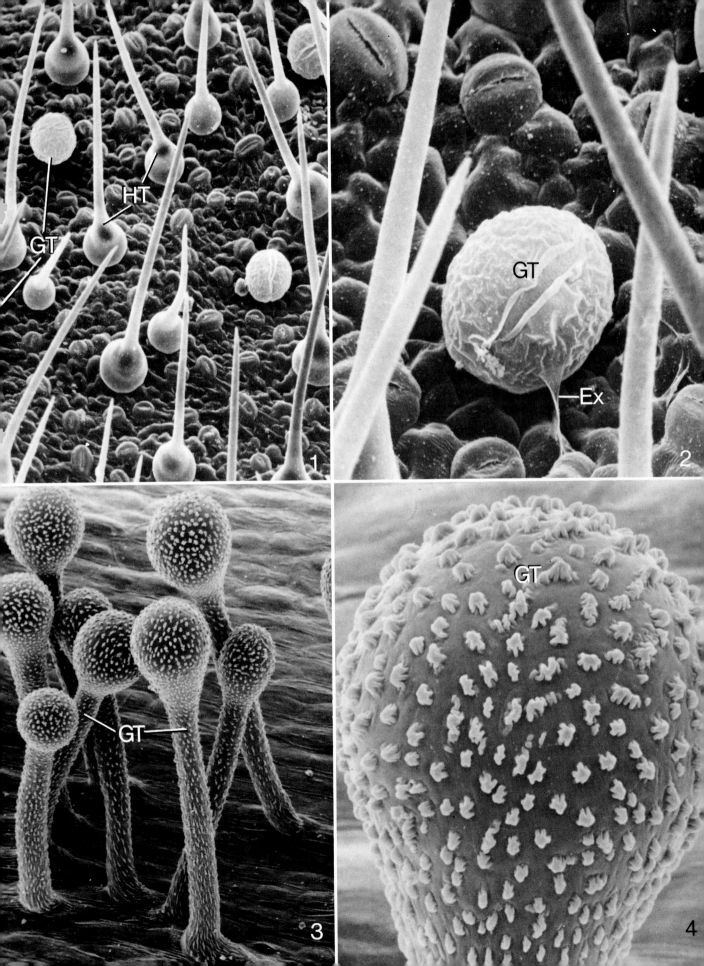

Peltate Scale on Leaf Surface
(Russian Olive and Spanish Moss)

AC   Absorbing cell
PS   Peltate scale

Fig. 1 (scales on under-leaf surface of Russian olive), ×140; Fig. 2 (an enlarged scale on the upper-leaf surface of Russian olive), ×515; Fig. 3 (absorbing scales on leaf surface of Spanish moss), ×140; Fig. 4 (an enlarged scale showing four absorbing cells), ×560.

The peltate scale represents another type of trichome associated with the plant surface. Each scale is a multicellular, umbrella-shaped structure. The cells comprising the scale are fused toward the center but separate at their margins (Fig. 2). The scale is attached to the plant by a central axis embedded in the epidermis. They may be so numerous that they overlap and form a thin silvery layer, as seen on the under-leaf surface of the Russian olive (*Elaeagnus angustifolia*) (Fig. 1). On the upper surface of the leaf, however, the scales are more dispersed and the epidermal cells on the leaf are visible (Fig. 2).

On the leaf surface of the Spanish moss (*Tillandsia usneoides*) (Fig. 3) the scales are modified into absorbing organs. The herbaceous plant exists in the form of loose pendulous tufts, which hang in long festoons from the branches of trees in tropical and subtropical American forests. They are also commonly observed attached to telephone wires. Each scale has four centrally placed absorbing cells (AC in Fig. 4). These cells can absorb water and nutrients for the plant from humid, polluted air. This plant has been used commercially as a stuffing and insulation material.

Rapid Movement of Leaves (Mimosa)

MC   Motor cells
Ll   Leaflet
Pe   Pulvinule
Pu   Pulvinus
Ra   Rachis

Fig. 1 (rachis), ×82; Fig. 2 (pulvinule), ×360; Fig. 3 (rachis and pulvinule cross-section), ×165; Fig. 4 (motor cell collapsed), ×33; Fig. 5 (motor cell fully turgid), ×48.

Mimosa is a sensitive plant which responds to touch. Normally the leaves of the plant are horizontal, but if one of them is lightly touched, all the leaflets (Ll in Fig. 1) fold and the rachis drops within 2 or 3 seconds. After a few minutes the leaves return to their original position. The *Mimosa* leaf is pinnately compound. Each leaflet has a cushion-like petiole named a pulvinule (Pe in Fig. 1), and the main petiole (called a rachis) (Ra in Fig. 1) in turn has a pulvinus (Pu) at its base (Fig. 4). When any part of the plant is touched, a series of visible successive responses of the pulvini can be observed, beginning at the site stimulated. The stimulus is conducted from cell to cell throughout the whole plant and the excitable cells may play the roles of both receptor and conductor.

Although it is well known that leaf movements are caused by a sudden loss of turgor in the motor cells (MC in Figs. 4, 5) of the pulvinus, the exact mechanism causing this turgor change is not understood. Histochemically, potassium salts are detectable in the motor cells before a response, but the large salt crystals appear in the intercellular spaces after a response. This indicates that the cell sap escapes to the exterior during movement and results in the decrease of turgor in the motor cells.

Other studies on fresh sections of pulvinules have demonstrated that the motor cells in *Mimosa* have contractile vacuoles whose activity cause the cell sap to be expelled from motor cells in a situation comparable to the contractile vacuoles in Protozoa. This activity may contribute to a decrease in turgor of the motor cells. The movement of the contractile vacuoles depends on a mechanochemical reaction related to an ATP-ATPase system.

All the motor organs that have been studied electrophysiologically exhibit the generation of an action potential prior to rapid movement. It is, therefore, to be expected that an action potential elicited in the motor cell triggers mechanical or chemical changes or both in its protoplasm as a secondary response. The triggering seems to occur through ionic changes in the plasma membrane. A direct effect of stimulus on the motor cells, or an effect of the propagated action potential which reaches them from the stimulated site, should cause the motor cell to generate a depolarization of the membrane and a resulting action potential. This may be the first response in the motor cells to a stimulus. The recovery process in the motor cells (Figs. 3, 5) may result from an active accumulation of the salts or ions much like the active uptake of potassium by guard cells.

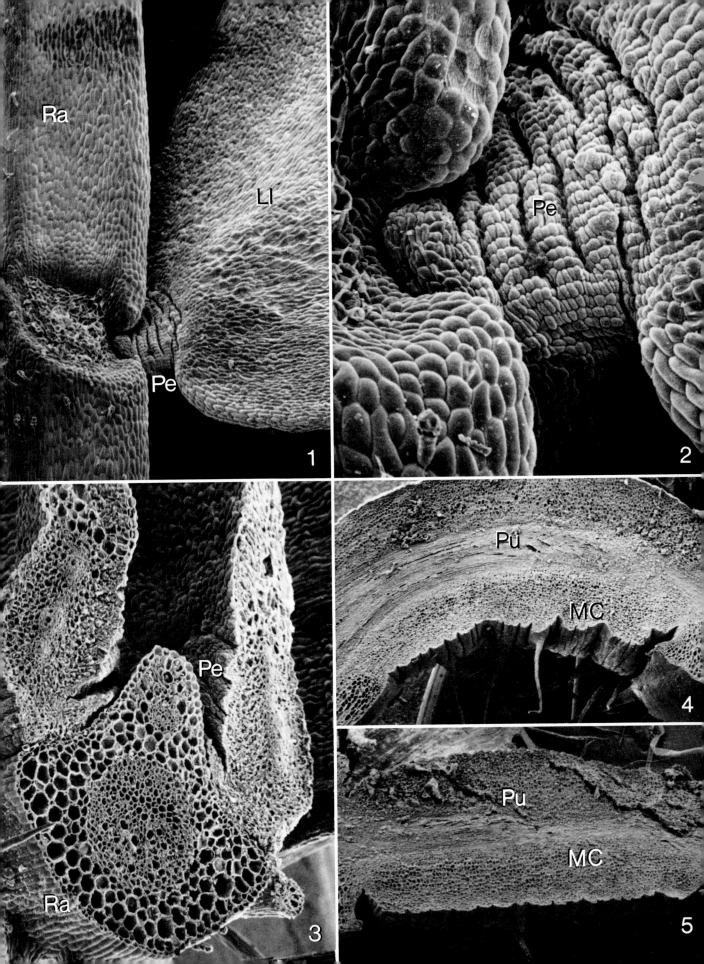

Venus Flytrap (Modified Leaf)

DG   Digestive gland
F   Fruit fly
R   Receptor
TH   Trigger hair

Fig. 1, ×270; Fig. 2, ×60.

Venus flytrap (*Dionaea muscipula*) is the most fascinating insect-catching plant. A native of the Carolina swamps, the plant traps and digests insects and other small animals to obtain part of its nitrogen and mineral requirements. Since the native habitat of the plant is nitrogen deficient, acquisition of the trapping mechanism is considered a necessary adaptation to the environment.

The plant leaf is divided into a blade portion and a prominent winged petiole. The blade consists of two identical semicircular lobes united by a mid-rib; it opens at an angle of about 50°. The outer margin of each lobe is provided with a comb of bristles that interlock when the trap is closed.

In the middle of the upper surface of each lobe are three trigger hairs which are very sensitive to touch. The presence of an insect on the leaf will trigger the hairs and stimulate the leaf blades to fold. Two separate stimuli, either to the same hair or to two different hairs, are required to spring the trap. After capture of the prey, the clamping movement continues slowly until the surfaces of the lobes press tightly against the prey. Meanwhile numerous digestive glands on the surface of the lobes become active and secrete an enzymatic sap. An insect is digested in about 10 days, after which the lobes reopen and the trap is ready for another catch.

The structural details on the leaf surface are shown in Fig. 1. A fruit fly (F) was placed on the leaf to illustrate the size of the prey in relation to the size of the trigger hairs (TH) and the digestive glands (DG) (Fig. 2). The cells at the base of the trigger hair are believed to act as sensory receptors (R in Fig. 1) since they produce receptor potentials when mechanically stimulated. The receptor potential spreads concentrically from the point of stimulation, and as it reaches the mid-rib, contraction of the abaxial cells occurs. Biochemical changes responsible for the closure of the mechanically stimulated trap are not well understood, though ATP appears to be a native source of energy for the closing process. The exact mechanism by which ATP produces the rapid movements is not clear, however.

References

JAFFE, M. J.: "The role of ATP in mechanical stimulated rapid closure of the Venus's-Flytrap." Am. J. Bot. **51**, 17—18 (1973).

TORIYAMA, H.: "Rapid movements in plants." Ann. Rev. Plant Physiol. **20**, 165—184 (1969).

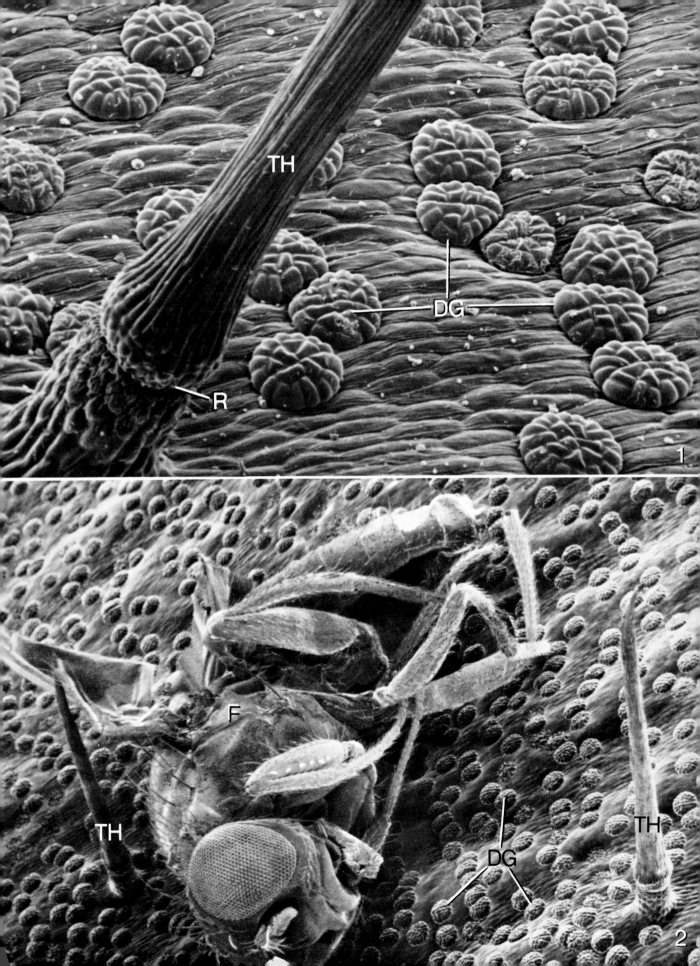

Tendrils (Modified Leaf)

DS Dorsal side
VS Ventral side

Fig. 1 (cucumber tendril), × 25; Fig. 2 (transverse section of a coiled cucumber tendril), × 75; Fig. 3 (squash tendril), × 85.

Some leaves or portions of leaves are so highly specialized that they lose their identity as leaves. A tendril is one type of specialized leaf which has a slender, elongate, thread-like structure. It is employed by less woody plants as an useful means of attaching themselves to external supports to maintain themselves in an erect position as they elongate.

An immature tendril is usually cylindrical or ribbon-like with a slight taper toward the apex. As the organ matures, it grows in length and diameter and eventually develops a hook at the tip. Tendrils normally attain maximum irritability after they have reached about three-fourths of their final length. When a tendril is touched with a stick, it will bend within five minutes and it completes one coil within 20 minutes (Fig. 1). If a mature tendril is never touched, it eventually becomes senescent. If the stick is withdrawn after touching, the tendril usually uncoils and becomes straight and irritable again. If the stick remains touching the tendril, it will tightly coil around the stick and thus anchor the plant (Fig. 3).

By histological means it was found that the early response of the tendril to rubbing is a contraction of the ventral surface accompanied by an expansion of dorsal surface. Subsequently, rapid expansion of the dorsal surface and a slower expansion of the ventral surface results in coiling. The mechanism of the primary response of the tendril to contact stimuli is not yet known. Somehow the stimulus generates an action potential that changes the permeability of membranes to ions, and as a consequence, a redistribution of ions between vacuoles and cytoplasm occurs. Water then flows rapidly out of those cells on the ventral side (VS) in response to osmotic gradients, while cells on the dorsal side (DS) swell (Fig. 2) by the uptake of water. ATP is believed to be involved in the ion pumping process and auxin is involved in plasticizing the dorsal cell walls.

References

JAFFE, M. J., and GALSTON, A. W.: "The physiology of tendrils." Ann. Rev. Plant Physiol. **19**, 417—434 (1968).
REINHOLD, L.: "Induction of coiling in tendrils by auxin and carbon dioxide." Science **158**, 791—793 (1967).

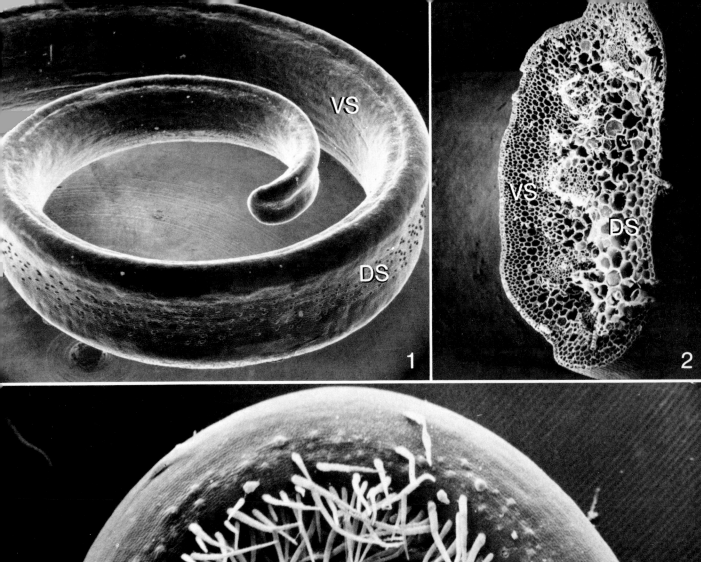

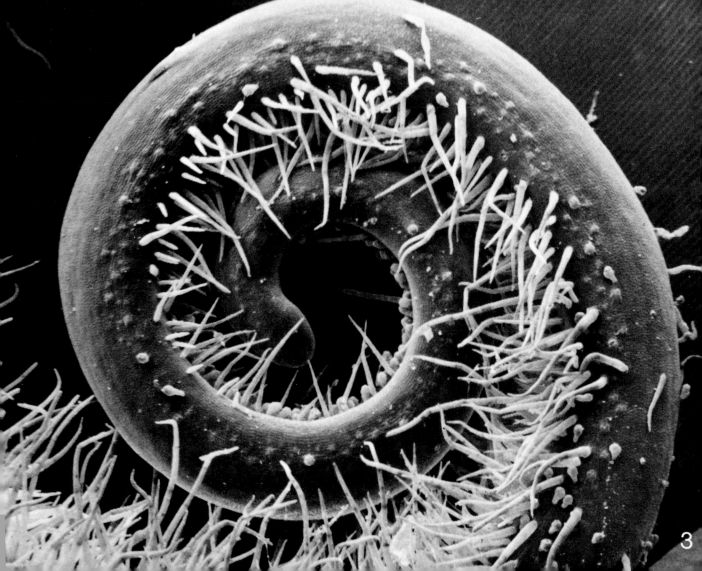

Flower

Photoperiodism and
Initiation of Flowering (Cocklebur)

Ap   Apex
Br   Bract
FP   Flower
LP   Leaf primordium

Fig. 1 (vegetative bud), × 240; Fig. 2 (staminate influorescence primordium), × 190.

For many plant species, day length is an important factor in the control of flowering, and the phenomenon is known as photoperiodism. Some plants flower only late in summer as the days shorten. They are called short-day plants; i.e., chrysanthemum and michaelmas daisies. Other plants flower only during the long days of summer and are called long-day plants; i.e., grasses, cereals, and clover. There are species in which day length does not affect flowering, and they are called neutral plants; i.e., the everbearing strawberry and cucumber.

Cocklebur (*Xanthium strumarium*) is one of the most widely studied short-day plants and is extremely sensitive to changes in day length. The plant requires only one short-day cycle for flowering. Once the plant is induced to flower by being kept in the dark for $8^1/_4$ hours or longer, it will continue to produce flowers even though the plant is returned to a long-day condition.

A remarkable feature of the dark period treatment is that it must be uninterrupted. With *Xanthium*, one minute of light at an intensity of 150 fc (1 500 lux) during a 9-hour dark period will suppress flowering. Since red light is found to be most effective in the night interruption and far red light can reverse the interruption, it is believed that the phytochrome system is involved in the process. Phytochrome is a protein pigment that controls a variety of plant physiological activities. When excited by red light (660 nm), phytochrome shifts from a red-absorbing form (Pr) to a far-red-absorbing form (Pfr). When illuminated with far red light (730 nm) or left in the dark, the reverse occurs.

The leaves usually perceive the photoperiodic condition. However, flower initiation itself takes place at the growing apex. Thus, a stimulus, called a florigen, apparently moves between leaves and apex to initiate flowering. This florigen (not yet isolated or characterized chemically) is synthesized after a certain minimum period of darkness, and the synthesis proceeds rapidly during the next few hours. It is apparently necessary for phytochrome to be present in the Pr form in order for flower hormone synthesis to proceed. Phytochrome is considered to be involved in a time-measuring process.

When a flowering stimulus arrives at the shoot apex, the vegetative apices (Fig. 1) cease producing leaves and buds and become converted into flowering apices. The first detectable change is an increase in cell division in the region of the corpus between the central mother cells and the rib meristem. It has been postulated that the florigen acts at the shoot apex by switching on the flowering gene. The exact nature of the gene-switching mechanism is still unknown.

Following cell division, the mantle gives rise to the bracts (Br in Fig. 2) and the flower primordia (FP in Fig. 2). Ultimately, the flower primordia extends over the whole surface of the apex (Fig. 2) so that all meristematic tissue becomes differentiated. Thus, a vegetative apex capable of unlimited growth becomes a determinate meristem of the influorescence.

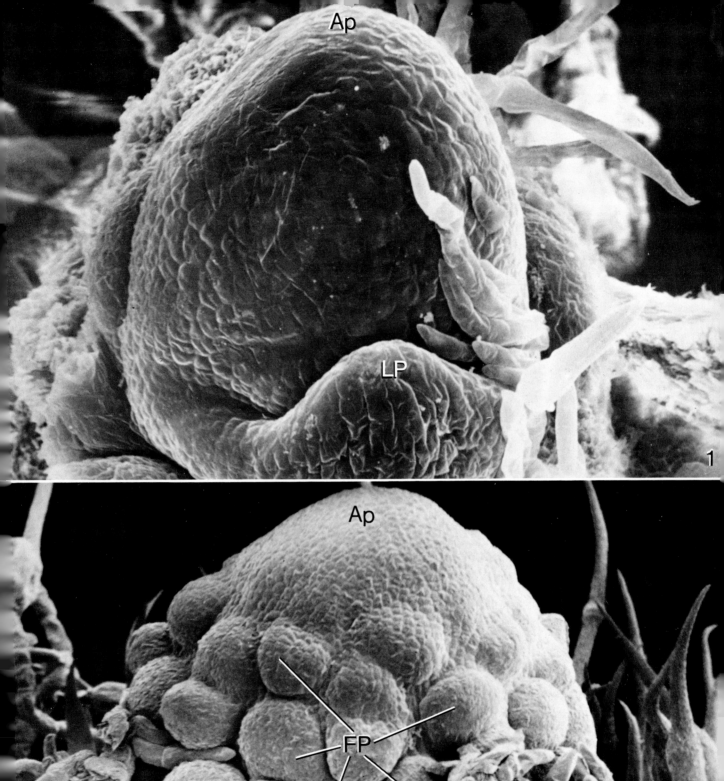

Development of Male Flowers (Corn)

Ca   Carpel
Fl   Floret
FP   Floret primordia
Le   Lemma
Lo   Lodicule
Pa   Palea
St   Stamen

Fig. 1, ×145; Fig. 2, ×92; Fig. 3, ×70; Fig. 4, ×28; Fig. 5, ×200; Fig. 6, ×163; Fig. 7, ×75.

Corn is a day-neutral plant whose flower initiation depends on the age of the plant rather than changes in day length. Corn has an imperfect flower. The staminate (male) flower is grouped at the top of the plant as a tassel, and the pistillate (female) flowers grow on the node and eventually develop into ears.

When the apex of a corn plant switches from vegetative to reproductive growth, the apical meristem becomes elongated and many ridges appear on the surface (Fig. 1). The ridges then protrude to become teeth-like floret primordia (FP) and a secondary ridge appears at the base of each primordium (Figs. 2 and 5). The bottom ridge later develops into two bracts, called a lemma (Le) and a palea (Pa) (Figs. 3 and 6), while the top portion of each floret (Fl) develops into three stamens (St), a carpel (Ca), and two lodicules (Lo) (Fig. 7). The carpel does not completely develop to maturity. Thus, each mature floret on the tassels consists of a sterile carpel, three stamens, two bracts, and two lodicules.

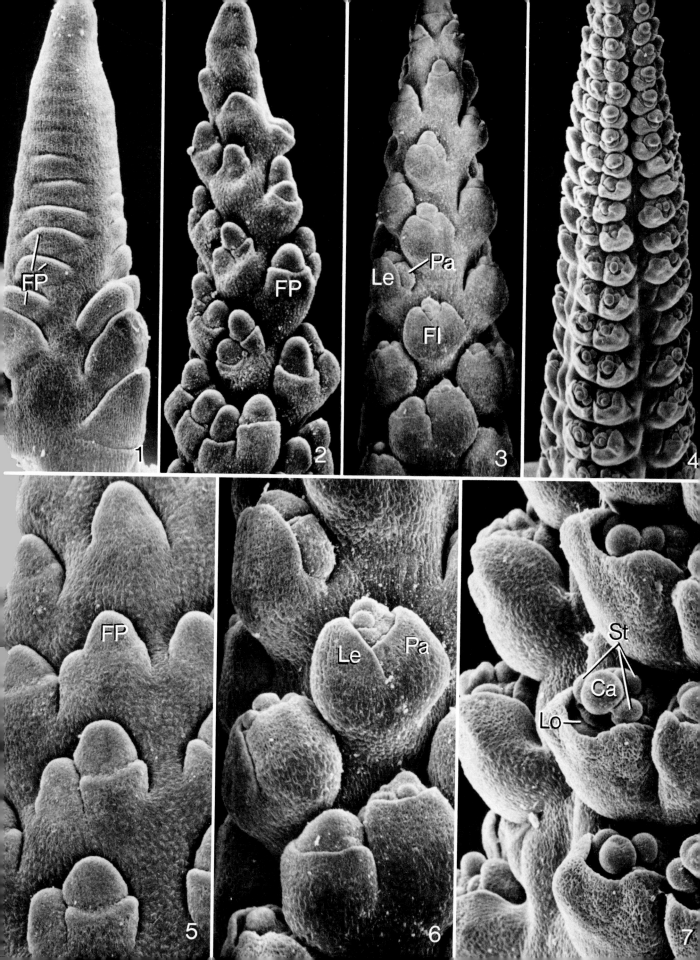

Morphology of Cactus Flower

Flowers are actually modified leaves. A typical flower is composed of four whorls consisting of (1) sepals, (2) petals, (3) stamens, and (4) carpels. All are attached to a receptacle. Sepals enclose the other flower parts during flower development and are usually green. Collectively, the sepals are called calyx. The petals (Pe), which are often showy and attractively colored, are collectively designated the corolla. The stamens form a whorl inside the corolla. Each stamen consists of a slender filament (Fi) and a terminal anther (An) which bears pollen (Po). Collectively, the stamens of a single flower are called an androecium. The carpels comprise the central whorl in a flower. Together, they are commonly called a gynoecium, or a pistil. The pistil consists of an expanded ovary, a slender stalk (style), and a stigma which receives pollen. A flower may not have all the indicated parts, but when it does, it is called a complete flower.

Mamminillaris chionocephala has a characteristic cactus flower structure: The sepals and petals are numerous and adjected basally, and carpels and styles are coalesced but stigmas are separate (Figs. 1 and 2). The pistil consists of an inferior ovary with a single chamber. Because of the scarcity of water in the desert, cactus flowers bloom only for a short time. Showy petals and numerous stamens are needed to attract insects to pollinate them effectively.

Of all characteristics of flowering plants, the flower is the least affected by environmental changes. The leaf shape is influenced by age, light, water, and nutrition, but flowers and fruits are not as affected as leaves. For this reason floral and fruit structures are used in plant classification.

Dandelion

Fig. 1 (floret), ×12; Fig. 2 (anther), ×30; Fig. 3 (stigma), ×41; Fig. 4 (pappus), ×100; Fig. 5 (achens), ×50.

The dandelion *(Taraxacum)* is a perennial herb belonging to the family Compositae. Members of this group are generally considered to be the most highly developed of the flowering plants. The Compositae is also one of the largest plant families with 12000 known species and it consists of many food plants such as artichokes, lettuce, and sunflowers as well as showy flower plants such as dahlia, chrysanthemum, aster, and zinnia.

Dandelions are worldwide in their distribution. The flowers bloom nearly throughout the year. Although dandelions serve to color nature golden, they are pestiferous weeds. The composite flower, as the name implies, consists of a group of many small flowers (florets) arranged to give the appearance of a single flower. This arrangement is useful in adaptation for pollination by insects. The packing of florets into heads ensures the pollination of a large number of flowers during a single insect visit.

The floret is epigynous (flower borne atop the ovary) (Fig. 1) and consists of a single petal (Pe). The calyx (Ca) is modified into a conspicuous tuft of hairs that enlarges during the fruiting stage to form a pappus (Pa in Fig. 4). There are two carpels as evidenced by the forked stigma (St in Fig. 3). The five stamens have separate filaments, but their anthers are coalesced into a tube (AT) around the style (Sy in Fig. 3). When the floret opens, the pair of stigmas are pressed together inside the anther (Fig. 2). The inside of the anther tube (or pollen sac) ruptures or splits so that pollen then fills the tube. The style gradually lengthens and carries the pollen out of the anther tube. Finally, the stigmas spread open, exposing the receptive surface (Fig. 3). If the flower is not pollinated by insects, the stigmas eventually curl back in order to touch any pollen grains of their own which remain on the style below. This action ensures self-pollination if cross-pollination does not occur. In the polyploid species of the genus *Taraxacum* no pollination is necessary in order to form an embryo. In these plants the first division of the megaspore mother cell is mitotic without reduction in the chromosome number, and the resulting egg cell can directly develop into an embryo. This development of a gamete into a new individual without fertilization is called parthenogenesis.

The inferiorly placed ovary contains one ovule and it ripens to form a dry, single-seeded fruit, a typical achene (Ac in Figs. 1 and 5). It is attached to the plumed pappus (Fig. 4), which renders the fruit sufficiently light to be dispersed by the wind.

Pollen Grains

Fig. 1 (geranium), ×400; Fig. 2 (ragweed), ×1180; Fig. 3 (poinsettia), ×920; Fig. 4 (begonia), ×2320; Fig. 5 (tobacco), ×1200; Fig. 6 (amaryllis), ×618.

Pollen grains are male gametophytes of plants. They are formed in the anther of flowers by meiosis and they carry haploid sperm cells. For protection, pollen grains typically develop a resistant exine layer outside the cellulose wall. The exine portion is a waxy or resinous layer which minimizes mechanical damages as well as desiccation.

Pollen grains from different plants vary widely in their size and shape, and in the number and arrangement of wall apertures and ornamentation of the exine. Several examples are included in the illustrations: the pollen grains of geranium (Fig. 1), ragweed (Fig. 2), poinsettia (Fig. 3), begonia (Fig. 4), tobacco (Fig. 5), and amaryllis (Fig. 6).

The durability of pollen grains together with their morphological variations is the information useful in various scientific works. The study of pollen grains, or palynology, is used by the plant taxonomist in plant classification, the geologist in oil exploration, the archeologist in paleontological work, the forester and agricultural researcher in breeding studies, the honey producers in improving production, and the aero-allergists in seeking the cause of allergies. Incidentally, it is the protein content, not the spiny shape, of the ragweed pollen grains that make many people sneeze.

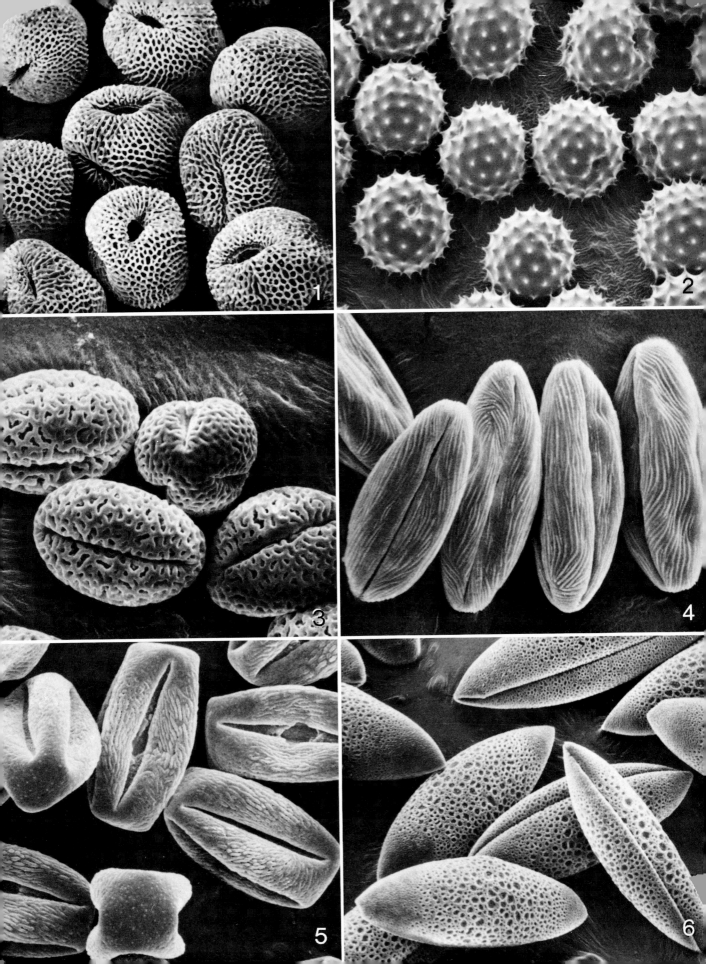

Stigma and Pollen Germination (*Vinca rosea*)

Pa   Papillae
Po   Pollen
PT   Pollen tube
SF   Stigma fluid
St   Stigma

Fig. 1 (stigma), ×100; Fig. 2 (pollen germination), ×100; Fig. 3 (pollen tube early growth), ×300; Fig. 4 (pollen tube late growth), ×600.

Pollination is the transfer of pollen grain from an anther to a stigma. Pollinating agents include wind, insects, and other animals. There are two types of pollination: self-pollination, which occurs in the same flower, and cross-pollination, which involves two plants with different genetic constitution.

Vinca is a self-pollinating plant. The anthers encircle the stigma (St) to ensure that the shed pollen (Po) will fall on the stigma. The stigma also produces a sugary secretion called stigma fluid (SF) to trap pollen (Fig. 1). When pollen grains fall on the stigma, they absorb water and swell (Fig. 2). The inner membrane extends through a pore in the outer wall (Figs. 3 and 4) of the pollen grain and forms a protoplast-linked tube. The pollen tube (PT) grows between the stigmatic papillae (Pa), penetrates the tissue (Fig. 3) of the stigma, grows down through the style, and enters the ovary. This kind of growth is believed to be chemically directed. Recent studies have shown that the ovules have a high calcium ion content to which the pollen tube is attracted.

Usually pollen consists of a generative cell and a tube cell when it is shed from the anther. Before or during germination the generative cell undergoes division and produces two sperm cells. These cells migrate to and remain at the tip of the pollen tube during its growth toward the ovary. The pollen of many plants will germinate in water or in a sugar solution. The *Vinca* pollen will germinate rapidly in 10% sucrose solution (Fig. 4), but the pollen of some species will not germinate in any artificial medium.

Germination of pollen is often the catalyst for ovarian growth since with successful pollination, there follows a burst of growth in the ovary and heavier pollinations usually result in increased fruit set. It is known that pollen is a rich source of auxin. An unknown type of stimulating factor has also been described to be present in pollen since pollination results in a stimulation of auxin formation in the ovary.

Seed

Cotton Seed and Bedstraw Fruit

Fi   Fiber
Ho Hook
SC Seed coat

Fig. 1 (cotton seed), $\times 210$; Fig. 2 (cotton fabric), $\times 135$; Fig. 3 (*Galium* fruit), $\times 40$; Fig. 4 (hooks), $\times 280$.

Each kind of seed or fruit has its own means of dispersal so that the plant can be propagated to a wide geographic area and to prevent excessive overcrowding. The chief agents in seed and fruit dispersal are wind, water, and animals, including man. Wind-dispersed seeds usually possess wings, tufts, or hairs to facilitate air travel. A good example is the cotton seed (Fig. 1), which possess many long fibers (Fi) on its seed coat (SC). This cotton fiber has been used by humans in making fabrics (Fig. 2) for so long that even in this synthetic-fiber era it is still a valuable fiber source.

Frequently seeds possess spines, beards, or a sticky secretion which enable them to adhere to the fur of mammals or the feathers of birds. Hooks (Ho) such as on the fruits of cocklebur or bedstraw (*Galium oporine*) (Figs. 3 and 4) are also effective in attaching the fruits to fur or cloth for dispersal. Some fruits such as the coconut float on water and can drift for thousands of miles. Some fruits such as berries may be eaten by animals, and the enclosed seeds may survive the digestive process and later become dispersed.

176

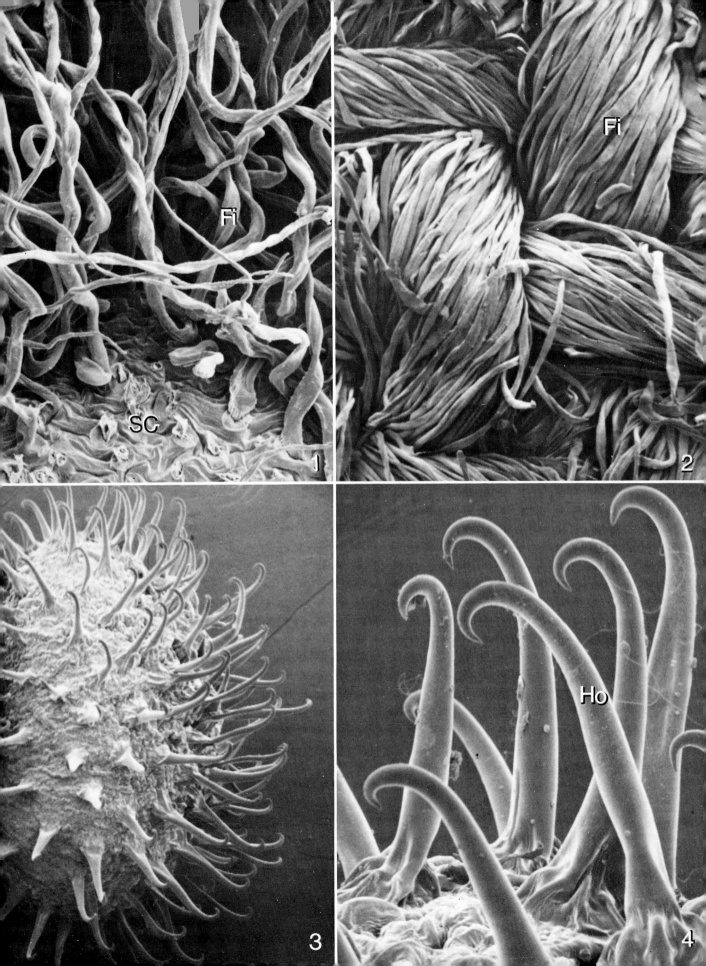

Germination (Barley Seed)

AL	Aleurone layer	Ra	Radicle
Ap	Apex	RC	Root cap
Cl	Coleoptile	Rt	Root
Cr	Coleorhiza	SA	Shoot apex
Em	Embryo	SC	Seed coat
En	Endosperm	Sc	Scutellum
Lv	Leaves	SE	Scutellar epithelium
Pl	Plumule	St	Starch

Fig. 1 (barley embryo), ×85; Fig. 2 (longitudinal section of a barley embryo), ×55; Fig. 3 (aleurone layer and scutellar epithelium), ×440; Fig. 4 (starch grains), ×4000; Fig. 5 (starch grains), ×3150.

A mature monocot seed typically consists of an embryo (Fig. 1) as well as a nutrient storage tissue called the endosperm (En in Figs. 2, 3). A longitudinal section of the embryo (Fig. 2) reveals that it is well differentiated internally. The shoot consists of an apex (Ap) and several immature leaves (Lv). The entire shoot is enclosed by a sheath called the coleoptile (Cl). The single cotyledon, called a scutellum (Sc), is firmly attached to the endosperm. At the lower end of the embryo axis, the root (Rt) and root cap (RC) are differentiated and enveloped by a tubular, thick sheath called the coleorhiza (Cr). Enlargement of the upper junction of the embryo and endosperm (Fig. 3) reveals the presence of an outer aleurone (Al) layer surrounding the endosperm portion of the seed. The aleurone layer is absent at the region where the endosperm and embryo join (arrow). In its place, a scutellar epithelium (SE) is present (Fig. 3).

Seeds may remain dormant for several weeks (for example, the silver maple) or for more than 1000 years (for example, the Indian lotus). After maturity they maintain a low metabolic activity during storage. Dryness and low temperature usually prolong seed life by slowing down its metabolism. With an adequate water supply and a favorable temperature, seeds will resume growth of the embryo. As water is absorbed there is an increase in seed volume and a general softening and weakening of the seed coat (SC). As a result, a developing embryo emerges (Fig. 1).

In its initial growth the seed depends on reserve nutrients. In the case of barley the nutrients are primarily in starch form (Figs. 4, 5); therefore, a mechanism is required to bring about a transformation of these reserves into a soluble form that can be transported to the growing regions of the seedling. Current research has demonstrated that during germination of barley seeds a plant hormone, gibberellin, is released by the embryo which induces the synthesis of amylase and other hydrolases in the aleurone layer (AL in Fig. 3).

These amylases, in turn, convert insoluble starch (St) into soluble sugars. The digestion of starch grains appears to involve local hydrolysis to create numerous pits initially (Fig. 4) followed by a progressive dissolution of successive layers along the pit margin (Fig. 5).

There is a dramatic increase in the rate of cell division in the meristematic region of the shoot apex and radicle during germination. The scutellum epithelium absorbs the digested food from the endosperm and transfers it to the growing regions. A seed germinates by firstly giving rise to a radicle (Ra in Fig. 1), followed closely by a plumule (Pl in Fig. 1), which is protected by a sheath, or coleoptile (Cl in Fig. 2). From the time that the radicle emerges from the seed to the time that it is capable of existing independent of stored food reserves in the seed, a plant is known as a seedling.

Reference

SHIH, C. Y.: "SEM studies of the internal organization of plant organs." Scanning Electron Microscopy/1974, O. JOHARI and I. CORVIN, Eds., IITRI 7, 343—349 (1974).

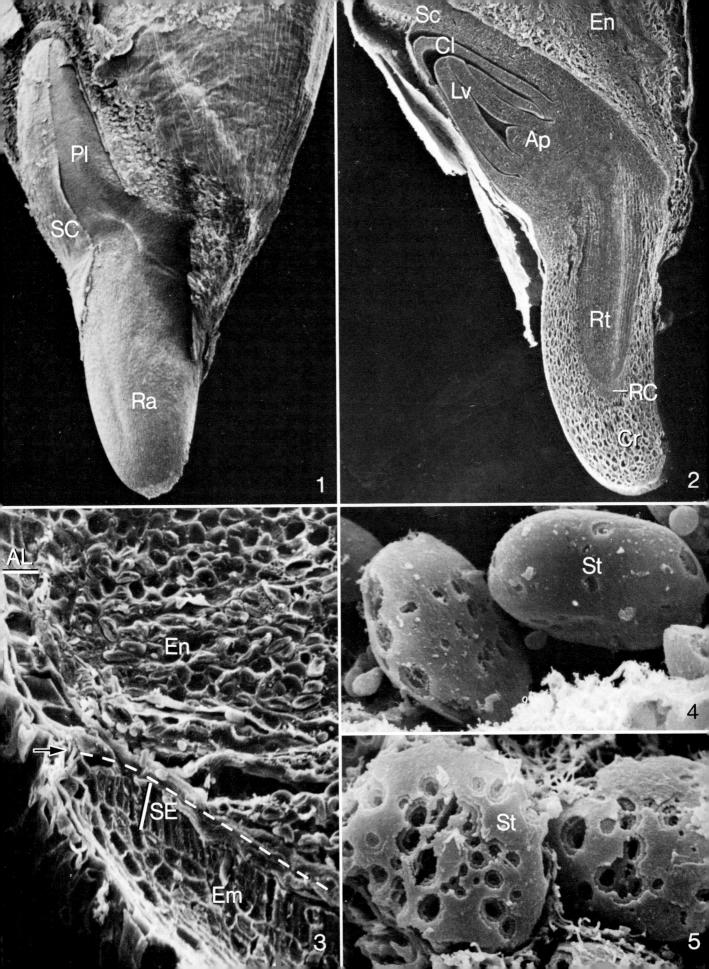

Chapter 8
Multicellular Animals

Sponges

Phylum Parazoa

Class Calcarea

Scypha

Ap	Apopyles or internal ostia
CC	Choanocyte (radial) canals
Ch	Choanocytes
IC	Incurrent canals
MS	Monaxon spicules
Os	Osculum
Pa	Paragaster (spongocoel)
Pr	Prosopyles
Sp	Calcareous spicules
TS	Triaxon spicules

Fig. 1, ×62; Fig. 2, ×70; Fig. 3, ×146; Fig. 4, ×30; Fig. 5, ×110; Fig. 6, ×297; Fig. 7, ×2330.

Sponges are primitive multicellular aggregates that are sessile in the adult state and illustrate radial symmetry or asymmetry. They are characterized by a system of pores through which a water current flows. The organism illustrated here is an example of a syconoid type of sponge and has a large exhalent opening called an osculum (Os in Figs. 1 and 2). At the margin of the osculum is a collar formed by the close apposition of many calcareous spicules (Sp in Figs. 1 through 3). The outer surface of the body has a number of incurrent canals (IC in Figs. 1, 3 through 5). Guarding these canals are numerous monaxon spicules (MS in Figs. 3, 5). The incurrent canals are continuous with choanocyte canals (CC in Figs. 4, 5) or radial canals by means of small pores called prosopyles (Pr in Fig. 6). The numerous choanocyte canals are lined with flagellated cells called choanocytes (Ch in Fig. 7). These cells serve to maintain a flow of water and ingest food from the water current. The choanocyte canals communicate with a large central cavity called a spongocoel or paragaster (Pa in Figs. 1, 2, 4) by openings called internal ostia or apopyles (Ap in Figs. 1, 2). The walls of the choanocyte canals are supported by monaxon spicules (MS) and triaxon (3 rays) spicules (TS in Fig. 6).

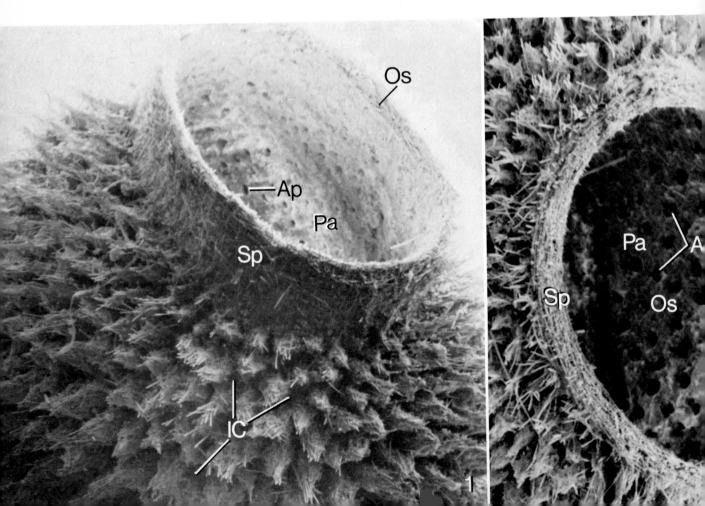

1

Hydra and Other Coelenterates

Phylum Cnidaria

Class Hydrozoa—Order Hydroida

Hydra oligactis

BD	Basal disc
Bo	Body or column
CM	Closed mouth
Hy	Hypodermis
Sp	Spermaries
Te	Tentacles

Fig. 1, ×70; Fig. 2, ×130.

Hydra belongs to a phylum considered to be the most primitive of the Eumetazoa. This organism has radial symmetry and an oral-aboral axis. It is covered by an epidermis derived from ectoderm and an inner gastrodermis derived from endoderm. Between the two layers is a noncellular layer, the mesogloea. There is a single internal cavity, the coelenteron, and a single opening termed the mouth, which is derived from the embryonic blastopore.

The major structural divisions of *Hydra* are illustrated in Figs. 1 and 2 and include the basal disc (BD), a column, or body (Bo), a hypostome (Hy) region at the end of which is a closed mouth (CM), and the arms or tentacles (Te) which surround the hypostome region. The bulbous structures associated with the body of the animal illustrated in Fig. 2 are spermaries (Sp) which contain the differentiating spermatozoa. The sexes are separate in *Hydra*.

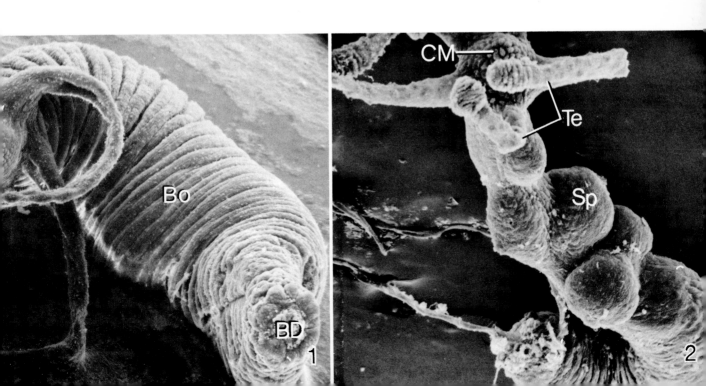

Phylum Cnidaria

Class Hydrozoa—Order Hydroida

Hydra oligactis (continued)

The mouth (CM in Fig. 3) is closed in this figure; it is located at the apex of the hypostome. The hypostome has a rough surface and is composed of several large epitheliomuscular cells (EC), whose free ends are bulbous in shape and which form a series of "lips" guarding the entrance to the mouth (Fig. 7).

The tentacles represent extensions of the body and possess an enormous capacity for contraction and extension. Defense and feeding involve the use of specialized and complex structures called nematocysts. Each represents a secretory product of an individual cell. Interstitial cells differentiate into cnidoblasts, which then secrete a nematocyst. They are more highly concentrated on the tentacles than elsewhere and they are grouped in recurring circular ridges of the tentacles where they are distributed into so-called batteries (NB in Fig. 4). A number of cnidoblasts, each with its secreted nematocyst, may be contained within a single epitheliomuscular cell, which is the primary cell type found in the epidermis. Nematocysts are independent effectors and when they are stimulated to discharge from the cell, a complex eversion process occurs. Nematocysts do not survive after discharge, and hence, are continually being formed as replacements for those discharged.

The cnidocil (Cn) is a trigger-like spine approximately 9 to 11 μm long which extends from the surface of the cnidoblast (Figs. 4, 6). The junction of the proximal portion of the cnidocil and the cell surface is characterized by a raised, ring-shaped papilla (arrows in Fig. 6), which would appear to allow for extensive movement of the cnidocil

Several different types of discharged nematocysts have been described in *Hydra* (i.e., stenotele (penetrant), streptoline glutinant (holotrichous isorhiza), desmoneme (volvent), and stereoline glutinant (atrichous isorhiza). The stenotele type is illustrated in Fig. 6 and consists of a capsule (Ca), everted shaft (ES), stylets (St), spines (Sp), and filament (Fi). This type, when discharged, penetrates the tissue of the prey and presumably injects a poison. The streptoline glutinant (SG) nematocyst possesses a relatively long cylindrical capsule and has a filament that bears numerous small spines or barbs (BF) directed toward the capsule (Fig. 5). The filament (SGF) of a stereoline glutinant nematocyst is also present in Fig. 5. This filament has a granular-appearing surface, a condition that may be due to the presence of an adhesive substance. A section through the body of *Hydra* demonstrates the surface of the gastrodermal cells (GC) which line the coelenteron (Fig. 8). The surface of these cells possesses numerous microvilli (Mv) and usually two flagella (Fl); the latter probably aid in the mixing of the contents of the coelenteron. Gastrodermal gland cells release digestive enzymes which participate in extracellular digestion in the coelenteron. The products of this activity are then taken up into the gastrodermal cells where the digestion process is completed intracellularly.

Hydra commonly is found attached to the substratum by its pedal disc. It can, however, glide on its pedal disc and it sometimes somersaults by using alternatively its tentacles and pedal disc. It can also float by secreting a gas bubble from the pedal disc cells. Although many cnidaria exhibit polymorphism (sometimes as colonies such as *Hydractinia*, to be illustrated next) with polyp and medusa stages, the medusa stage of *Hydra* is minute. *Hydra* is a polyp form capable of both asexual reproduction (by budding) and sexual reproduction, in which egg or sperm are differentiated from interstitial cells in specific regions of the body called spermaries or ovaries.

Reference

BEAMS, H.W., KESSEL, R.G., and SHIH, C.Y.: "The surface features of *Hydra* as revealed by scanning electron microscopy", Trans. Am. Micros. Soc. **92**, 161—175 (1973).

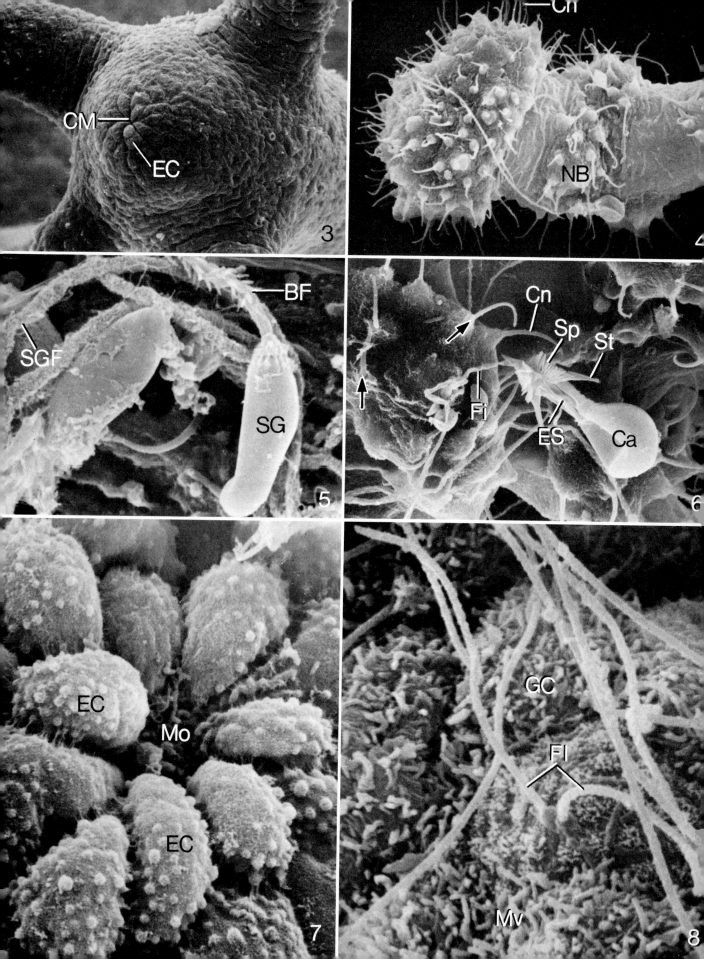

Phylum Cnidaria

Class Hydrozoa—Order Hydroida

Hydractinia

Cn   Cnidocil (on tip of dactylozooid tentacle)
CT   Capitate tentacles
Da   Dactylozooid
FT   Filiform tentacles (of gastrozooid)
Ga   Gastrozooid
Go   Gonozooid (or Gonophore)
Kn   Knobs (on distal end of dactylozooid)
Pe   Perisarc
St   Stolons

Arrows: Cnidocil spines in Fig. 2

Fig. 1 (*Hydractinia*), × 360; Fig. 2 (*Hydractinia*), × 300; Fig. 3 (*Hydractinia*), × 370; Fig. 4 (*Pennaria*), × 25; Fig. 5 (*Pennaria*), × 60.

Hydractinia is a marine colonial hydroid which grows as an encrusting mat, especially on the surface of shells occupied by hermit crabs. Several types of hydroids or "individuals" comprise the colony and all are interconnected by a system of stolons. Stolons are cellular tubes, consisting of a layer of gastrodermis and epidermis (coenosarc), which surround a cavity continuous with the coelenteron of all zooids. A secretion of the epidermal cells forms a nonliving covering over the stolons. This perisarc serves as protection and also for attaching the colony to a substratum.

The stolonic system is not illustrated here, but is hidden from view by the numerous zooids that attach to it. The *Hydractinia* colony thus exhibits polymorphism with dactylozooids distributed among gastrozooids and gonozooids. The feeding hydranth or gastrozooid (Ga in Figs. 1, 2) has a body with a roughened surface and is surrounded distally by a row of filiform tentacles (FT), the tips of which are covered by cnidocil spines (arrows) of nematocysts contained in the cells of the epidermal layer (Fig. 2). The mouth of the gastrozooid in Fig. 2 is obscured by the tentacles. Those zooids or individuals in the colony concerned with sexual reproduction are called gonozooids, or gonophores (Go in Fig. 1). These gonophores produce a minute medusa stage by sexual reproduction. Another member of the colony, more limited in distribution, is the spiral zooid, or dactylozooid (Da), which is specialized for defense. The example shown in Fig. 3 is, however, preserved in an extended (rather than coiled) condition, and it can be seen that it is quite elongated and bears knobs (Kn) at its distal end. These knobs contain numerous nematocysts, the distribution of which is apparent from the position of the cnidocils (Cn).

Pennaria tiarella

This marine colonial hydroid has a free medusa stage, unlike *Hydractinia*. The hydranths shown in Figs. 4 and 5 have capitate tentacles (CT) and a basal whorl of filiform tentacles (FT). The gonophores (Go) arise from the hydranth just distal to the filiform tentacles. The hydranths are connected by coenosarc tubes which are surrounded by a noncellular, protective perisarc (Pe). The stalks that support the hydranths are called stolons (St), and this system is collectively called the hydrorhiza.

188

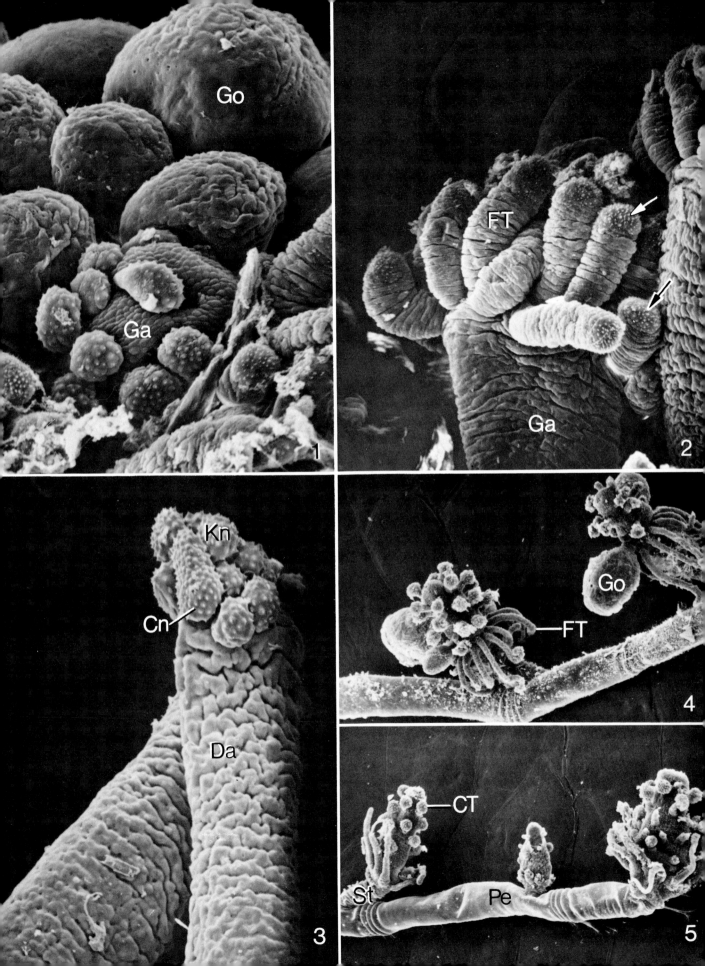

Flatworms

Phylum Platyhelminthes

Class Turbellaria—Order Tricladida

Dugesia (Planaria)

Au   Auricle
Br   Sensory bristles
CA   Cilia on auricles
Mo   Mouth
Ph   Pharynx
Rh   Rhabdites

Fig. 1 (dorsal view of entire planaria), × 15; Fig. 2 (ventral view of entire planaria), × 40; Fig. 3 (everted pharynx), × 195; Fig. 4 (surface of pharynx), × 3850; Fig. 5 (auricle), × 720.

The flatworms represent the most primitive of the acoelomate bilateria. The most widely studied example of free-living flatworms is the planaria. This worm is dorso-ventrally flattened and elongated along the primary body axis (Fig. 1). The only cavity present internally is that associated with the digestive tract; it has only one opening—a mouth located on the mid-ventral surface and derived from the embryonic blastopore. A muscular pharynx (Ph in Figs. 2, 3) representing the initial part of the digestive system can be everted through the mouth (Mo in Fig. 3) and food can be sucked up through the pharynx to the remainder of the digestive system by the muscular activity in the pharynx wall.

An initial stage in cephalization is illustrated in *Dugesia*. The "head" end is extended laterally in the form of paired auricles (Au in Fig. 5) and two eyespots are present in an area closely associated with a cerebral ganglion (or primitive "brain").

An epithelial layer encloses the animal externally; it is unique in several ways. On the ventral side ciliated epidermal cells are abundant, and both the number and length of these cilia are conveniently illustrated in the SEM (CE in Fig. 6). Ciliated epidermal cells are not as numerous on the dorsal side of the animal (Figs. 7, 9).

Many small turbellaria move largely by means of the beat of cilia associated with the ventral epidermal cells. Gland cells on the ventral surface secrete a slime which, among other functions, provides a resistance against which the cilia can thrust. In order to display the ciliated cells clearly in *Dugesia*, the organisms were treated to remove the obscuring slime coat. Larger turbellaria use cilia for locomotion, but in addition, they have muscle layers (circular, longitudinal, diagonal, parenchymal) internal to the epidermis which can produce body undulations effective in locomotion.

190

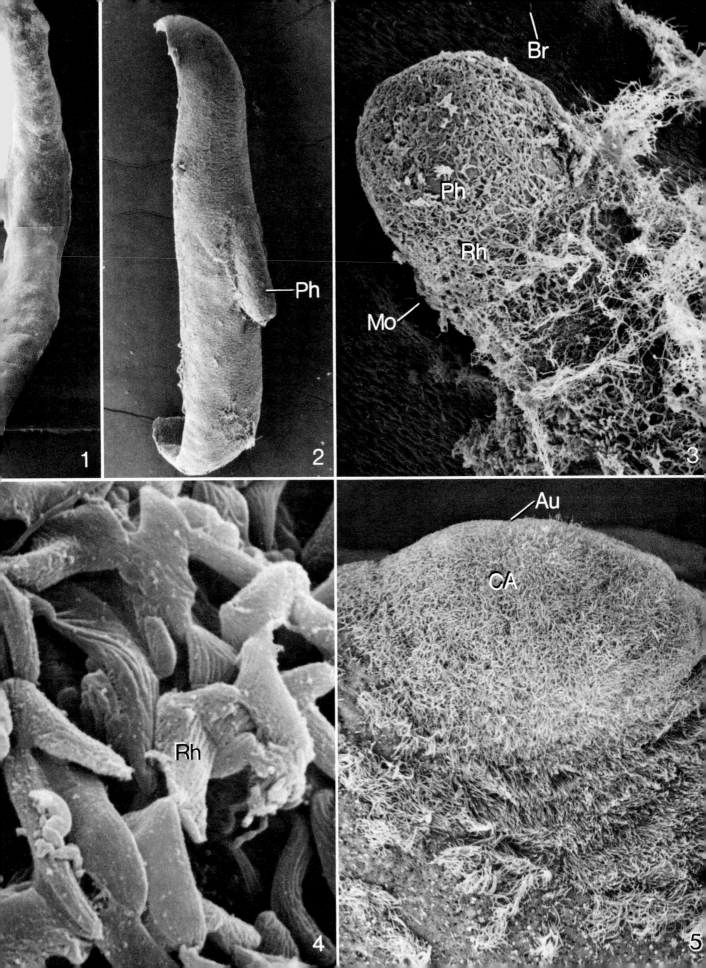

Phylum Platyhelminthes

Class Turbellaria—Order Tricladida

Dugesia (Planaria)
(continued)

Br	Sensory bristle	Po	Pores in epidermal
CE	Ciliated epidermal		cells
	cells	Ph	Pharynx
Mi	Microvilli on epi-	Rh	Rhabdites
	dermal cells		

Arrows: Discharge of rhabdites

Fig. 6 (ventral surface), ×2215; Fig. 7 (anterior dorsal surface), ×2020; Fig. 8 (ventral surface), ×3765; Fig. 9 (dorsal surface), ×2625.

Many of the epithelial cells on both the dorsal and ventral sides of the body have small projections or microvilli (Mi in Figs. 6, 9) associated with their free surface. In addition, numerous pores or apertures about 1 to 1.5 μm in diameter are associated with the surface of the epithelial cells (Po in Figs. 6, 7). They are more prevalent on the dorsal side of the organism than on the ventral surface. It is well known that the epidermal cells of turbellaria secrete and store numerous elliptical bodies called rhabdites. The chemical constitution of these bodies is not clear, but they are known to be released from the surface of epidermal cells and to dissolve rapidly in water when extruded. It is thought that these secretory bodies serve to protect the flatworm against irritating chemicals. Stages in the release of rhabdites (Ra in Figs. 7 through 9) or rhabdoids are particularly numerous in the micrographs (arrows, Figs. 7, 9). The great frequency of discharge in these animals probably represents a response to the glutaraldehyde fixative that appears to preserve their structure. Some of the pores associated with the epidermal cells may represent surface areas of the epidermal cells through which the rhabdites have been discharged, for they are similar in size to those openings through which the elliptical rhabdites appear to be in the process of emerging. The rhabdites range from about 6 to 9 μm in length and 2 to 3 μm in width. They are much more numerous in epithelial cells on the dorsal surface than in similar cells on the ventral surface. Therefore, the discharge of rhabdites from epithelial cells is more frequently observed on the dorsal surface in the scanning electron micrographs.

There is only a rather primitive development of sense organs in the flatworms. As a result of active locomotion and the acquisition of an anterior-posterior axis, there is some achievement with respect to cephalization. In addition to the anterior, paired, pigment-spot ocellus-type photoreceptors, the paired auricles are known to contain numerous chemoreceptors. The chemoreceptors frequently take the form of ciliated pits or grooves containing glandular and chemoreceptor cells. The dorsal side of the auricles is covered with an entangled meshwork of cilia (CA in Fig. 5), which obscure the chemoreceptor and tactile cells associated with the auricles. It is thought that the heavily ciliated cells on the auricles serve to circulate water over the chemoreceptor cells and perhaps provide some directional information by this action. While chemoreceptors are thought to be concentrated on the auricles, tactile cells are generally distributed over the body surface, although they may be concentrated anteriorly and tend to be grouped in clusters. Both chemoreceptor cells and tactile cells have been described as having one or more external " bristles" associated with their free surface. The touch receptor is a neurosensory cell with one or more bristles extending from one pole between the epidermal cells. What may represent the surface bristles (Br) of tactile neurosensory cells are illustrated from the dorsal anterior (Fig. 9) as well as ventral surface (Fig. 3) of *Dugesia*. The precise identification of these structures as sensory bristles versus a cilium or flagellum must be tentative until correlated transmission electron microscope studies of sections from this area are made.

The parenchyma (mesoderm) consists of cells packing the area between the body wall and digestive system. In it are differentiated a primitive excertory (osmoregulatory) system and a reproductive system.

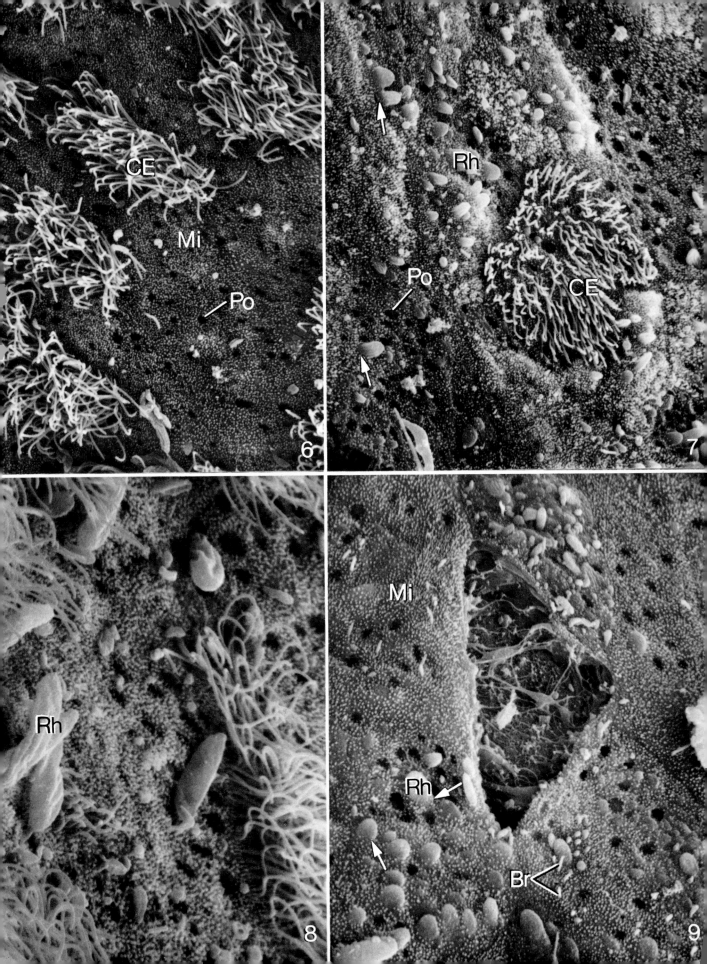

Phylum Platyhelminthes

Class Trematoda—Order Digenea

Schistosoma mansoni (Fluke)

Schistosoma mansoni constitutes a major world health problem because these digenetic trematodes produce a serious parasitic disease called schistosomiasis. The adult male and female (dioecious) flukes have an interesting association, which is illustrated in Fig. 1. The male (♂) is larger than the female (♀), and the body of the male is curved so as to form a ventral groove, called the gynecophoric canal, in which the female resides. The organisms illustrated were obtained from the mesenteric veins of a mouse. The adult worms typically reside close to sources of nutriments and they can be observed with the light microscope to move through these veins in an undulating manner. As shown in Figs. 1 and 3, the male is covered by numerous surface papillae or tubercles. Many of these tubercles are covered with spines (ST) that are 1 to 2 µm long (Figs. 3, 5, black arrows). These surface specializations may assist the male in moving through the portal vein. A few of the surface tubercles, however, lack spines (Fig. 3, SmT). Further, single short projections are distributed on the flat portion of the body surface (Fig. 3, white arrows). The papillae are not found on the anterior portion of the male, nor are they present on the surface of the gynecophoric canal. Two suckers are located in an anterior position on the ventral surface of both the male and female (Fig. 1). These suckers, called anterior (AS) and ventral suckers (VS) (Fig. 1), are uniformly covered by numerous spines (SS), which are also about 1 to 2 mµ long and which appear to project through ring-shaped orifices in the sucker surface (Figs. 2, 4).

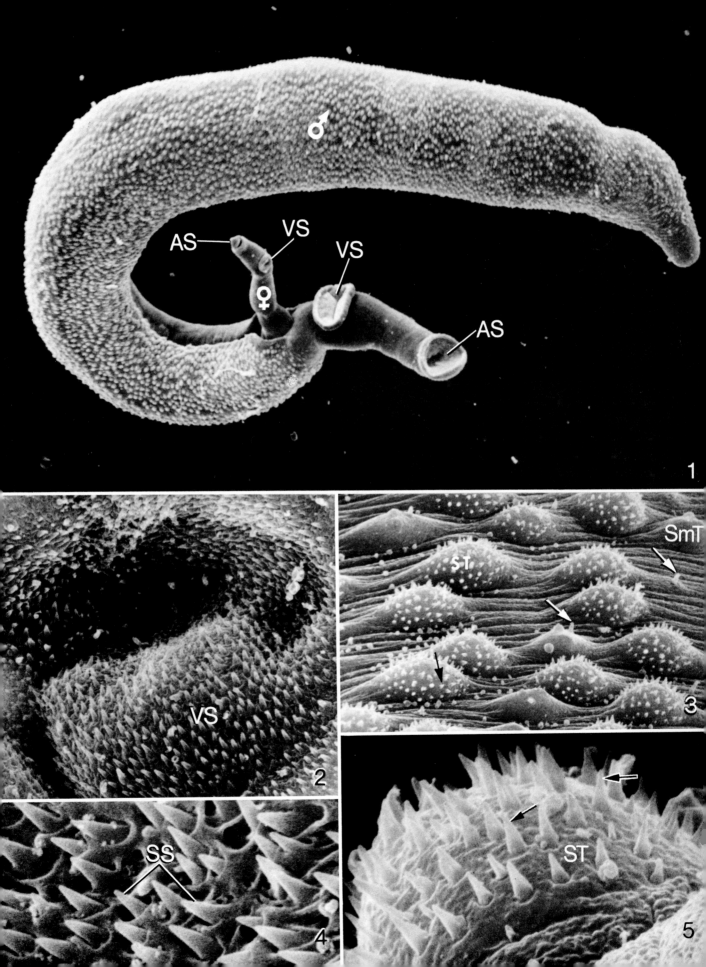

Phylum Platyhelminthes

Class Trematoda—Order Digenea

Mesocercaria of *Allaria canis*

A mesocercaria stage in the life history of this fluke is illustrated in the scanning electron micrographs on the adjacent page. The adult flukes measure about 2.5 to 4.5 mm in length and live in the intestine of canids including dogs. The eggs laid by the adult flukes are voided in the feces, and after about two weeks of development, the embryos hatch in the form of ciliated miracidia larvae which penetrate an appropriate species of snail. Inside the snail the miracidia develop into sporocysts and subsequently into cercariae larvae. The cercariae leave the snail and attach to tadpoles of, for example, *Rana pipiens*. Once the cercariae have entered through the integument of the tadpole, they develop into mesocercariae. If a tadpole or frog is eaten by a canid, the mesocercariae are liberated from the tadpole tissues and eventually the adult worms come to reside in the intestine. A ventral view of a mesocercaria is illustrated in Fig. 1. An oral sucker (OS) is located anteriorly and the ventral sucker, or acetabulum (Ac), takes the form of a triangular opening in the ventral surface of this organism. This appearance seems to be due to the withdrawal of the sucker during fixation (Fig. 3). Several circular folds enclose the oral sucker (Fig. 2), and the body is abundantly provided with numerous pointed spines (Sp) which extend from elevated tubercles (Tu) covering the body surface including that associated with the suckers (Fig.4). The small opening on the ventral surface (arrow in Fig. 1) probably represents the excretory pore.

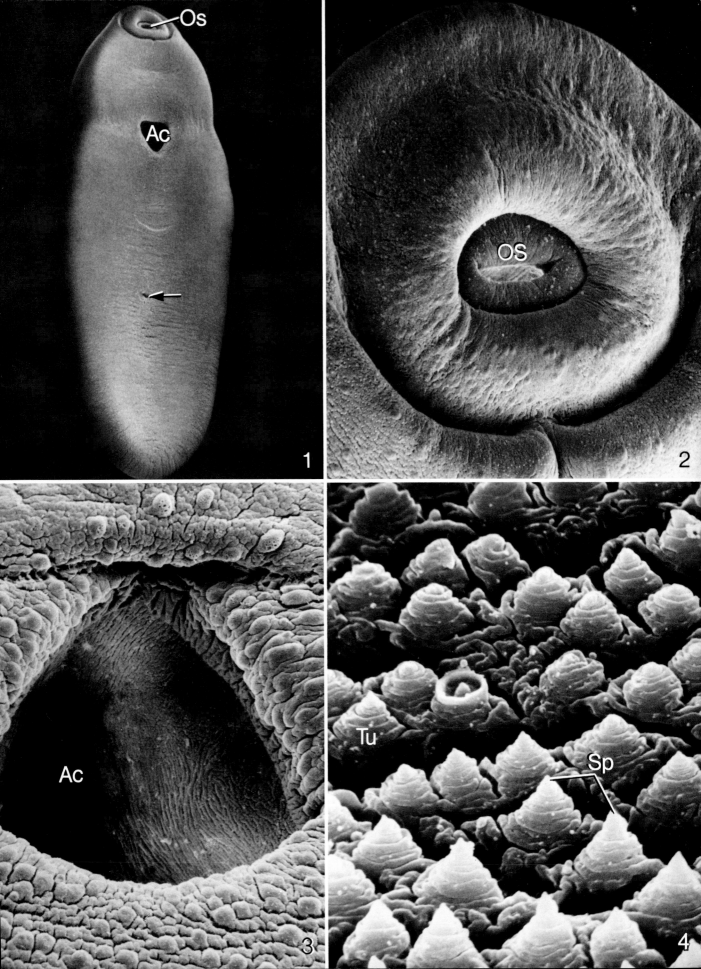

Phylum Platyhelminthes

Class Trematoda—Order Digenea

Prosthodendrium volaticum sp.n.

Ac	Acetabulum
Ci	Cirrus
EP	Excretory pore
GO	Genital opening
GP	Genital papilla
Mo	Mouth
Sp	Spines

Fig. 1, $\times 270$; Fig. 2, $\times 270$; Fig. 3, $\times 1020$; Fig. 4, $\times 7500$.

This adult fluke typically resides in the intestine of the red and big brown bat. The organisms illustrated here are dorsoventrally flattened and have a fleshy body. The posterior end is rounded and the anterior end is broadly pointed. A dorsal (Fig. 1) and a ventral (Fig. 2) view are included. On the ventral side the anterior mouth (Mo) is apparent; the excretory pore (EP) is located at the posterior end of the animal. Near the mid-ventral region of the body is a ventral sucker complex or acetabulum (Ac), which is withdrawn in Fig. 2. The genital opening is favorably illustrated in Fig. 3. The extended components include a portion of the cirrus (Ci) projecting from a genital opening (GO) at the end of an extended genital papilla (GP) While this species has been described to be aspinose, it is apparent in Fig. 4 that numerous tongue-shaped spines (Sp) do project from the body surface. Further, the pointed end of these curved spines, which measure about 0.5 to 0.6 µm at their widest point, all appear to project posteriorly.

Reference

BLANKESPOOR, H.D., and ULMER, M.J.: "*Prostodendrium volaticum* sp. n. (Trematoda: Lecithodendriidae) from two species of Iowa bats", Proc. Helminth. Soc. Washington **39**, 224—226, 1972.

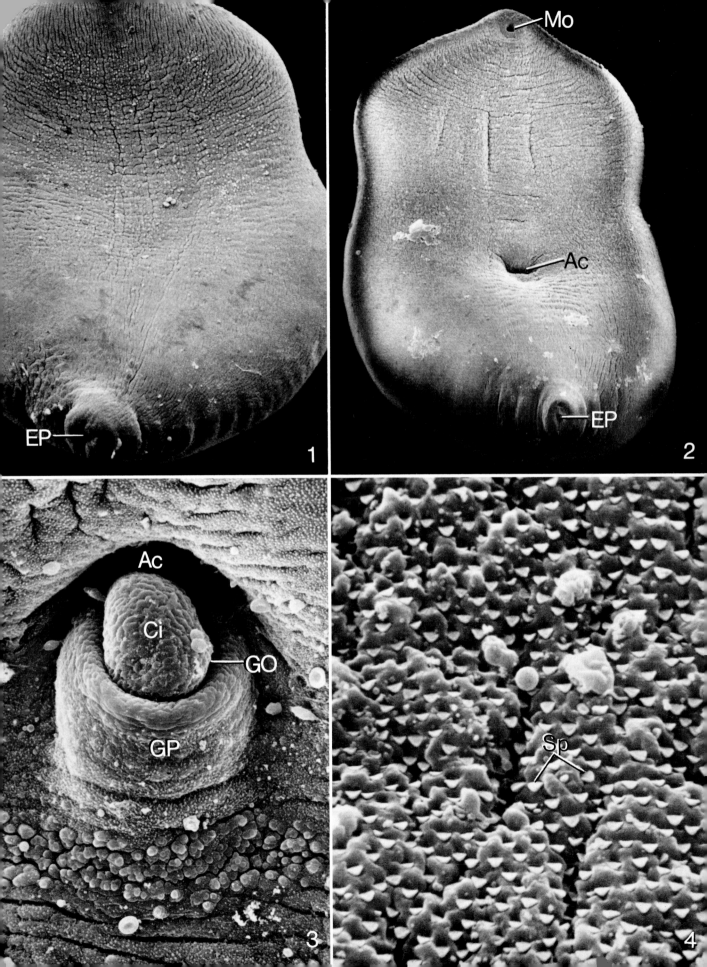

Phylum Platyhelminthes

Class Cestoidea—Order Cyclophyllidea

Taenia pisiformis
and *Hymenolepis diminuta* (Tapeworms)

The tapeworms comprise one group of acoelomate bilateria and have a parasitic existence with a complex life cycle involving several hosts. The anterior end of the tapeworm has a well-developed head or scolex (Figs. 1, 3, 4) which bears four cup-shaped suckers (Su in Figs. 1, 3, 4). The scolex of *Taenia pisiformis* is illustrated in Fig. 1. In this cyclophyllidean cestode four cup-shaped suckers (Su in Fig. 1) are associated with the scolex. In additon, there is present a rounded apical rostellum (Ro in Fig. 1) which is armed with numerous rostellar hooks (RH in Figs. 1 and 2). The rostellar hooks are recurved posteriorly (Figs. 1 and 2) and are arranged in two rings around the rostellum. The 20 hooks located in the more apical band are longer than those located posteriorly and measure approximately 130 μm in length. Those hooks arranged in a circle just posteriorly are slightly shorter and are set in between the hooks in the apical circle (Fig. 2).

The scolex of *Hymenolepis diminuta* is illustrated in Fig. 3. The scolex of this tapeworm has suckers but does not possess a rostellum. In the tapeworm scolex illustrated in Fig. 4 the rostellum (Ro) is retracted. The tapeworm body is a ribbon composed of a series of asexual buds called proglottids (Pr in Fig. 3). The proglottids are flattened, and those located just posterior to the scolex are smaller and immature compared to the larger, mature proglottids located in a more posterior position. New proglottids are produced immediately behind the scolex (NP in Fig. 3). The tapeworm attaches itself to the intestinal mucosa by means of the suckers on the scolex (and rostellum if present), and digested food from the host's alimentary canal is absorbed through the body wall.

The posterior proglottids are massive reproductive units which produce both eggs and sperm (hermaphroditic, self-reproducing units), and they detach from the posterior end of the worm and degenerate. The mature tapeworm eggs (Eg in Fig. 5) measure about 50 μm in diameter. A portion of the proglottid section contained in Fig. 5 is enlarged in Fig. 6 to illustrate the sperm. The small anterior head (He in Fig. 6) and the long flagella (Fl in Fig. 6) comprising the sperm are apparent. The noncellular tapeworm surface (cuticle) is provided with numerous but extremely small projections (Pro in Fig. 7).

Phylum Platyhelminthes

Class Cestoidea—Order Cyclophyllidea

Taenia pisiformis
and *Hymenolepis diminuta* (Tapeworms)
(continued)

Fig. 1 (scolex of *T. pisiformis*), ×105; Fig. 2 (rostellum of *T. pisiformis*), ×670; Fig. 3 (anterior end of *H. diminuta*), ×195; Fig. 4 (scolex), ×735; Fig. 5 (proglottid section with eggs), ×90; Fig. 6 (sperm), ×2550; Fig. 7 (cuticle projections), ×16,050.

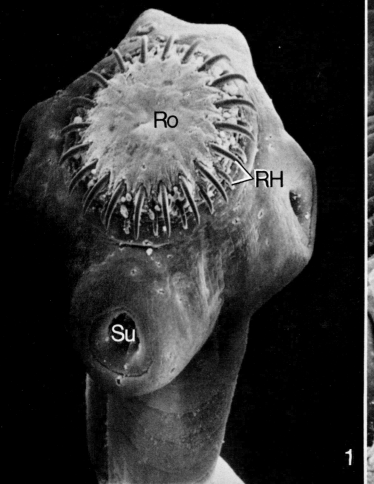

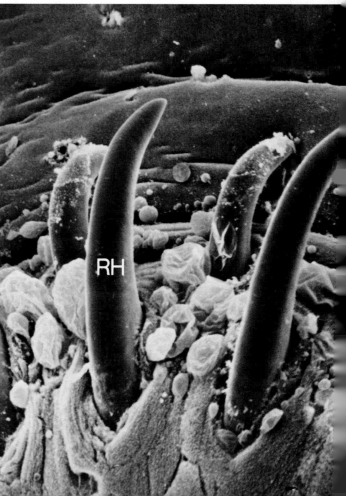

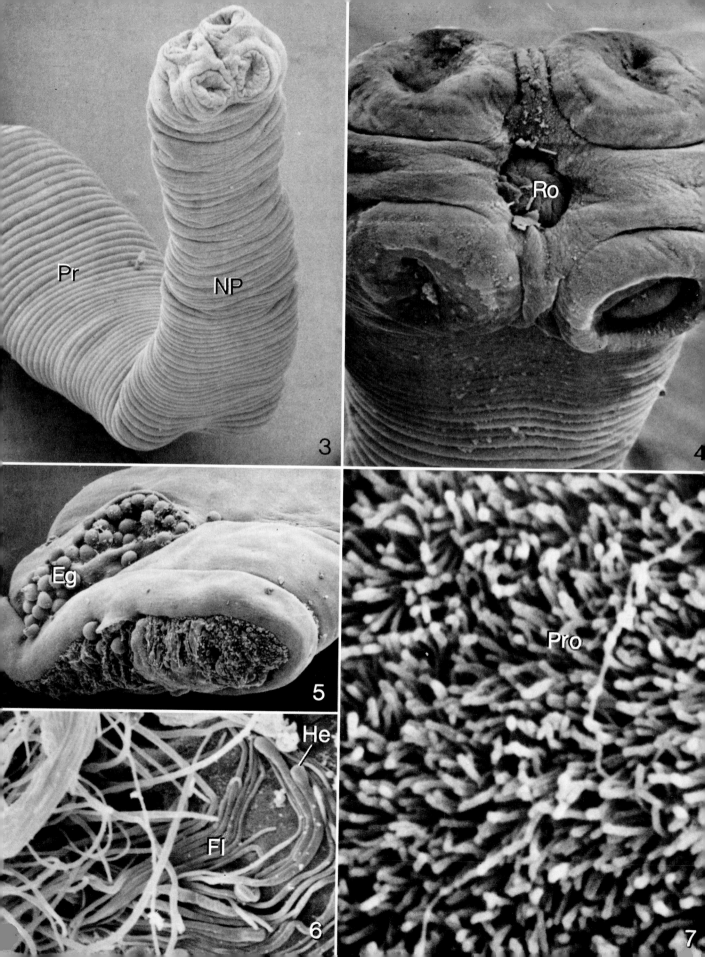

Spiny-headed Worms

Superphylum Aschelminthes— Phylum Acanthocephala

Prosthorhynchus gracilis
and *Macracanthorhynchus hirudinaceus*

This group of worm-like pseudocoelomates consists of only about 500 species, which are parasitic. They are characterized by having an anterior proboscis equipped with spines; hence the name spiny-headed worm. These parasites require two hosts to complete their life cycle. The juveniles typically parasitize insects and crustaceans and the adults reside in the vertebrate digestive tract. The major body regions include the retractable proboscis (Pr) and neck (Ne) as well as a trunk (Tr in Fig. 1). The entire proboscis can be retracted with muscles into a proboscis receptacle. The worms have no digestive system since digested food from the host digestive system is absorbed directly through the body wall.

The proboscis of *Prosthorhynchus gracilis* from the starling is illustrated in Figs. 1 through 3. The proboscis is covered with rounded papillae (Pa) in the center of which a spine (Sp in Fig. 2) emerges. This spine is recurved posteriorly and its form and size are illustrated in Fig. 2. The spine is about 25 to 30 µm in length and their number is species specific. The spiny proboscis is used to attach the worms to the host's intestinal mucosa and it can cause considerable damage. Therefore, it is frequently difficult to obtain parasites with a clean and undamaged proboscis. Not all the papillae covering the proboscis appear to be armed with spines, for those at the posterior end of the proboscis (SP in Fig. 3) appear to be filled with a material which may represent a secretory product. Sense organs are not well developed in these worms, but a sensory pit (SeP in Fig. 1) is located at the tip of the proboscis. The anterior end of another Acanthocephala, *Macracanthorhynchus hirudinaceus*, is illustrated in Figs. 4 and 5. This organism parasitizes the digestive tract of the pig. The proboscis (Pr in Fig. 4) is partially retracted, but the distribution of the pointed spines (Sp in Figs. 4, 5) is evident. A section through the body of the female worm, such as included in Fig. 6, illustrates both immature eggs (IE) and more mature ones (ME). The mature shelled eggs measure about 2.5 µm in length. The surface of the trunk of the worm illustrated in Fig. 7 is covered with rod-shaped bacteria. Note the numerous filaments (Fi in Fig. 7) that extend between the bacteria and between the bacteria and the body surface.

The Acanthocephala are of general biological interest because they are almost exclusively parasitic, have degenerate internal organs, and demonstrate in other ways doubtful affinities to other animal groups.

Superphylum Aschelminthes— Phylum Acanthocephala

Prosthorhynchus gracilis
and *Macracanthorhynchus hirudinaceus*

(continued)

Fi   Filaments (associated with bacteria)
IE   Immature eggs
ME   Mature eggs
Ne   Neck
Pa   Papillae (on proboscis)
Pr   Proboscis
SeP Sensory pit
Sp   Spines (on proboscis)
SP   Papillae (lacking spines)
Tr   Trunk

Fig. 1, ×136; Fig. 2, ×940; Fig. 3, ×940; Fig. 4, ×330; Fig. 5, ×925; Fig. 6, ×400; Fig. 7, ×10125.

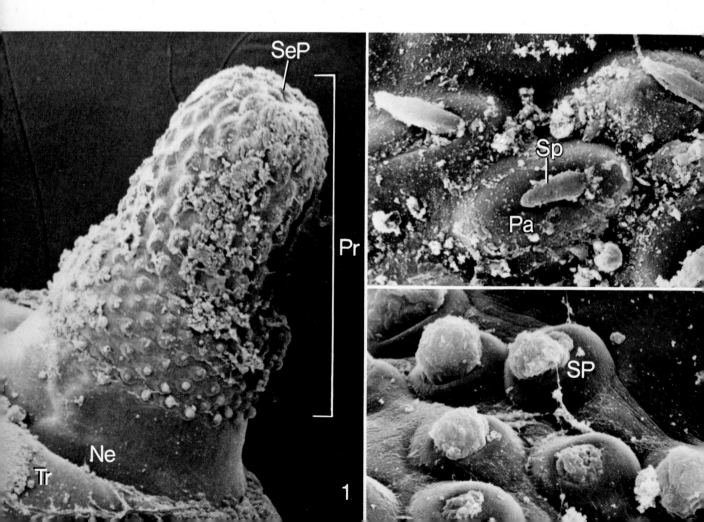

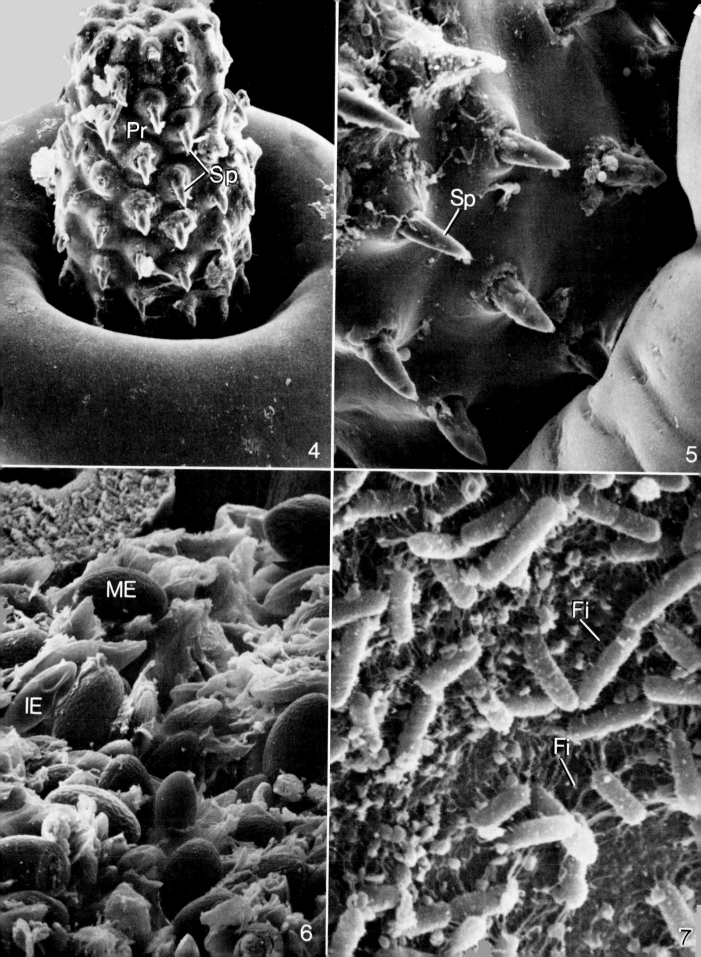

Nematodes

Superphylum Aschelminthes— Phylum Nemotoda

Class Secernentia—Order Strongylida

Nematospiroides dubius

CB	Copulatory bursa
LR	Longitudinal ridges
TA	Transverse annulations
Vi	Villi (of intestine)
♀	Female worm

Fig. 1, ×136; Fig. 2, ×385; Fig. 3, ×3188; Fig. 4, ×188.

This heligmosomid nematode inhabits the gut of rodents for a part of its life cycle and is characterized by a tightly spiraled form in the adult. Newly attached larvae become closely associated with the villi (Vi in Fig. 1) of the mouse intestinal mucosa within 24 hours after ingestion. Twenty-four to 48 hours after ingestion they penetrate the mucosal surface and take up residence near the longitudinal muscle layer close to the exterior surface of the organ. After undergoing two molts in this position the larval parasites return to the lumen, where they undergo a final molt and achieve reproductive maturity. It is after the final larval molt that the organisms assume their coiled form (Fig. 1). The adults become entangled in the mucosal villi and mucus, which apparently provide some measure of protection against dislodgement from the intestine (Fig. 1).

The tightly coiled adult female measures about 18 to 21 mm in length; the male is generally smaller (8 to 10 mm in length). The posterior end of the male is modified into a bursa (CB in Fig. 4), which during copulation surrounds the posterior end of the female (♀ in Fig. 4) at the genital aperture. A number of longitudinally oriented ridges occur in the cuticle (LR in Figs. 2, 3), and the arrangement of these ridges is used by taxonomists for species identification of heligmosomid parasites in rodents. Transverse annulations (TA in Fig. 3) or striations in the cuticle are also present, but on close inspection in the SEM they appear to be grooves in the cuticular surface extending between and perpendicular to the longitudinal ridges. These annulations are associated with the characteristic dorso-ventral undulatory movement of nematodes, and their presence undoubtedly permits dorso-ventral flexures in the cuticle during locomotion.

References

BIRD, A.F.: The Structure of Nematodes. New York: Academic Press, pp. 74—80 (1971).

EHRENFORD, F.A.: "The life cycle of *N. dubius* baylis". J. Parasitol. **40**, 480 (1954).

FAHAMY, M.A.: "An investigation of the life cycle of *N. dubius* with special reference to the free living stages." Z. Parasitenke **17**, 394—399 (1956).

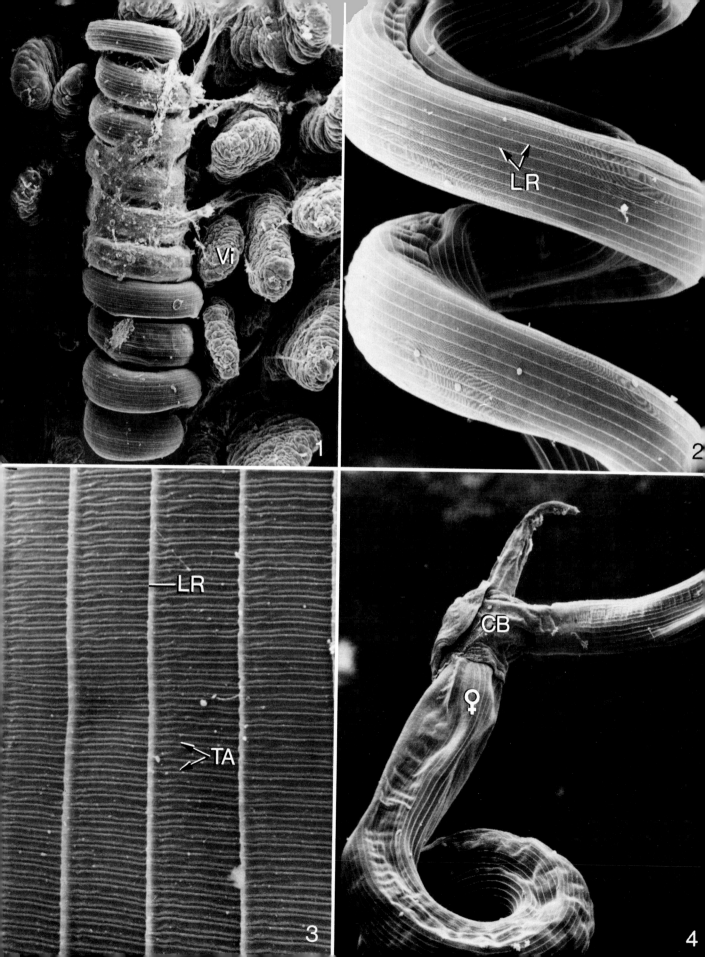

Superphylum Aschelminthes—Phylum Nematoda

Order Spirurida—Superfamily Filarioidea

Foleyella sp.

AE   Anterior end
Al   Alae
An   Anus
CP   Caudal papillae
DS   Dorsal surface
Mo   Mouth
PE   Posterior end
Pr   Projections on ventral surface

Fig. 1, × 665; Fig. 2, × 130; Fig. 3, × 1375.

Nematodes are elongate, unsegmented worms with an elastic cuticle made of protein. The species of this genus commonly live as parasites in adult form in the mesentary and body cavity of the frog. They have also been described in the subcutaneous tissues and muscle of saurians and other amphibians in addition to the frog. The adult worms (filarioids) produce microfiliariae which circulate in the blood stream. Mosquitoes represent the intermediate host in the life cycle.

Small adult male filarioids are illustrated in Figs. 1 and 2. A mouth (Mo) is present at the anterior tip of the worm (AE), while the anus (An) is located on the ventral side at the posterior end (PE) of the worm (Figs. 2 and 3). In this species four pairs of rounded projections called caudal papillae (CP), are associated with the anus (Fig. 3). The dorsal surface (DS) of the organism appears smooth and the ventral side is covered with numerous but small projections (Pr in Figs. 1 and 3). Adults have lateral alae (Al) extending nearly the entire length of their bodies (Figs. 1 and 3).

210

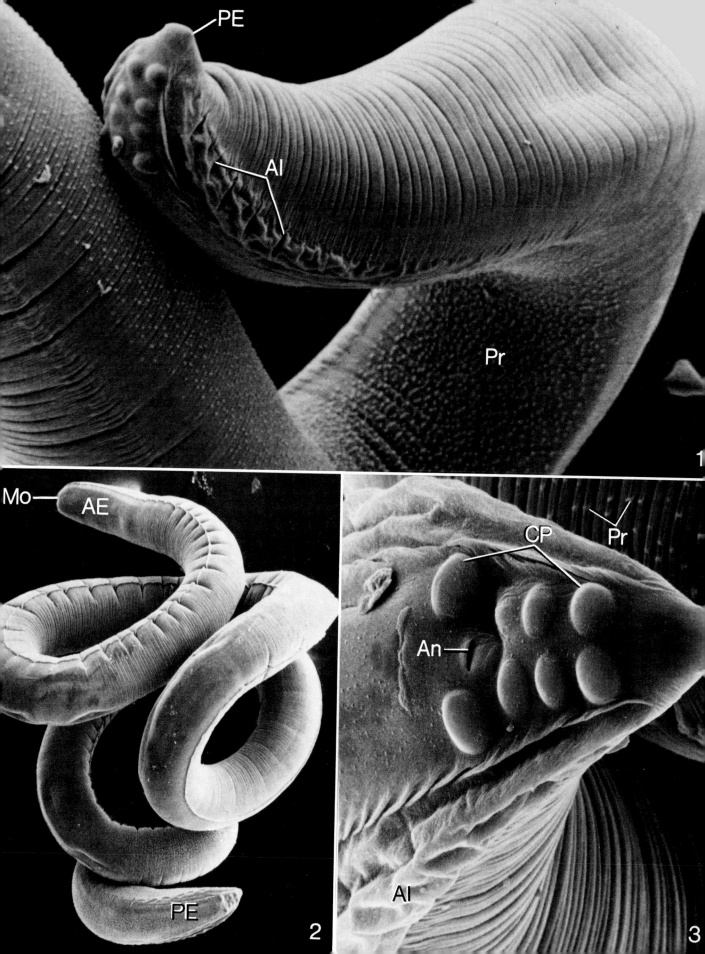

Ectoprocts or Lophophorate Coelomates

Phylum Ectoprocta

Class Gymnolaemata—Order Cheilostomata

Bugula

Bugula is a marine lophophorate coelomate that is sessile and colonial (Fig. 1). Erect branching colonies are formed which superficially resemble seaweed. The individual members (zooids) of a colony (zooarium) are all small and of several types. The zooids of a colony intercommunicate with each other through pores. Thus, the coelom of the entire colony is continuous. The branching in the colony is dichotomous and the entire colony is enclosed by an exoskeleton of calcified chitin called a zooecium (Zo in Fig. 2), which serves a protective and a supportive function. Feeding zooids (autozooids) can project through apertures in the exoskeleton or they can be retracted. Each autozooid is characterized by a ring of tentacles (Te in Figs. 2, 3) surrounding the mouth. The retractable tentacles are collectively called a lophophore (Lo in Fig. 2). The lophophore tentacles of an autozooid are illustrated in Figs. 3 and 4. Note that each tentacle tip contains many cilia (Ci in Figs. 3, 4). In Fig. 6 the zooids are retracted into the zooecium. Note the expanded frontal window (FW in Fig. 6) of the zooecium and the spines (Sp). The zooecium is uncalcified in the region of the frontal window. In the cheilostomes (e.g. *Bugula*) the zooecium aperture is closed by an operculum (Op in Fig. 6). When the lophophore is extruded by coelomic fluid pressure (hydraulic mechanism), the coelomic space within the tentacles of the lophophore inflates the tentacles.

Another type of zooid (heterozooid), called an avicularium (Av in Fig. 2), is smaller than the autozooid and has a greatly reduced internal structure. It is modified into a movable jaw. Avicularia may be sessile or stalked and are used to capture the larvae of small organisms for food.

Heterozooids may also be modified into reproductive individuals. Most ectoprocts are hermaphroditic, and in *Bugula*, eggs are brooded within a special chamber called an ovicell (Ov in Figs. 2 and 5). The eggs undergo radial cleavage and develop into trocho-phore larvae. The larvae then escape from the ovicell. Reproduction by budding can also occur in the ectoprocts. Another type of individual is a vibracula (Vi in Fig. 6). The vibracula are simply long bristles that can move and so serve to "sweep" the colony clean. There are no circulatory, respiratory, or excretory systems in the ectoprocts.

Phylum Ectoprocta

Class Gymnolaemata—Order Cheilostomata

Bugula (continued)

Av   Avicularium
Ci   Cilia
FW   Frontal window
Lo   Lophophore
Op   Operculum
Ov   Ovicell
Sp   Spines
Te   Tentacles
Vi   Vibracula
Zo   Zooecium

Fig. 1, ×30; Fig. 2, ×190; Fig. 3, ×695; Fig. 4, ×5025; Fig. 5, ×885; Fig. 6, ×275.

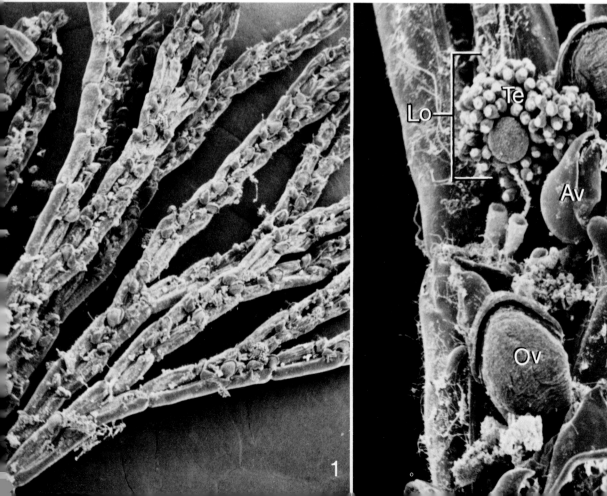

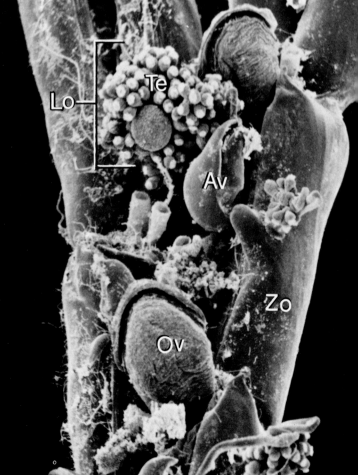

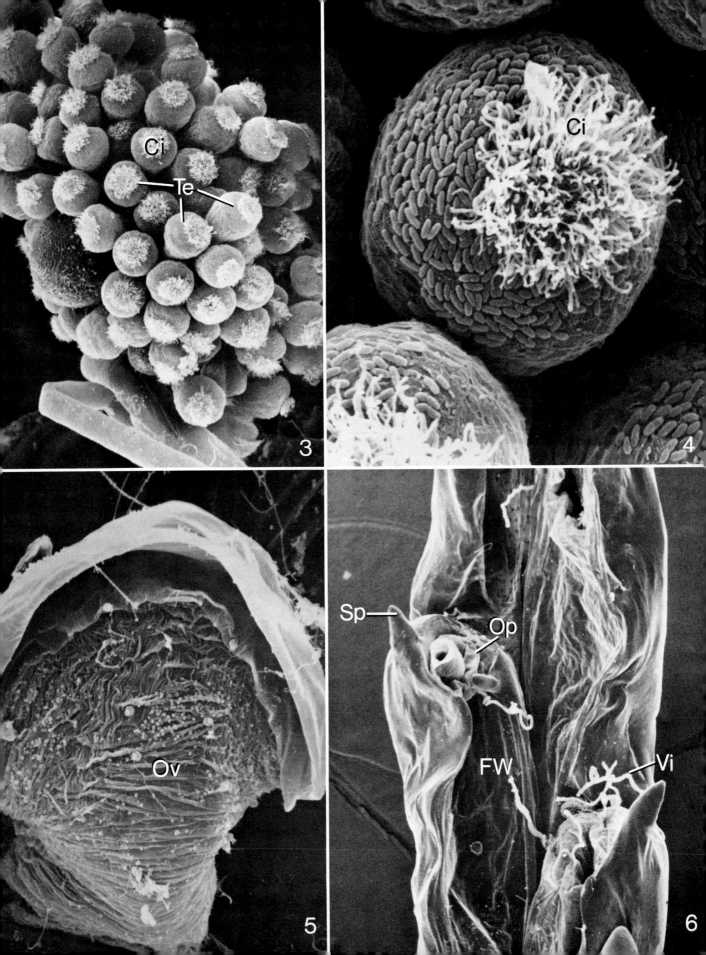

Annelid Worms

Phylum Annelida

Class Oligochaeta

Lumbricus terrestris (Earthworm)

An	Anus	OO	Openings of oviducts
Cu	Cuticle	Pe	Peristomium
DP	Dorsal pores	Ph	Pharynx (partially
Ds	Dorsolateral pair of		everted)
	setae	Pr	Prostomium
LE	Large epidermal cells	VS	Ventrolateral pair of
MG	Male genital openings		setae

Fig. 1, ×45; Fig. 2, ×60; Fig. 3, ×1215; Fig. 4, ×14000; Fig. 5, ×410; Fig. 6, ×53; Fig. 7, ×40.

Epidermal cells in the body wall secrete a thin cuticle (albuminoid material) which covers the animal. Large epidermal cells (LE) which project above the adjacent cuticle (Cu) are widely distributed over the surface of the earthworm (Fig. 3). They are, however, especially numerous at the anterior end of the animal (Fig. 1, arrows). The free surface of these cells is usually adorned with numerous, finger-like projections, or microvilli (Fig. 3). Pores or apertures are sometimes observed to be associated with the surface of these cells, as are secretory droplets. These cells may, therefore, represent secretory (mucous or albuminous) gland cells in the epidermis. Still other cells on the surface of the earthworm contain openings that are surrounded by several projections of variable length (Fig. 4). These cells, which are more restricted in number, may represent the surface of sensory receptor cells.

The body segments of *Lumbricus,* except for the first and last, have bristles, or setae, which represent vestiges of parapodia. There are eight setae per segment arranged in four groups of two. On each side of a segment are two dorsolateral (DS) and two ventrolateral (VS) setae (Fig. 1). The dorsolateral setae probably represent a vestige of the notopodium and the ventrolateral setae are remnants of the neuropodium. The tips of the setae are illustrated projecting through the body wall in Fig. 5. These bristles can be extended or retracted by means of extrinsic muscles located inside the body.

A pair of dorsal pores per segment connects the coelomic compartments with the exterior of the earthworm (DP in Fig. 6). Through these apertures coelomic fluid can be extruded onto the surface of the worm. Earthworms are hemaphroditic, and the external openings of the female and male genital systems are located on the ventral side of specific segments. The paired external openings of the male genital system (MG) are located on segment 15 (Fig. 7) and are larger than the paired external openings of the oviduct (OO), which occur on segment 14 (Fig. 7).

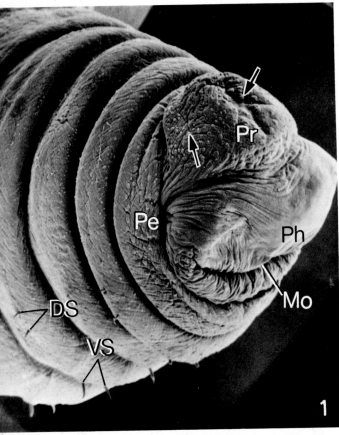

This segmented annelid worm has a burrowing existence, and as a consequence, its cephalic sense organs and external appendages are not well developed. The mouth (Mo) is located on the ventral portion of the first segment (Fig. 1). The first segment is composed of two portions: a dorsal projection called the prostomium (Pr) and the area surrounding the mouth, which is called a peristomium (Pe). The earthworm is herbivorous and feeds on detritus in the soil. Food enters the digestive system by a pumping action of the pharynx (Ph), which may be partially everted during the process (Fig. 1). An anus (An) is located in the center of the terminal posterior segment (Fig. 2).

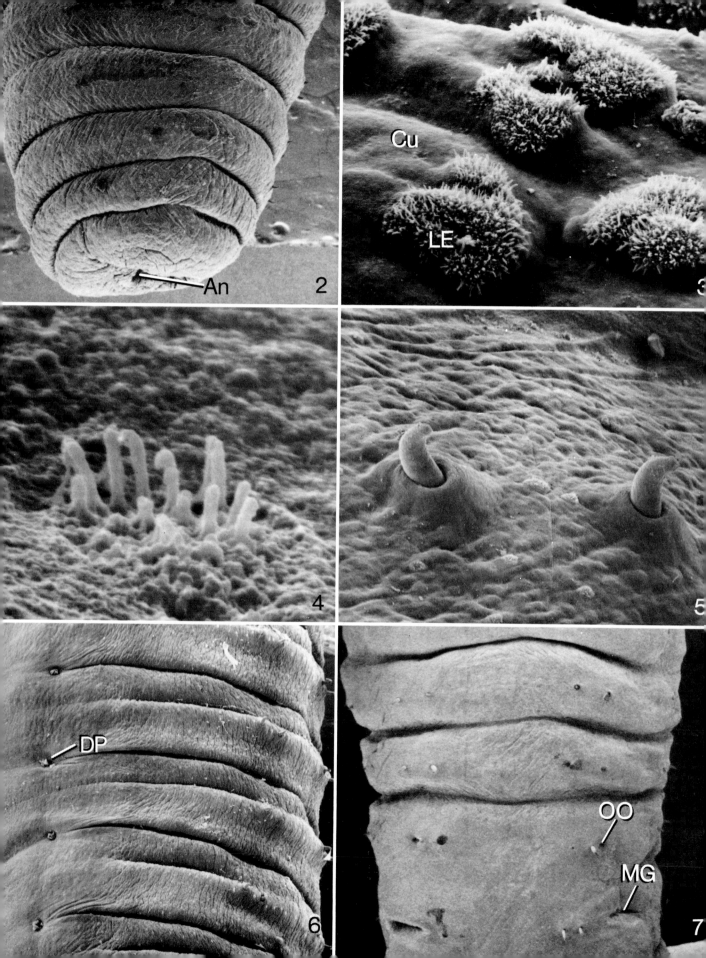

Phylum Annelida

Class Polychaeta—Subclass Errantia

Nereis limbata

DC	Dorsal cirrus	PeC	Peristomial cirrus
EP	Everted pharynx	PP	Prostomial palps
IL	Inferior ligula	Pr	Prostomium
Ja	Jaws	PT	Prostomial tenacles
Ne	Neuropodium	Se	Setae
No	Notopodium	SL	Superior ligula
Pa	Parapodium	VC	Ventral cirrus
Pe	Peristomium		

Fig. 1, ×24; Fig. 2, ×25; Fig. 3, ×30.

In contrast to the burrowing annelids represented by *Lumbricus*, the errant polychaetes as represented by *Nereis* have well-developed sensory appendages associated with the head and prominent segmental appeandages used primarily in locomotion.

The dorsal anterior portion of *Nereis* is illustrated in Fig. 1. Note that the pharynx is partially everted (EP) so that a pair of pharyngeal teeth or jaws (Ja) are visible. The prostomium (Pr) is modified into prostomial tentacles (PT) and palps (PP). Segments 1 and 2, represented by the peristomium (Pe), have two pairs of tentacles or cirri (superior peristomial cirri and inferior peristomial cirri) associated with each side (PeC in Fig. 2). Beginning with segment 3, each segment has a pair of lateral paddlelike, locomotory projections which are called parapodia (Pa). Details of the parapodia from the mid-region of the body are illustrated in Fig. 3. The appendages are divided into a dorsal notopodium (No) and a ventral neuropodium (Ne). Each of these, in turn, is divided into a superior and an inferior ligula, (SL) and (IL). A short projection is associated with the notopodium (a dorsal cirrus, DC) and the neuropodium (a ventral cirrus, VC). Numerous and long setae (Se) project from the notodium and neuropodium.

218

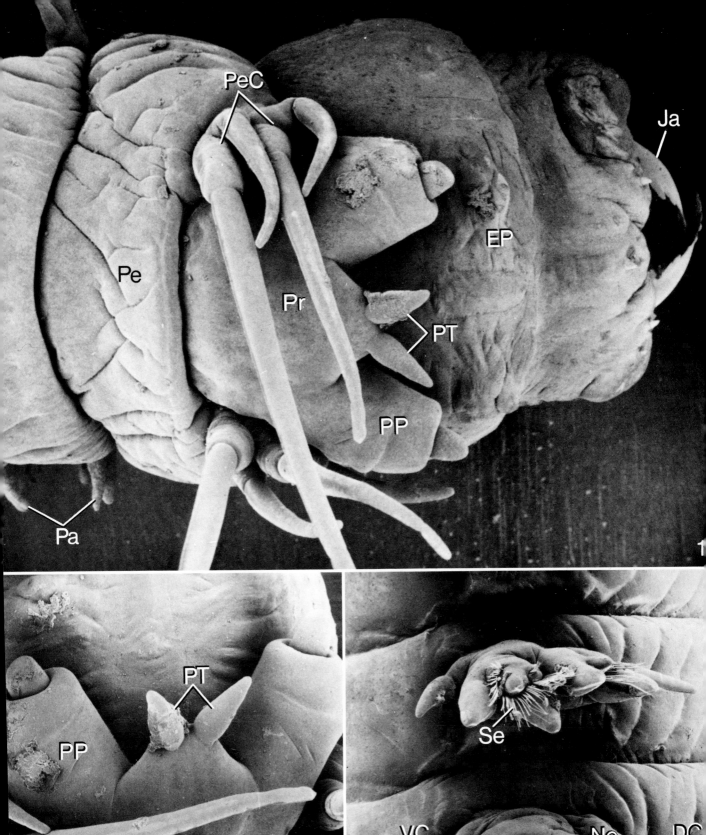

Phylum Annelida

Class Polychaeta

Polychaetes

The external structure of polychaetes, especially that of the head and parapodia, is quite variable and dependent on habitat. The burrowing, marine polychaete illustrated here shows still another major structural variation in the annelid appendage or parapodium. The anterior end of the organism is illustrated in Fig. 1. The tentacles (Te) associated with the prostomium (Pr) and peristomium (Pe) are apparent. The parapodia (Pa) associated with the most anterior body segments consist of a dorsal notopodium, which is long and narrow, and a ventral neuropodium, which possesses setae. The appendages associated with the main part of the worm, however, are highly modified. In this region (Figs. 2 and 3), the notopodia (No) have a number of filamentous branches (NF) extending from them. As many as 10 to 12 notopodial filaments (NF in Fig. 3) may be associated with each appendage. The ventral neuropodia (Ne in Figs. 2 and 3) have setae projecting from them (Se in Figs. 2 and 3). The notopodial filaments are characterized by possessing many cilia (Ci in Fig. 4) extending from their lateral surfaces. These parapodia are structurally modified for aeration and respiration, thus enabling the maintenance of a water current over the burrowing worm. Gas exchange occurs between the water and the blood vessels in the thin notopodial filaments. The presence of large numbers of cilia on the notopodial filaments undoubtedly represents an additional mechanism for ensuring water movements over the filaments.

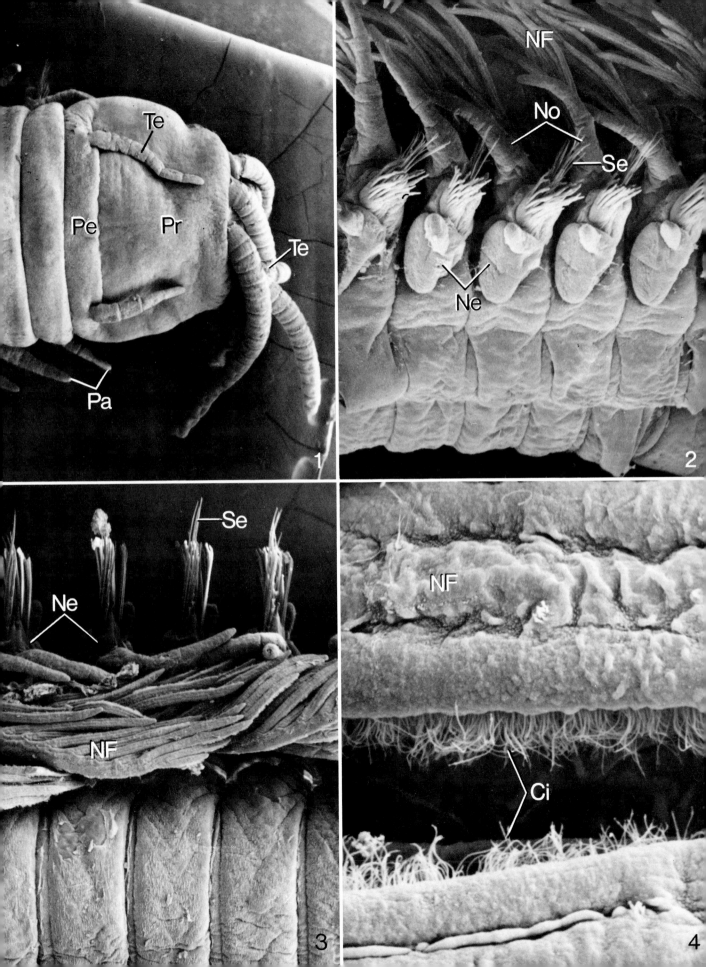

Molluscs

Phylum Mollusca

Class Gastropoda—Subclass Pulmonata

Snail

The snails illustrate one typical molluscan feature well; namely, the thickening of the ventral side of the body so as to produce a flattened muscular foot (Fo) used in locomotion (Fig. 1). There is also a definite head (He) with a mouth (Mo) and cephalic tentacles (CT) bearing eyes (Fig. 1). The ventral mouth is surrounded by two lateral lips (LL) and an anterior lip (AL) and leads into a buccal cavity (BC) containing a specialized region called a radula sac (Fig. 2). The radula sac forms a belt on which teeth are held and it is called a radula (Ra in Fig. 2). The radula can be projected from the mouth and by alternate movements can scrape algae from surfaces over which the animal glides.

Gland cells, predominantly mucous, are particularly abundant in the molluscan epidermis. This material is used in lubricating and attaching the foot during locomotion, in entrapping food, and in keeping the body surface moist. Mucous epidermal cells are particularly abundant on the head and foot and especially on the lips surrounding the mouth. These cells frequently have microvilli (Mi) associated with their free surfaces (Fig. 8). The mucus (Mu) slime associated with the surface of one of the lips is illustrated in Fig. 8.

Many epidermal cells also have large numbers of long cilia (Ci) associated with their free surfaces. These cilia aid in maintaining water currents around the mouth and foot and over the tentacles, and can move food particles entrapped in mucus. The tentacles are especially well ciliated not only at their tips but along their length (Figs. 3 through 5). In addition, large numbers of ciliated cells are present in the foot (Fig. 6). The body surface around the head is also ciliated (Fig. 7).

Phylum Mollusca

Class Gastropoda—Subclass Pulmonata

Snail (continued)

Fig. 1 (ventral view of head and foot), ×38; Fig. 2 (mouth), ×160; Fig. 3 (tentacle), ×165; Fig. 4 (tip of tentacle), ×1635; Fig. 5 (side of tentacle), ×1575; Fig. 6 (surface of foot), ×5400; Fig. 7 (dorsal side of head), ×3150; Fig. 8 (lip), ×8700.

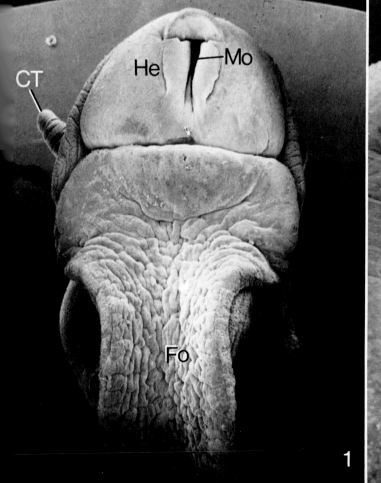

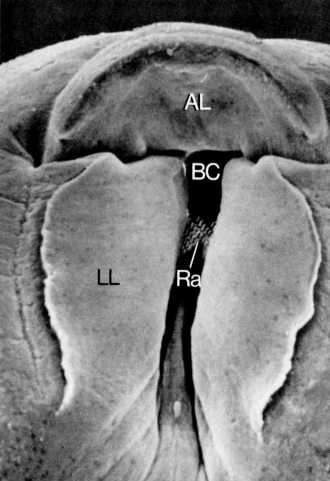

Phylum Mollusca

Class Gastropoda—Subclass Pulmonata

The Gastropod Radula

The radula is a characteristic molluscan feeding device which is particularly well developed in the gastropods. It consists of a belt or ribbon of connective tissue on which are placed many curved teeth which project backward (Fig. 1). The radula is contained within a radula sac and is supported by a rod called the odontophore. Protractor and retractor muscles attach to the ends of the tooth belt. When used in feeding, the odontophore is extended from the mouth by muscle action so that the radula is applied to the surface to be scrapped. The belt is then moved back and forth over the odontophore by muscles attached to the belt. New teeth are constantly being differentiated in the posterior end of the radula sac to replace those lost or worn off the anterior part of the radula ribbon. The number of rows of radula teeth is highly species specific. The radula illustrated in Fig. 1 shows the central tooth row (CT), the broader teeth in the lateral tooth rows (LT), and the more pointed teeth comprising the marginal tooth (MT) rows. The teeth in the central and lateral tooth rows are enlarged in Fig. 2. The number of radula teeth may vary in the molluscs from only one to more than one-half million.

Arthropods

Phylum Arthropoda

Class Insecta—Order Diptera

Two-Winged Flies

The true flies, or Diptera, feed mainly on nectar, but some prey on other insects, and still others have taken to sucking the blood of man and other animals. As the name implies, they retain only the first pair of wings; the second pair is reduced to little knob-like sense organs of balance, called halteres. The variety of mouthparts found in members of this group demonstrates divergence from the primitive biting type. There is always a proboscis formed principally from the elongated labium, ending in a pair of lobes, the labella. The labium may function to support and guide the remaining mouthparts which are enclosed within it.

The head of the housefly is shown in Fig. 1. The mandibles are absent and the feeding apparatus consists of a proboscis (Pr), which is hinged to the head. The basal section, or rostrum (Ro), is actually part of the head, and the clypeus (Cl) is situated on its anterior section. The paired maxillary palps (MP),

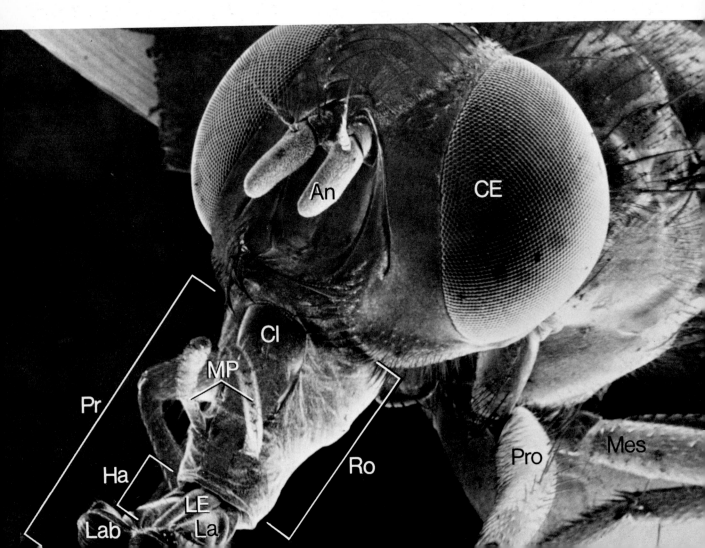

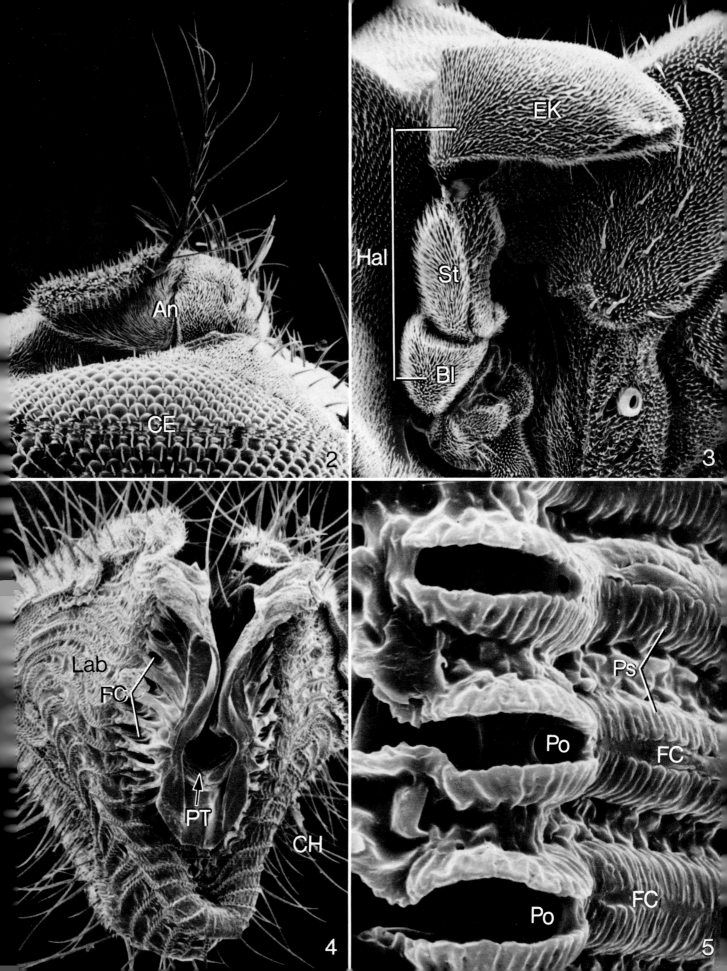

representing remains of the maxillae, are located laterally on the rostrum. The middle region of the proboscis, called the haustellum (Ha), consists of a highly modifed labium (La) and a labrum-epipharynx (LE); the latter covers a deep labial groove in which lies the blade-like hypopharynx (not shown). The hinged arrangement of the proboscis permits it to be folded under the head when not in use. The distal portion of the proboscis is composed of two fleshy lobes, the labella (Lab), which represent the termination of the labium. Note also the antennae (An), the compound eyes (CE), and a portion of the pro- and mesothoracic legs (Pro and Mes).

The tri-segmented antenna and compound eye (CE) of *Drosophila* (the fruit fly) are shown in Fig. 2. Olfactory pits, supplied with sensory neurons, are present on the antennae of both flies. The antennae are extended during flight so as to expose the olfactory pits to the air stream flowing over the fly's head.

The hind wings of the Diptera are modified to form halteres. Each haltere (Hal in Fig. 3) consists of a basal lobe (Bl), a stalk (St), and an end knob (EK) which projects backward from the end of the stalk so that its center of gravity is also behind the stalk. The entire structure is rigid except for flexibility in the ventral surface near the base which permits some freedom of movement. The halteres articulate with the thorax in a manner similar to the wings. During flight they vibrate up and down at a frequency identical to the wings.

The labella (Lab) of a blowfly are shown in Fig. 4. They are ringed with a variety of chemosensory hairs (CH) and contain a series of fine food channels (FC), or straws, which are better seen in Fig. 5. The food channels are strengthened by rings of chitin and are referred to as pseudotrachea (Ps). The food channels open externally via tiny pores (Po), through which liquid food and fine food particles can be imbibed. In the groove between the labella are a series of prestomal teeth (PT), which can be everted and used for rasping.

There are many chemosensory hairs around the periphery of the labella as well as on the legs of flies. In each hair are neurons that aid the fly in discriminating between salts, sugars, amino acids, and water. Also present is a neuron that responds to mechanical stress. Upon contact of a fly's tarsal hairs with a favorable food, the proboscis is immediately everted. A dorsal view of the tip of a fly's leg (tarsus) is shown in Fig. 6. Note the tarsal hairs (TH), tarsal claws (TC), and pulvilli (Pu). If the surface is too smooth for the claws to grip, the fly makes use of adhesive organs on the

An	Antenna	Mi	Microtrichia
Bl	Basal lobe	Om	Ommatidia
CH	Chemosensory hairs	Po	Pores
Cl	Clypeus	PT	Prestoma teeth
CE	Compound eye	Pr	Proboscis
EK	End knob	Pro	Prothoracic leg
FC	Food channel	Ps	Pseudotrachea
Ha	Haustellum	Pu	Pulvilli
Hai	Hairs	Ro	Rostrum
Hal	Haltere	St	Stalk
Lab	Labella	TC	Tarsal claws
La	Labium	TH	Tarsal hairs
LE	Labrum-epipharynx	TeH	Tenent hairs
MP	Maxillary palps	TS	Trichoid sensilla
Mes	Mesothoracic leg	Ve	Vein

Fig. 1 (head), ×52; Fig. 2 (antenna), ×300; Fig. 3 (haltere), ×480; Fig. 4 (labellum), ×207; Fig. 5 (labellum), ×2665; Fig. 6 (tarsus), ×260; Fig. 7 (eye), ×1500; Fig. 8 (eye), ×7500; Fig. 9 (wing), ×870.

pulvilli, or tarsal or tibial pads. These organs generally consist of dense collections of tubular hairs (tenent hairs, TeH) with delicate expanded tips that are moistened by a glandular secretion.

Most adult insects have a pair of compound eyes (CE in Fig. 1), one on either side of the head, which bulge to a greater or lesser extent to give a wide field of vision. Each compound eye is an aggregation of similar units known as ommatidia (Om), the number of which varies tremendously. When a large number of ommatidia are present, the facets that they present to the outside are packed close together and assume a hexagonal form (see Figs. 7 and 8, *Drosophila*). In *Drosophila* there are large hairs (Hai) inserted into sockets between the facets (Figs. 7 and 8). The function of these hairs may be mechanosensory. Note the small but numerous papillae covering each facet (arrows, Fig. 8).

The success of insects as terrestrial animals is at least partly due to their ability to fly. The surface of the Dipteran wing membrane (Fig. 9) is often covered with small non-innervated spines called microtrichia (Mi). Larger hooked spines, called trichoid sensilla (TS), are inserted into sockets along the wing veins (Ve). These sensory hairs are probably mechanoreceptors which respond to touch and possibly the flow of air over the wings during flight.

230

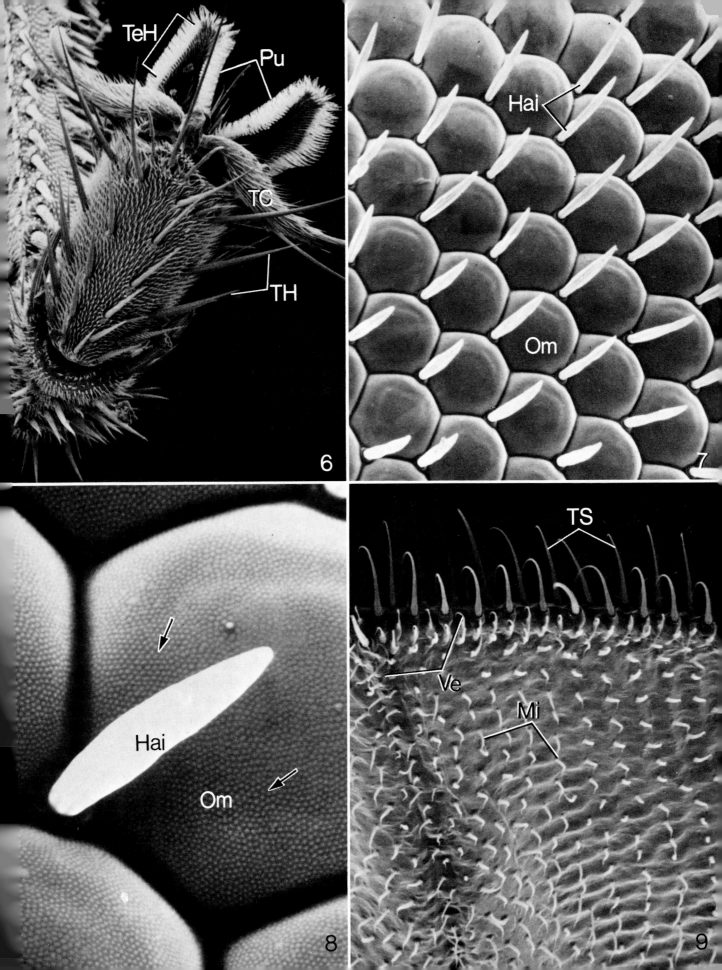

Phylum Arthropoda

Class Insecta—Order Lepidoptera

Butterflies

The butterflies and moths have two pairs of wings that are completely covered with scales as well as mouthparts modified for sucking (siphon-tube type). The mandibles are rudimentary but the paired maxillae are enormously elongated to form a long sucking tube, or proboscis (Pr), which is well adapted for securing the nectar of flowers (Figs. 1 and 2). When not in use, the proboscis is coiled below the thorax. The maxillae (Max in Fig. 2) that form the proboscis are grooved on their medial surface and fit together to form a tightly constructed coilable tube. In feeding, the proboscis is extended by blood pressure. Fluids are sucked into the proboscis by muscular activity in the pharynx and the buccal cavity. The sensory labial palps (LP) are found at the base of the proboscis (Fig. 1). Also illustrated in this figure are antennae (An), a compound eye (Ce), and the many hairs (Ha) covering the head of this specimen.

The wings of Lepidoptera are completely covered with scales which vary in form from typical hair-like structures to flat plates and usually cover the entire body as well as the wings. Note the overlapping serrated scales (Sc in Fig. 3) found on the wing of a butterfly. A flattened scale in transverse section consists of two lamellae with an intervening airspace. The two lamellae are supported by internal struts called trabeculae (not shown here). The scales are inserted in sockets of the wing membrane so that they are inclined to the surface and overlap each other to form a complete covering. Pigments in the scales are responsible for the color of many Lepidoptera. The pigment is located in the wall or the cavities of the scale. A portion of a scale is enlarged in Fig. 4. The scales consist of overlapping plates (OP) which form ridges. Crossbridges (Cr) extend between each ridge of the scale.

232

Phylum Arthropoda

Class Insecta—Order Lepidoptera

Butterflies (continued)

AS   Antennal scales
Pa   Patches
SP   Sunken pegs
Tw   Thick-walled pegs

Fig. 5 (antenna), $\times 110$; Fig. 6 (antenna), $\times 195$; Fig. 7 (antenna), $\times 495$; Fig. 8 (antenna), $\times 2500$.

The antennae of Lepidoptera, as in the case for all insects (Fig. 5), have a wide variety of sensilla distributed over their entire surface. A sensillum is defined as a specialized area of the integument consisting of formative cells, sensory nerve cells, and in some cases auxillary cells. Many different types of sensilla can be seen distributed over the surface of the butterfly antennae (refer to Figs. 5 through 8). These sensilla serve a variety of functions including olfaction, chemoreception, and mechanoreception. Several types of sensilla can be seen among the antennal scales (AS). Long, tapered, thin-walled pegs occur in patches (Pa) on the inner medial antennal surface of this specimen (Figs. 5 through 7). Larger, thick-walled (Tw) hairs (Figs. 6 and 8) are inserted in sockets in the cuticle and are probably mechanoreceptors and may serve also as contact chemoreceptors. Also shown here are sunken pegs (SP), the function of which is not known (Fig. 8).

A widespread phenomenon in the Lepidoptera is the sexual dimorphism of antennae, which is sometimes associated with the occurrence of different sense organs. Usually, the antennae of males are more complex. This dimorphism is probably the result of the male's necessity to find a mate that releases a sex pheromone (sexual attractant). Sex pheromones are interspecific odorous molecules (signals) which are detected by receptors, usually specific sensilla on the antennae, and result in very complicated and stereotyped behavioral responses by the male. The silkworm moth, *Bombyx*, possesses very large antennae with numerous hairs containing many thousands of olfactory receptor cells, about 30000 of which are specialized for the detection of the female sex pheromone.

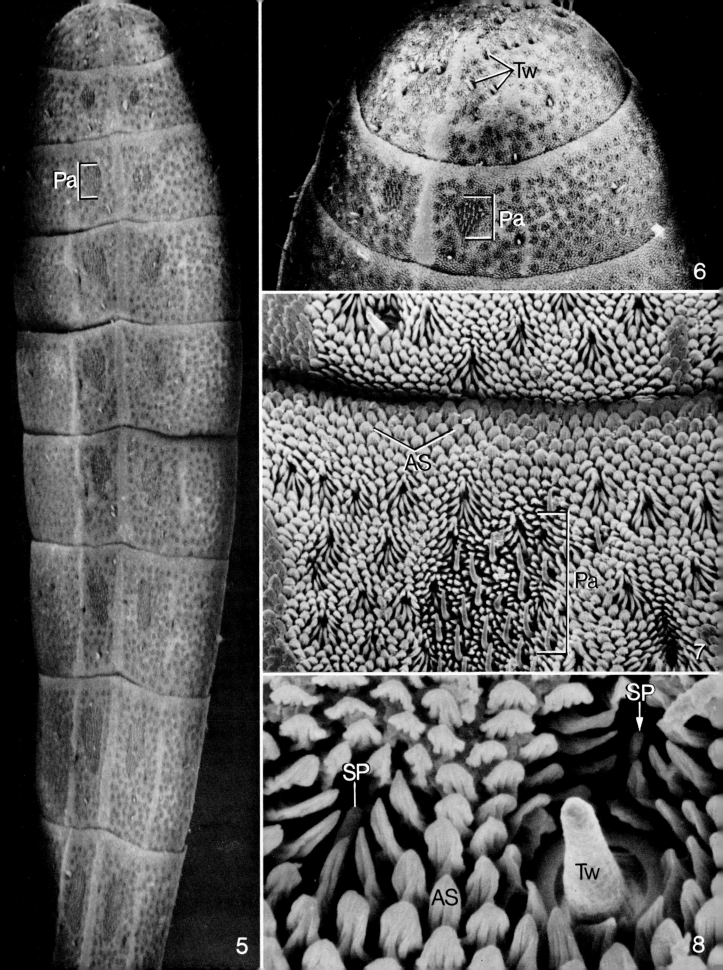

Phylum Arthropoda

Class Insecta—Order Hymenoptera

Honeybee

The head and mouthparts of a worker bee are shown in Figs. 3 and 4. The worker bee has large compound eyes (CE) with many small, long hairs protruding from their surface. The antennae (An) are used in the detection of food and the location of the hive; the distal end of one antenna is illustrated in Fig. 3. The antennae arise close together from the face between the lower halves of the compound eyes, where each is inserted into a small, circular membranous socket (So in Fig. 5). Note a portion of the scape (Sc) of the antenna, which consists of a single, long joint inserted into the antennal socket of the head by a prominent knob (Kn). The knob is covered with a variety of hairs (Ha) that serve a mechanosensory function when bent. This articulating knob is attached to the rim (Ri) of the socket by a circle of membrane, but it is also pivoted on a slender peg-like process projecting upward from the lower edge of the socket. Hence, the antenna is free to move in any direction, but at the same time it is held firmly in position by the pivot. The typical plumose hairs (PH) that cover the surface of the bee's head are also illustrated in Fig. 5.

The mouthparts of the bee are adapted for biting and licking (Figs. 3 and 4). The labrum (La) is a wide, transverse flap at the lower edge of the face. The paired mandibles (Ma) are hinged to the lateral part of the head below the compound eyes (Fig. 3). The mouth opens between the bases of the mandibles. Originating behind the mandibles is a composite structure, with its terminal parts elongated to form a proboscis. The proboscis consists of the maxillae (Max) and labium (Lab), united at their proximal end. The central labium is a multiple structure including the labial palps (LP), the internal short paraglossae (Pa in Fig. 3), and the long glossa (Gl in Figs. 3 and 4) or tongue (actually formed from two glossae), on the tip of which is a small lobe, the labellum. Nectar and honey flow up a ventral groove in the glossa and eventually into the mouth. The nectar collected by the labium (tongue) from flowers is stored in the crop, where the action of saliva changes it to honey.

The antennae (Fig. 6) are covered with a variety of sensilla. The tenth (10) and terminal (11) segments of the antennal flagellum are illustrated. The sensilla trichodea (ST) of the last eight joints of the flagellum of a worker are slender and curved toward the tip of the antenna. Thicker and more blunt hairs are abundant near the tip of the antenna; they are called sensory pegs or sensilla basiconica (SB). In addition, there are numerous and characteristic small elliptical plates called sense plates or sensilla placodea (SP). They are slightly depressed below the surface of the antenna and are delicately hinged to the encircling cuticle.

The hind legs (metathoracic) of the worker bee are of particular interest because they bear structures by which the bee transports pollen from flowers to the hive. The outer surface of the hind tibia of the worker is smooth, slightly concave, and fringed on both edges by long hairs that curve outward (Fig. 1). The entire structure is called the pollen basket (PB). The inner surface of the large basal segment, or planta (Pl in Fig. 2), of the hind tarsus is covered with sharp, stiff spines closely arranged in about 10 transverse rows. These spines or brushes (Br in Fig. 2) are used to clean the insect, and as the two brushes of the hind legs are rubbed against each other, the pollen is collected into a ball, which is then placed in the pollen basket. Also shown here is a row of spines (Sp in Fig. 2) on the inner surface of the tibia which constitute the comb.

237

Phylum Arthropoda

Class Insecta—Order Hymenoptera

Honeybee (continued)

An	Antenna	Pl	Planta
Br	Brushes	PH	Plumose hairs
CE	Compound eye	PB	Pollen basket
Gl	Glossa	Ri	Rim
Ha	Hairs	Sc	Scape
Kn	Knob	SB	Sensilla basiconica
LP	Labial palps	SP	Sensilla placodea
Lab	Labium	ST	Sensilla trichodea
La	Labrum	So	Socket
Ma	Mandibles	Sp	Spines
Max	Maxillae	10	Tenth segment
Pa	Paraglossa	11	Terminal segment

Fig. 1 (outer surface of tibia), ×48; Fig. 2 (inner surface of planta), ×45; Fig. 3 (head and mouth parts), ×48; Fig. 4 (mouth parts), ×36; Fig. 5 (antenna), ×330; Fig. 6 (detail of antenna), ×3505.

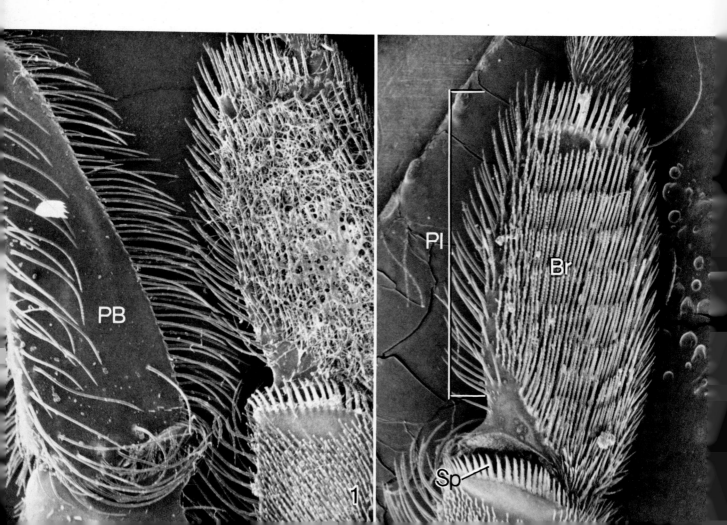

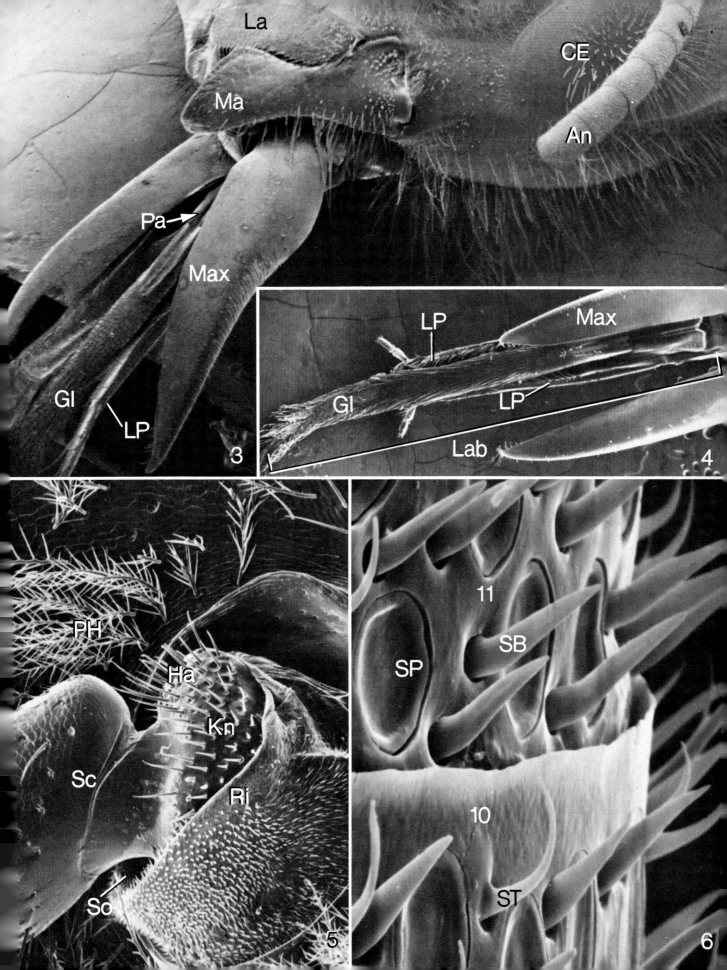

Phylum Arthropoda

Class Insecta—Order Diptera

Mosquito

An	Antenna	Max	Maxillae
Cla	Claws	MP	Maxillary palps
Cl	Clypeus	Mi	Microtrichia
CE	Compound eye	Sca	Scales
Fl	Flagellum	Sc	Scape
Lab	Labella	SG	Serrated galea
La	Labium	Se	Setae
LE	Labrum-epipharynx	To	Torus

Fig. 1 (head and mouthparts), ×125; Fig. 2 (antenna), ×1655; Fig. 3 (mouth parts), ×475; Fig. 4 (tip of tarsus), ×1015; Fig. 5 (wing), ×660.

Mosquitoes are blood-sucking Dipterans in which the mouth parts are well adapted for piercing the integument. The typical food of both sexes is nectar and other plant juices, but the female has mouth parts modified for obtaining blood meals. Mosquitoes serve as vectors of several diseases, including malaria and yellow fever. The most prominent features of the head include the large compound eyes (CE), which almost touch each other (Fig. 1). The antennae (An) are located anteriorly, and each has two large basal segments including a narrow scape (Sc) and a globular torus (To). The remainder of the antenna consists of a multi-segmented flagellum (Fl), each segment of which bears a ring of setae (Se in Fig. 2). There are a variety of antennal sensilla (Fig. 2). Many of these are probably olfactory and mechanosensory and aid the mosquito in locating a suitable place to pierce its prey. The clypeus (Cl) extends anteriorly from between the eyes and articulates with the mouth parts.

The mouth parts are illustrated in Figs. 1 and 3. The long labium (La in Fig. 1) is covered with scales and terminates in paired labella (Lab in Fig. 3), from which several long hairs extend. The labella represent reduced labial palps which serve a tactile function. A dorsal groove (arrow in Fig. 1) extends the length of the labium. The remaining mouth parts are long, slender, and closely apposed. They include the labrum-epipharynx, paired mandibles, paired galeae of the maxillae, and hypopharynx. The close association of these mouth parts makes identification difficult. However, a portion of the labrum-epipharynx (LE) and the maxilla (Max) can be identified in Fig. 1. The food channel through which material is moved to the mouth is formed dorsally by the labrum-epipharynx and ventrally by the hypopharynx. The long, pointed hypopharynx has a fine longitudinal groove along which saliva flows to prevent coagulation of blood.

The female mosquito has two long mandibles which are extremely fine, needle-like structures with distal serrated edges. It also has two long, but thicker galeae (highly modified distal joints of the maxillae) which also have serrated edges at their distal ends (SG in Fig. 3). The mandible and galeae function together to pierce the skin. In contrast, the male, which does not suck blood, has much shorter mandibles and galeae. In addition, the functionless hypopharynx is fused to the labium. The remaining mouthparts, the maxillary palps (MP in Fig. 1), are sensory in function.

The claws (Cla in Fig. 4) of a mosquito are located on the terminal segment of the tarsus (pretarsus) and aid the animal in obtaining purchase on its intended prey. The rod-like scales (Sca) on the mosquito wing are illustrated in Fig. 5. In addition, the nonsensory spines, or microtrichia (Mi), cover the surface of the wing.

240

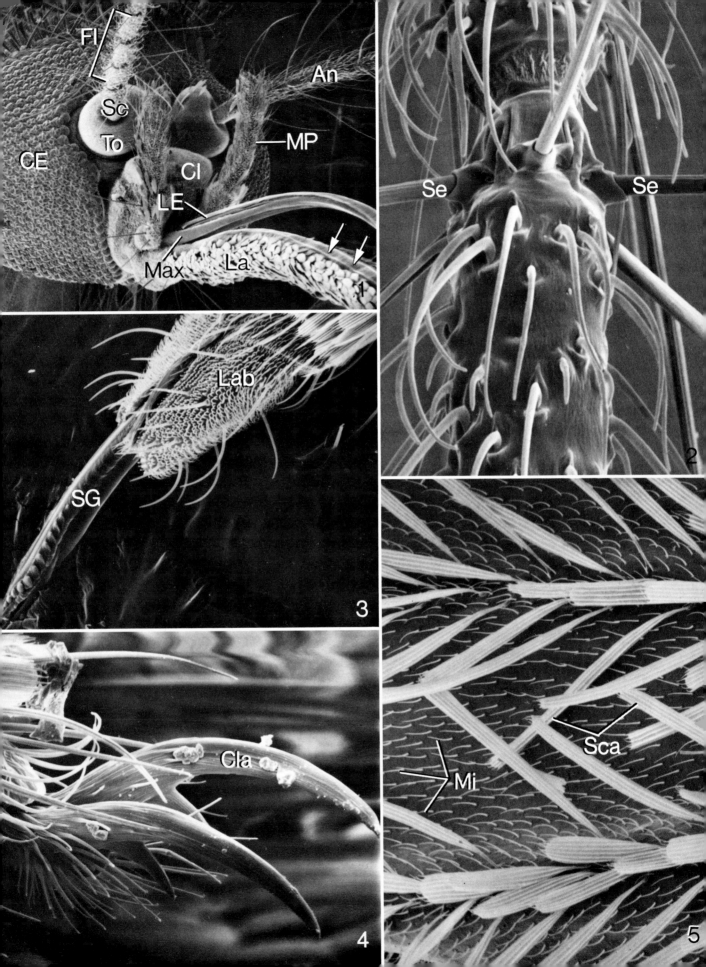

Phylum Arthropoda

Class Insecta—Order Hymenoptera

Ants

Ants are related to bees and wasps, but are readily distinguished from other members of the Order Hymenoptera by a series of characters. Perhaps the most striking feature of ants is the differentiation of the abdomen into two strongly marked regions: a slender one- or two-jointed, highly mobile pedicel that allows the ant to bend in all directions, and a larger, more compact terminal portion, the gaster (Fig. 1). Another distinguishing characteristic is the elbowed antennae (An in Fig. 2). The first segment of the antenna is called a scape (Sc) and is highly elongated (Fig. 2). The compound eye (CE), clypeus (Cl), and labrum (La) are shown in Fig. 2 as are a portion of the thorax (Th), and the coxa (Co) and trochanter (Tr) of the legs.

A wide variety of sensilla are distributed over the surface of the flagellum (Fl in Fig. 2) of the antenna. The number of sensilla occurring at the distal end of the flagellum is much greater than on the scape portion of the antenna (compare Figs. 3 and 4). At the base of the scape (Sc) is located a knob which is covered with spines and articulates with a socket (So), permitting great freedom of movement in the antennae (Fig. 3). The facets (Fa) of the compound eye and the sparsely distributed hairs (arrows) are well illustrated in Fig. 5.

Ants are among or most efficient and tireless runners and walkers in the insect world. They often walk long distances and are able to climb a perpendicular surface, such as a wall or a tree trunk, with the greatest of ease. They are also strong and can carry or drag objects much larger and heavier than themselves. The ventral side of the distal end of a leg (Fig. 6) consists of a fifth (terminal) tarsal segment (TS) which terminates in the form of two claws (Cla) and an adhesive pad (Pa) with bristles. These structural adaptations aid the ant in adhering to the substratum.

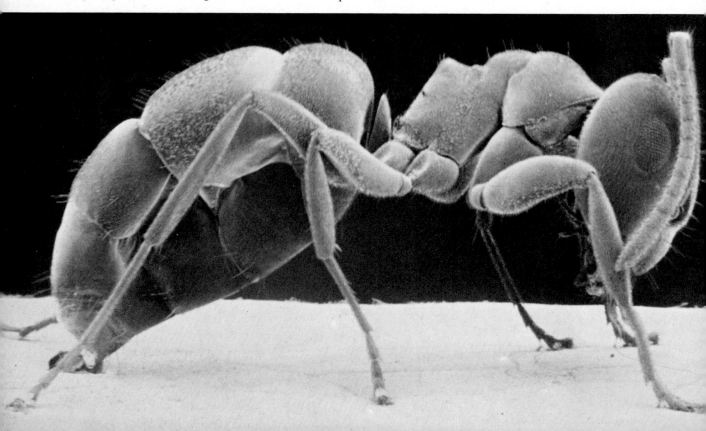

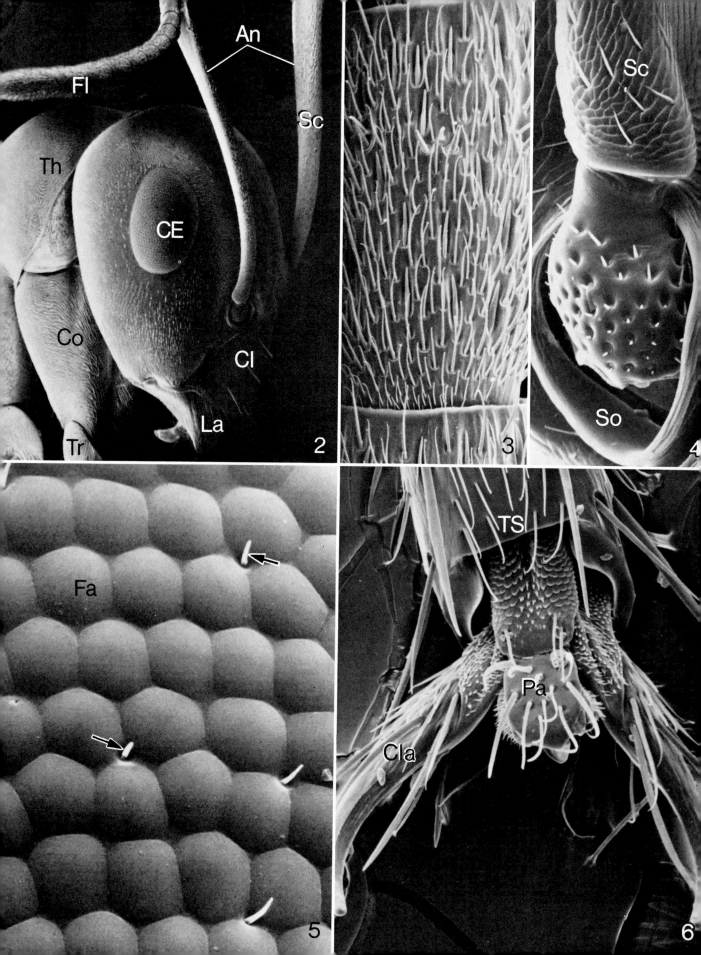

Phylum Arthropoda

Class Insecta—Order Isoptera

Termites

Termites are social insects in which a division of labor occurs within the castes, and the morphology of individuals varies with their function. Members include sterile workers and soldiers that do not have wings and sexual males and females that do possess wings. There are also intermediate castes among termites, and the number of castes is much greater than in other social insects such as bees and ants.

The soldier members of a caste include mandibulate and nasute forms which have a large, eyeless head (He) which is strongly chitinized, a wingless thorax (Th), and an abdomen (Ab). Mandibulate soldiers (Fig. 2) have larger mandibles (Ma) than nasute soldiers, and they are used for both offensive and defensive purposes. A frontal pore (FP) opens on the head behind the antennae (An). A viscid secretion used in repelling enemies is released from this pore. The labrum (La), maxillary palps (MP), preclypeus (Pr), and postclypeus (Po) portions of the head are identified (Fig. 4). The anterior, paired antenna are of the monoliform type and originate from the first postoral segment (Figs. 1 and 2). The antenna typically consists of a basal segment or scape (Sc) associated with an articulating socket (ArS), an adjacent small pedicel, and a many segmented flagellum (Figs. 4 and 5). Insect antennae typically have chemical and mechanical sensory receptors for tactile, olfactory, and auditory stimuli. Note the sensory spines (SS) which form a ring around the distal portion of each of the antennal segments in the worker (Fig. 5).

The worker caste illustrated in Figs. 1 and 3 through 5 performs many functions for the colony including forming tunnels, chewing wood, feeding and cleaning other castes, cleaning the nest, and attending to the young. The cellulose present in the wood eaten by termites is digested in many forms by Protozoa (symbionts) such as *Trichonympha* which live in the intestine, and additional protein is obtained by eating dead members of the caste as well as fecal ma-

terial. Nymphs are fed fungus and vegetable matter which are partially predigested by adults. In this manner the nymphs become infected with symbiotic bacteria necessary to digest cellulose.

The head (He) of a worker is eyeless, but much smaller than that of the soldiers. The anterior sclerites including the postclypeus (Po), preclypeus (Pr), and labrum (La) are apparent in Fig. 4. Also note the antennae (An), mandibles (Ma), maxillary palps (MP), and labial palps (LP) in Figs. 1 and 4. The three segments of the thorax (Th) with prominent pronota are distinct (Fig. 1). The three pairs of similar legs (Le) have strong coxas. The tarsi are composed of three short basal joints with which the claws (Cl) articulate (Figs. 1 and 4). The abdomen (Ab) has ten segments (the eleventh being vestigal in the termite). Note the tergum (Te), sternum (St), and pleural membrane (PM) portions of an abdominal segment (Fig. 1). Associated with the posterior end of the male soldier termite (Fig. 3) are anal styles (AS) and anal cerci (AC). The workers and soldiers (sterile castes) differ from the reproductive forms in their smaller size, absence of compound eyes, lack of wings, and lesser degree of chitinization of the thorax and abdomen.

245

Phylum Arthropoda

Class Insecta—Order Isoptera

Termites (continued)

Ab   Abdomen
AC   Anal cerci
An   Antenna
ArS Articulating antennal socket
AS   Anal styles
Cl   Claws
FP   Frontal pores
He   Head
La   Labrum
Le   Legs
LP   Labial palps
Ma   Mandibles
MP   Maxillary palps
PM   Pleural membrane
Po   Postclypeus
Pr   Preclypeus
Sc   Scape
SS   Sensory spines
St   Sternum
Te   Tergum
Th   Thorax

Fig. 1 (worker), ×50; Fig. 2 (mandibulate soldier), ×55; Fig. 3 (posterior end of male worker), ×185; Fig. 4 (head of worker), ×120; Fig. 5 (antenna of worker), ×415.

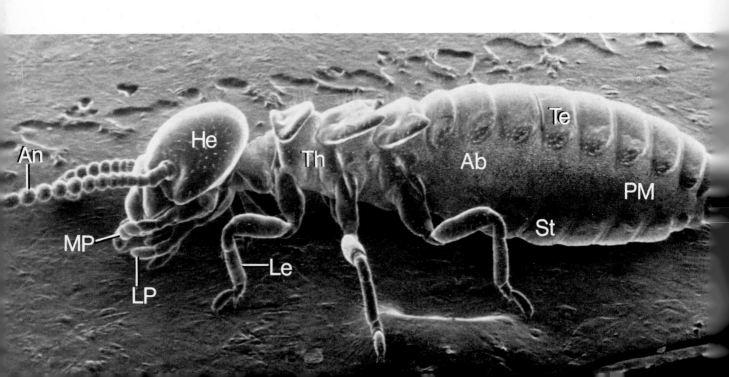

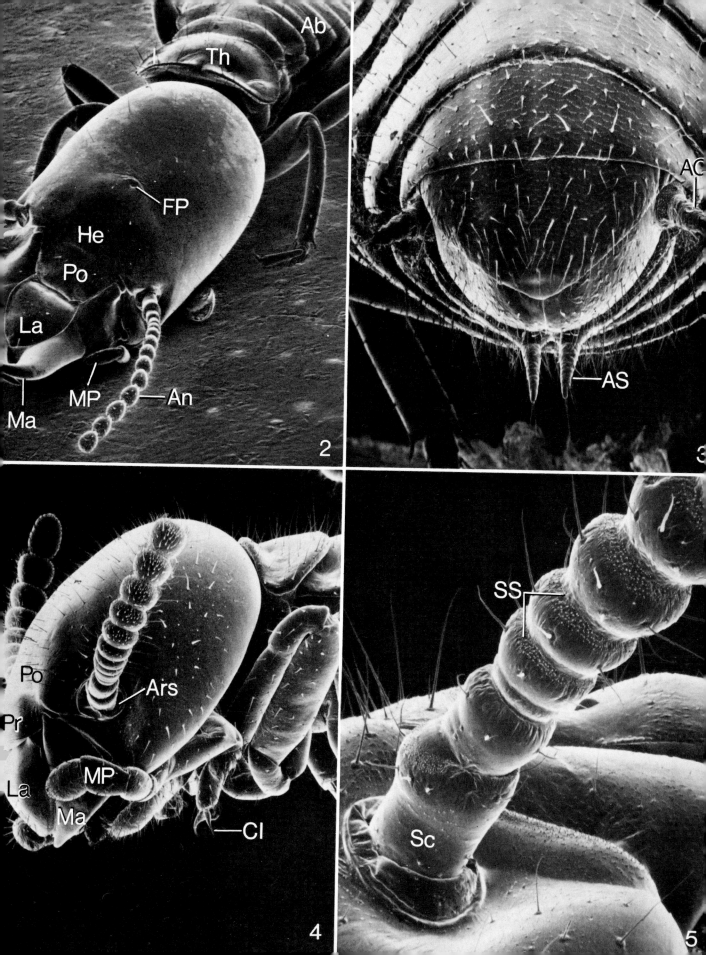

Phylum Arthropoda

Class Chilopoda

Terrestrial Mandibulates—Centipedes

An   Antenna
Ma   Mandibles
Max   Maxillae
Oc   Ocelli
PC   Poison claws
Sp   Spriacles
St   Sternite
Te   Teeth
Ter   Tergite

Fig. 1, ×52; Fig. 2, ×45; Fig. 3, ×41.

Centipedes are dorso-ventrally flattened and divided into a head and body region. The head (Figs. 1 and 2) is provided with one pair of antennae (An). Note the variety of circularly arranged hairs and spines (sensilla) associated with each of the antennal segments. The mouth parts include one pair of mandibles (Ma in Fig. 1) and two pairs of maxillae (Max in Fig. 1). A number of ocelli (Oc in Figs. 1 and 2) comprise the paired eyes of this organism, but are absent in some centipedes. The body consists of a variable number of identical body segments, most of which have a single pair of legs (each with seven joints).

The first body segment has limbs modified as poison claws (maxillipedes) (PC in Fig. 1), while the last two segments usually lack appendages. Centipedes are carnivorous and their food consists of small animals killed by means of the poison claws. The powerful poison claws are directed forward, curve inward, and end in sharp points (each is pierced by the opening of a poison gland). Each milliped has four free joints and a basal joint fused to the sternum of the first segment so as to form a large plate which bears teeth (Te in Fig. 1).

The exoskeletal plates covering the body consist of dorsal tergites (Ter in Fig. 2) and ventral sternites (St in Fig. 3), both of which have hairs extending from them (see Fig. 3). In the chilopod illustrated, the prominent spiracles of the tracheal system (respiratory function) are located ventrally and laterally on the posterior four body segments (Sp in Fig. 3).

248

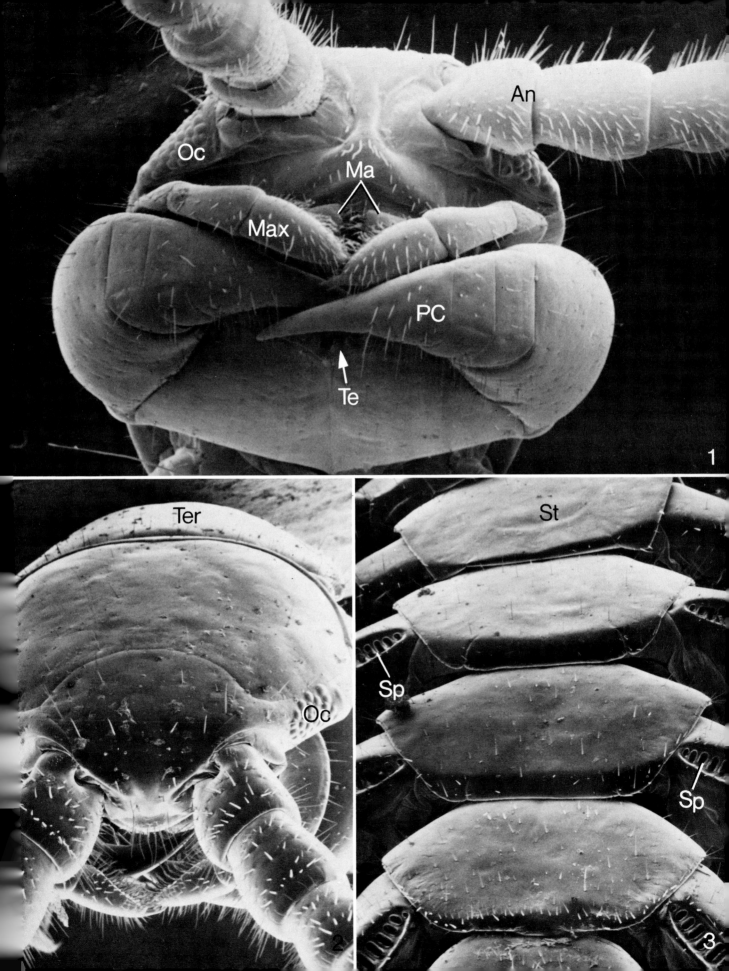

Phylum Arthropoda

Class Arachnida

Chelicerates (Mites)

Ca	Caruncle	Oc	Ocelli
Ch	Chelicerae	Op	Opisthosoma
Cl	Claws	Pe	Pedipalps
Co	Coxa	Pr	Propodosoma
Fe	Femur	Ta	Tarsus
Ge	Genu	Te	Tectum
Gn	Gnathosoma	Ti	Tibia
Me	Metapodosoma	Tr	Trochanter

Fig. 1 (mite), ×116; Fig. 2 (head and mouth parts), ×255; Fig. 3 (tip of foot), ×2425.

The order Acarina includes both ticks and mites, many of which are microscopic in size. The majority of mites are terrestrial and live as scavengers or as ectoparasites on plants and animals. The Acarina are so morphologically diverse and display so few unifying features that acarologists consider the order to be polyphyletic (having several evolutionary origins). A most striking characteristic is the apparent lack of body divisions. Abdominal segmentation has disappeared, and the abdomen has fused imperceptibly with the prosoma. Another general feature of the group is the change that has taken place in the head region which contains the mouth parts; it is called the capitulum. The dorsal body wall projects forward to form a rostrum, or tectum.

The Acarina have been given divisional landmarks by taxonomists. The gnathosoma (Gn in Fig. 1), or capitulum, includes the oral opening, mouth parts, chelicerae (Ch in Fig. 2), and pedipalps (Pe in Fig. 2). The chelicerae and pedipalps are variable in structure depending on their function. They are usually composed of two or three segments and may terminate in pincers, or the distal piece may fold back on the base to form a crushing and grasping appendage. The propodosoma (Pr in Fig. 1) includes that area of body containing legs 1 and 2. A pair of ocelli (Oc) can be seen on each side of the propodosoma. The metapodosoma (Me in Fig. 2) includes the body region of legs 3 and 4. Together the gnathosoma, propodosoma, and metapodosoma constitute the prosoma. The posterior region on which is found the anus is called the opisthosoma (Fig. 1, Op). Note the tectum (Fig. 2, Te) and the row of long hairs protruding from the upper part of capitulum.

In many forms the four pairs of legs are each six-segmented and are composed of a coxa (Co), trochanter (Tr), femur (Fe), genu (Ge), tibia (Ti), and tarsus (Ta) (Fig. 1). The tarsus often bears a pair of claws (Cl in Fig. 3). These appendages can be modified for such functions as walking, swimming, and jumping.

250

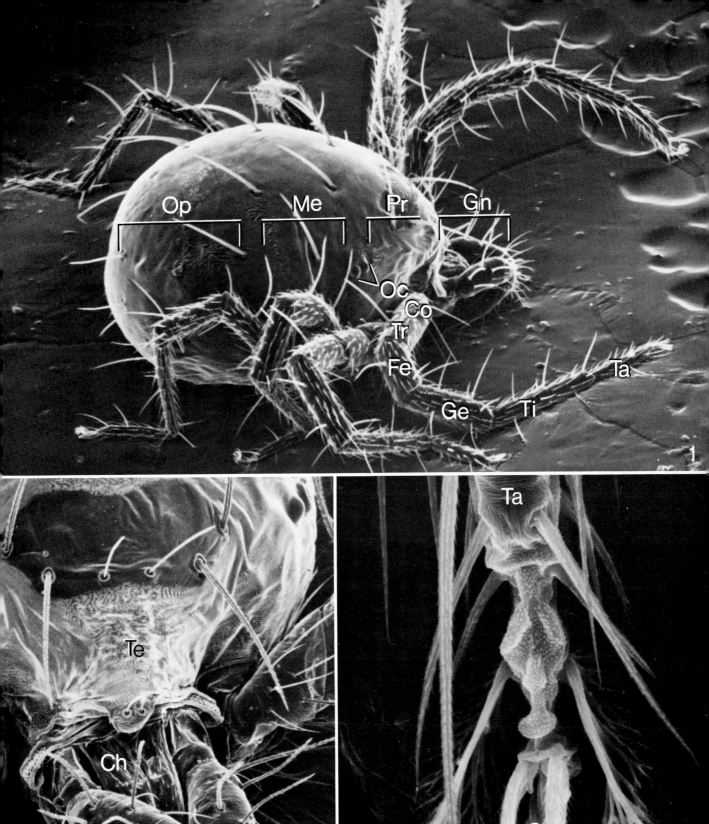

Phylum Arthropoda

Class Arachnida

Chelicerates (Mites)

(continued)

Cl   Claws
Se   Setae
So   Socket
Sp   Spatulate setae
Ta   Tarsus
TH   Tenent hairs

Fig. 4 (mite), ×95; Fig. 5 (body suface), ×2915; Fig. 6 (tarsus), ×1100; Fig. 7 (tenent hairs), ×2915.

To illustrate the diversity displayed by mites, a second organism is included for comparison (Figs. 4 through 7). The mite illustrated in Fig. 4 is covered with many more setae (Se) than the first mite. As in other arthropods, many of these setae are sensory and assume a variety of shapes. Note the spatulate (Sp) setae illustrated in Fig. 5. They are inserted in sockets (So in Fig. 5) and are probably mechanosensory. The nature and position of the setae in mites is an extremely important diagnostic characteristic in the classification and identification of species. Note also the sculpturing displayed in the exoskeletal covering of the mite in Fig. 5.

The tarsus (Ta) of two different mites can be compared in Figs. 3 and 6. Both bear a pair of claws (Cl), but the form of the tarsus is variable. Both are supplied with a variety of hairs. In additon to the claws, a delicate sucker-like structure, the caruncle (Ca), is present on the tarsus of one mite (Fig. 3), whereas on the other mite are numerous hairs specialized for walking on smooth surfaces. These hairs are called tenent hairs (TH in Figs. 6 and 7). Note that in Fig. 7 the tenent hairs have expanded filiform tips and thus form a very delicate feather carpet of hairs. The tips of the hairs are provided with a glandular secretion which allows adhesion to smooth surfaces.

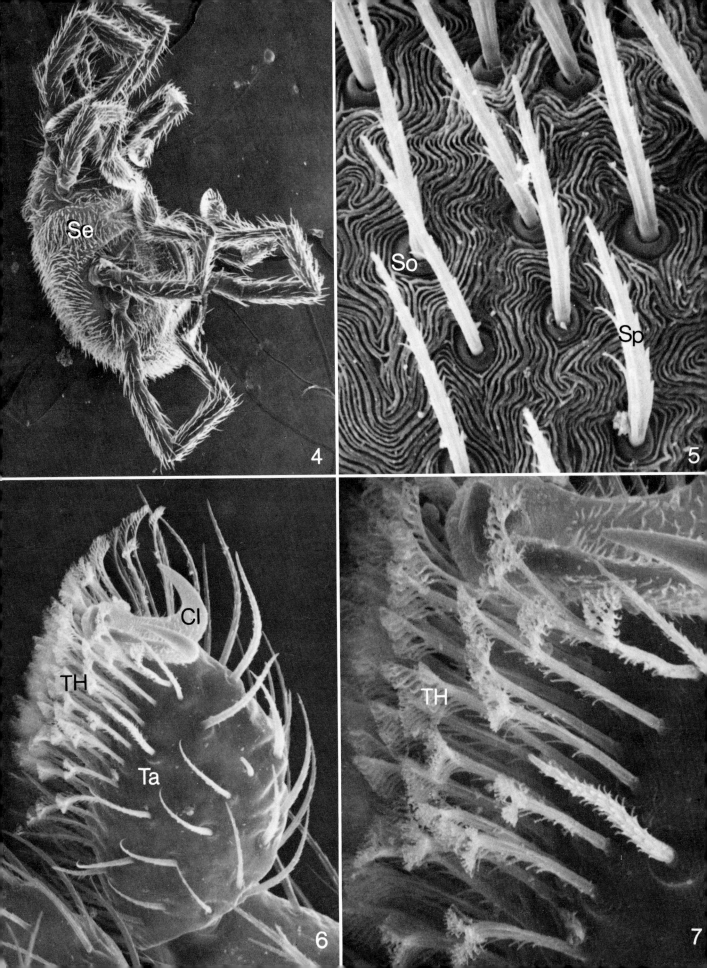

Phylum Arthropoda

Class Crustacea—Subclass Branchiopoda—
Order Cladocera

Daphnia (Water Flea)

Daphnia, sometimes called the water flea, is a free-living, freshwater crustacean. The body of *Daphnia* is enclosed by a carapace (Ca in Figs. 1, 2) which is joined dorsally but open ventrally (OC, Fig. 1). The carapace is nearly closed in the organism illustrated in Fig. 1, but is open in the organism illustrated in Fig. 2. The carapace is thus a hinged structure, shaped much like a bivalve shell which serves to enclose the laterally compressed thorax. The surface of the carapace is marked by a series of crossing ridges (Figs. 1, 6). Numerous small spines (Sp in Fig. 2) extend from the margin of the posterior portion of the carapace. A single large spine (PS in Fig. 1) extends from the posterior end of the carapace. A portion of the posterior spine is illustrated in Fig. 5. It can be seen from this figure that a number of closely packed, longitudinally arranged plates comprise the spine. Many of the plates, in turn, terminate in the form of long tapered spines (SP in Fig. 7). These spines consist of a number of small, closely packed rods which extend the length of the spine (arrows in Fig. 7).

The carapace is attached anteriorly and dorsally to the head and to the first two thoracic segments. The head (He) is not clearly separated from the thorax, but the junction is marked by a groove. The covering exoskeleton of the head is bent vertrally so that the pointed rostrum (Ro) projects downward (Figs. 1, 2). The position of a single, large compound eye (Ey) is indicated in Fig. 2. Under normal conditions many of the head appendages are hidden from view by the exoskeleton of the head and the carapace. The first appendages (segment 2) are called the antennules (An) and are identified in Figs. 4 and 8. The antennule is small, unjointed, and terminates in a group of sensory bristles (SB, Fig. 4) called olfactory setae. In Fig. 4 the distal portion of the antennule is observed to be composed of rod-shaped structures about 40 to 50 μm

long. At their proximal attachment are several needle-like processes which are much thinner and shorter than the rods (Fig. 4). The base of the antennule also has a small spine associated with it (arrow in Fig. 4). The second pair of appendages, called the antennae (Ant), are very large and extend beyond the exoskeleton (Figs. 1, 5). Each antenna is a biramous appendage having a basal joint (BJ) extending distally as two rami (Ra). The rami end distally in numerous, feather-like bristles (Br in Fig. 5), which are called natatory setae. These prominent antennae in *Daphnia* are highly modified for swimming, and the large number of setae associated with them improves their effectiveness in locomotion. The large mandibles (third appendages) are located close to the mouth. This tapered structure is covered with fine hairs (Ma in Fig. 8). The maxillae (fourth appendages) are located below the mandible in Fig. 8.

The carapace covers the thorax, which has 5 pairs of thoracic appendages called phyllopodia (TP in Figs. 2, 8). The thorax ends as a telson which has two unjointed rami whose distal ends have bristles or setae (Te in Fig. 6). The thoracic appendages are covered with a delicate cuticle, but are sufficiently flexible so as to move without requiring joints or subdivision into segments. The gnathobases of the filtering thoracic appendages are enlarged in *Daphnia* so as to accommodate large numbers of bristles (filter fringes) used in feeding. Structural variations exist in the thoracic phyllopodia. In general, they consist of a corm which supports endites (endopodites), exites (exopodites), an epipodite, sometimes a proepipodite. These appendage components have large numbers of prominent bristles associated with them. A portion of the endopodite of the second thoracic appendage is illustrated in Fig. 3. Note the terminal extensions of the apical lobe of this endopodite (ALE), which are co-

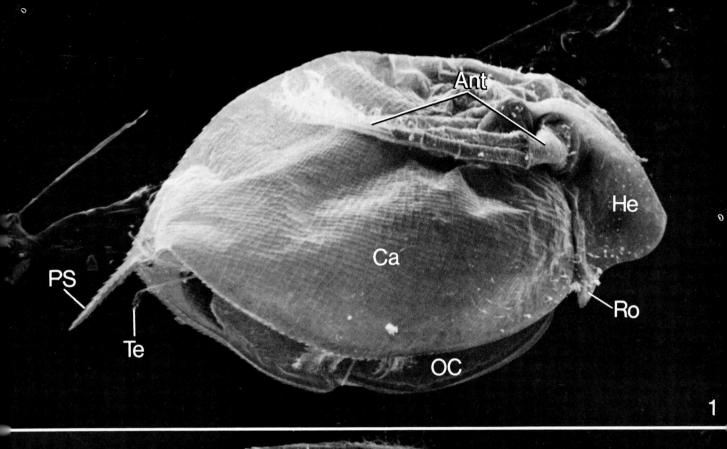

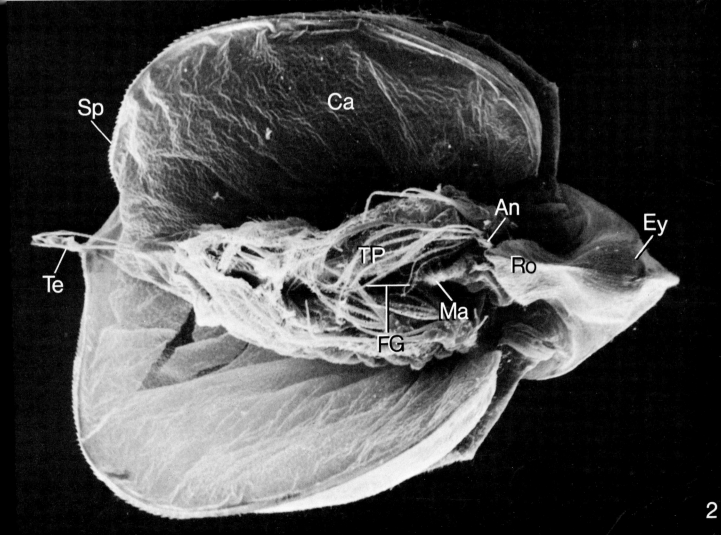

vered with numerous bristles (BrE in Fig. 3). *Daphnia* is a filter feeder in which the thoracic phyllopodia play an important role. Movements of the appendages create water currents, and food particles are entrapped by the bristles. The food particles (e.g., algae) are then moved anteriorly in a mid-ventral food groove (FG in Figs. 2, 8) to the mouth.

ALE   Apical lobe of endopodite
An   Antennules
BJ   Basal joint of antenna
Br   Bristles (antenna)
BrE   Bristles of endopodite
Ca   Carapace
FG   Food groove
Ma   Mandible
PS   Posterior spine
Ra   Rami (of antenna)
Ro   Rostrum
SB   Sensory bristle (antennule)
Sp   Spines (carapace)
SP   Spinous projections of posterior spine
Te   Telson
TP   Thoracic phyllopodia

Arrows: Multiple structure of projections from posterior spine

Fig. 3, ×825; Fig. 4, ×1500; Fig. 5, ×165; Fig. 6, ×395; Fig. 7, ×2750; Fig. 8, ×170.

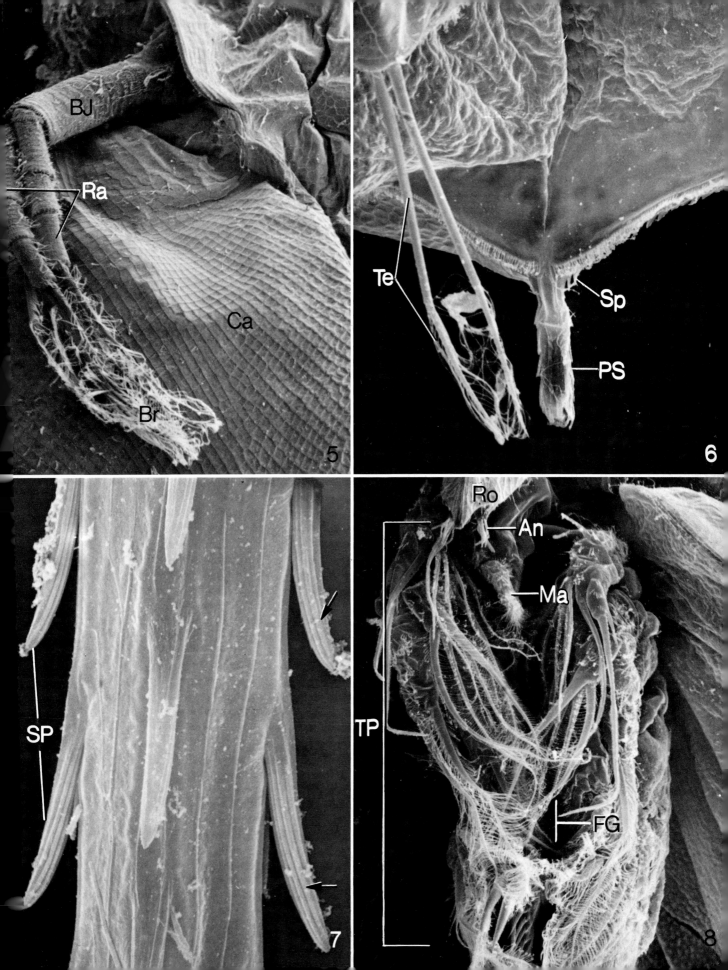

Phylum Arthropoda

Class Crustacea

Nauplius larva of *Artemia salina* (Brine-shrimp)

The characteristic initial larval stage of many Crustacea is called the nauplius. The nauplius larva of the brine-shrimp after hatching from the egg, but prior to the first moult, is illustrated in Figs. 1 and 2. The larva has three pairs of appendages which possess filaments or setae. The first pair of appendages are called antennules (An in Figs. 1, 2), and the tip of each appendage has two long setae and a shorter setal rudiment (AS in Figs. 1, 2). The second pair of appendages, the antennae, are the major locomotory structures in the larva. The biramous antennae (Ant in Figs. 1, 2) consist of a long protopod. Note the single seta associated with the basal enditic process of the antenna (ES in Figs. 2, 4). A single protopodal seta (PS in Figs. 2, 4) is also apparent. The remaining swimming setae (SS in Figs. 2, 4) associated with the expod and endopod of the antenna vary in number from 11—13.

The third pair of appendages, the mandibles (Ma in Figs. 1—3), are short and uniramous. Each of the mandibles has three terminal setae arising from the expod (ExS in Fig. 3). A second pair which represents the endopodal setae (EnS in Fig. 3) are located close to the terminal setae. Another pair of setae, the protopodal setae (PrS in Fig. 3) extend medially near the base of the mandibles. In later developmental stages of the nauplius larva, the number of setae change, some of the setae develop terminal setules and many of the antennal setae become hinged.

The mouth of the larva is covered by a flap-like structure called the labrum (La in Figs. 2, 3). No segmentation is yet evident externally in the trunk (Tr in Figs. 1, 2). The nauplius larva may be followed by other larval stages in many of the Crustacea as the result of moulting, addition of segments, and changes in the structure of the appendages.

Phylum Arthropoda

Class Crustacea

Nauplius larva of *Artemia salina* (Brine-shrimp)

(continued)

An	Antennules
AS	Antennule setae
Ant	Antennae
EnS	Endopodal setae (of mandible)
ES	Enditic seta (of antenna)
ExS	Exopodal setae (of mandible)
La	Labrum
Ma	Mandible
PS	Protopodal seta (of antenna)
PrS	Protopodal setae (of mandible)
SS	Swimming setae (of antenna)
Tr	Trunk

Fig. 1 (lateral view) ×410; Fig. 2 (ventral view) ×360; Fig. 3, ×965; Fig. 4, ×300.

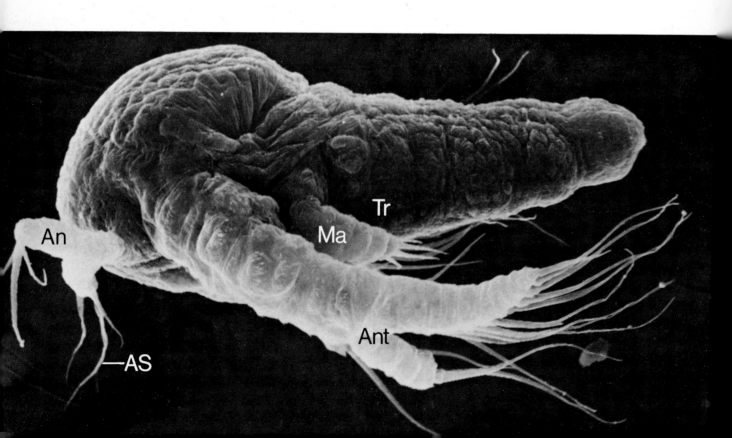

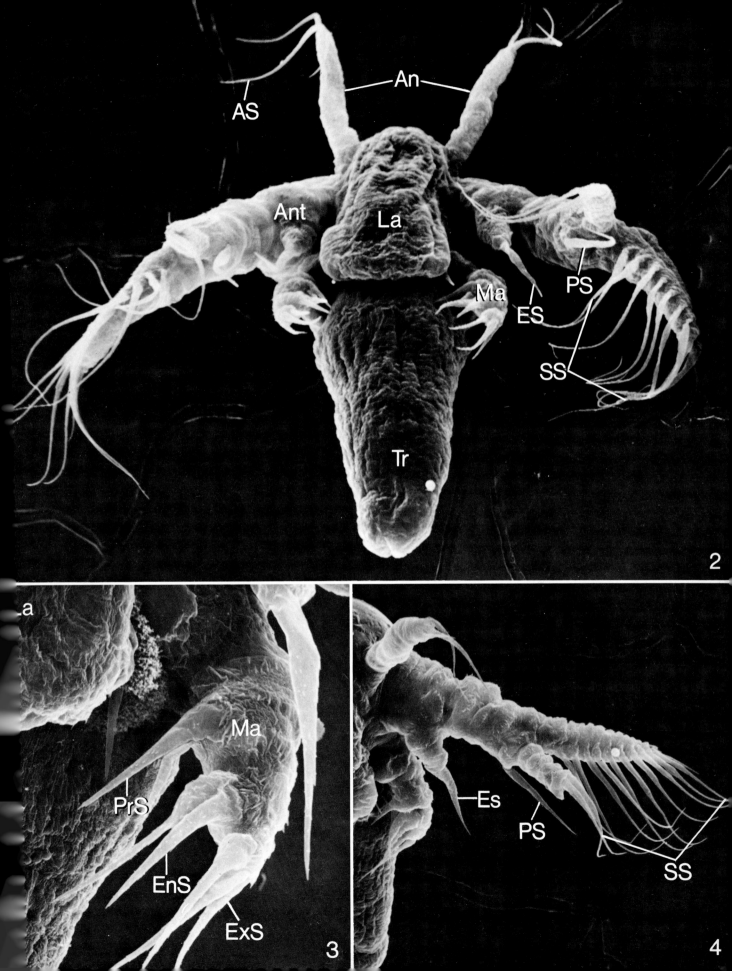

Chapter 9
Tissue and Organ Systems of Animals

Blood

Red Blood Corpuscle and Platelets

Fig. 1 (RBS's, light microscope preparation), ×2000; Fig. 2 (SEM of thrombus in rat lung), ×4580; Fig. 3 (SEM of human RBC's), ×7020; Fig. 4 (SEM of human platelets), ×10600.

The human red blood corpuscle (RBC) represents a highly differentiated structure adapted to serve in combining with, transporting, and releasing oxygen and carbon dioxide under appropriate conditions which involve the partial pressure of these gases in different regions of the body. The RBC is formed principally in the bone marrow and the stem cells have large numbers of polyribosomes for synthesizing hemoglobin, a complex protein-containing iron, which is its eventual major constituent. When fully formed, the RBC no longer has a nucleus or those organelles typical of other cells. In human males there are approximately 5.5 million RBC's per cubic millimeter of blood, and each has a life span in the circulating blood of about four months. The appearance of RBC's in normal blood smears stained with the Giesma method is illustrated in Fig. 1. From such images, it is difficult to determine that the natural shape of the RBC in mammals is a biconcave disc. This shape, however, is clearly demonstrated in the SEM (Figs. 2 and 3). RBC's average about 7.75 µm in diameter by about 1.9 µm in width at the edge. RBC's, or erythrocytes, are extremely flexible and very sensitive to mechanical stress. They are also sensitive to osmotic changes in the blood plasma. Under normal condition the RBC and blood plasma are in osmotic equilibrium. Under hypotonic conditions, in which the molecular concentration of the blood plasma becomes reduced due to increased water content, the water tends to enter the RBC and it swells. As a result, hemoglobin may escape from the RBC into the plasma under acute conditions. If, however, the molecular concentration of the blood plasma is increased through a corresponding decrease in water content of the plasma, water tends then to exit from the RBC, and it becomes irregularly shrunken or crenated in shape. RBC's have a strong tendency to adhere to one another at their flat surfaces and may exist in the form of stacks; such a condition is termed rouleaux formation.

Blood platelets are variable in shape, but often appear as round or oval biconvex discs (Fig. 4). Their diameter is also variable, but averages about 3 µm. There are approximately 300000 platelets per cubic millimeter of blood. Platelets are formed from giant megakaryocytes in bone marrow in an unusual manner which involves the fragmentation of the cytoplasm of these cells. Thus, platelets are cytoplasmic derivatives of megakaryocytes and do not have a nucleus. They survive in the circulating blood for about 10 to 12 days. When the continuity of a blood vessel lining is interrupted, platelets seem to be involved in causing the vessel to constrict so as to reduce blood loss. Further, platelets have adhesive surfaces and tend to agglutinate at the site of blood vessel injury so as to assist in closing the defect. Platelets also participate in the production of thromboplastin, which is involved in the process of blood clotting. Thromboplastin is produced in response to substances derived from platelets and locally in the injured tissue and is necessary, in part, for the conversion of prothrombin into thrombin. Calcium ions and other factors are also involved in the process of blood clotting. Thromboplastin is produced in response to substances derived from platelets and locally in the injured tissue and is necessary, in part, for the conversion of prothrombin into thrombin. Calcium ions and other factors are also involved. Thrombin is, in turn, involved in the conversion of fibrinogen into fibrin. Fibrin exists in the form of thin fibers (Fi in Fig. 2). With the TEM these fibers have an axial periodicity of about 250 Å. The formation of fibrin threads in a blood clot is, therefore, a three-step process.

The SEM has been useful in recent studies investigating the effects of different drugs and chemicals on changes in platelet shape (WALSH and BARNHART, (1973).

References

WALSH, R.T., and BARNHART, M.I.: "Blood platelet surfaces in 3-dimensions." In Scanning Electron Microscopy/1973, O. JOHARI and I. CORVIN, eds. IITRI 6, 481—488 (1973).

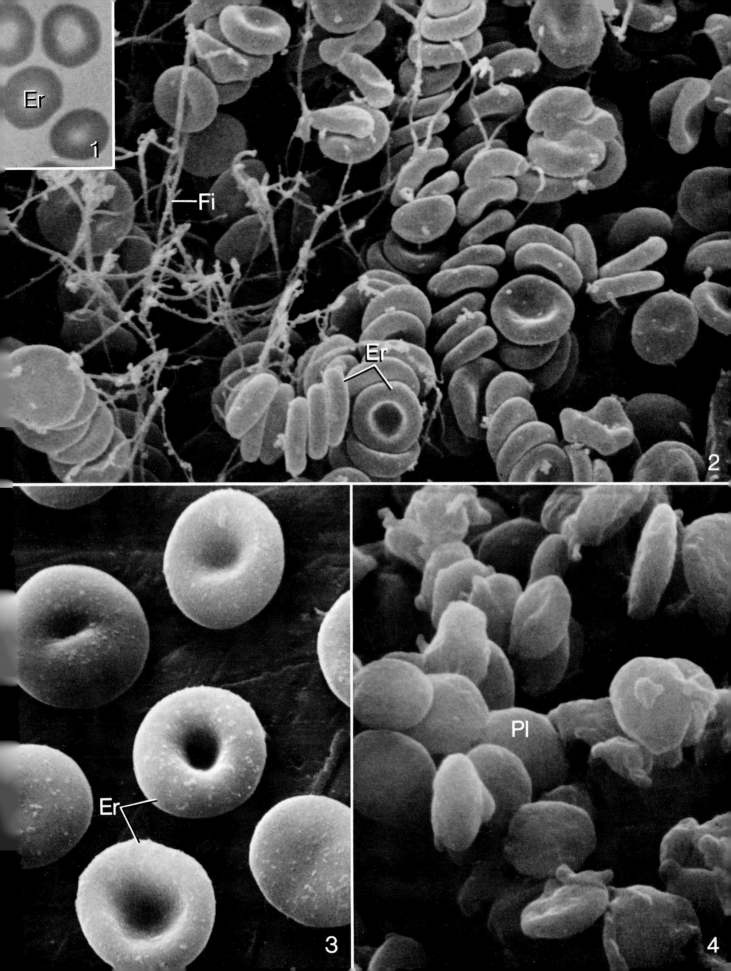

Er

Fi

Er

1

2

Er

3

Pl

4

Leukocytes

The circulating blood contains several types of colorless cells called leukocytes, which in normal human blood average 5000 to 9000 per milliliter. The number of leukocytes, however, varies widely during disease states of the body. Leukocyte types (neutrophils, lymphocytes, monocytes, eosinophils and basophils) are easily identified in slides of blood smears treated with Giemsa's or Wright's blood stains. In such preparations, classification of the cells is made on the basis of (1) size, (2) staining reaction and size of cytoplasmic granules if present, and (3) size, shape, and staining reaction of the cell nucleus. It has been demonstrated recently that the SEM can be usefully used to characterize the overall shape of leukocytes as well as the surface specializations that may be associated with these cells. Because leukocytes can be highly motile cells, it is not surprising that most of the studies on leukocytes with the SEM have demonstrated that considerable variation in cell form and surface specializations are possible. In view of this feature, it is necessary to use correlative measures to precisely identify a particular type of leukocyte based on cell morphology (WETZEL et al. 1973). In the study by WETZEL et al. (1973), human leukocytes that had settled from suspension on a coverslip were initially fixed and then stained with Giemsa's method. The preparations were then examined as a wet mount and specific cells identified with the light microscope. Then, the same cells were dried by the critical point method using carbon dioxide and examined in the SEM. This study demonstrated (1) a wide variation in form and surface morphology even among each of the leukocyte types, (2) striking similarities among leukocytes of different series in the SEM, and (3) that morphologic variants of cells in a given series may resemble corresponding forms in another series (WETZEL et al., 1973). Therefore, at least at this time, independent methods are necessary for specifically identifying individual cells observed in the SEM. Furthermore, there is evidence that such factors as temperature and length of incubation result in even greater variation in surface morphology of the leukocytes. That is to say, cell form and surface appendages can vary markedly during the attachment process of a cell to the substrate and during its migratory activity. It will be necessary and of interest in future studies to define the possible variations in form and nature of surface specializations of each class of leukocyte. SEM studies on purified populations of leukocyte types may be of value as well.

Several types of leukocytes (Le) are illustrated in Figs. 1 through 4. The buffy coat of blood obtained from normal individuals was prepared by centrifugation of anti-coagulated venous blood at 1000 rpm for 10 minutes at room temperature. A small portion of the buffy coat was resuspended in 0.1 ml cold isotonic saline (pH 7.4) and placed on 22 mm square plastic coverslips. The cells were allowed to settle onto the coverslips for 6 hours in a chamber saturated with saline to prevent drying. The preparation was then flooded with half-strength Karnovsky's paraformaldehyde-glutaraldehyde fixative in phosphate buffer (pH 7.2) and fixed for approximately 4 hours. The coverslips were then washed in several changes of phosphate buffer for a duration of 12 hours, dehydrated in ethanol and dried by the critical point method using liquid carbon dioxide. The cells were then coated with gold-palladium and carbon and examined in the SEM. The leukocytes (Le) illustrated in Fig. 1 vary in their size and surface complexity. The surface modifications of the leukocytes illustrated in Fig. 1 include blebs (Bl) and finger-like projections or microvilli (Mi). Both types of specializations may be associated with a single leukocyte, while in other cases only one form of surface specialization predominates on a leukocyte. Other studies have shown that ruffles (lamellipodia) and filopodia may also be associated with the surface of leukocytes depending on such factors as degree of attachment and motility when prepared for study (See also Chapter 1, Figures 19—22).

It has been reported that some of the circulating lymphocytes present in the blood have a rather smooth surface and these are thought to be derived from the thymus. However, other lymphocytes have many surface projections and these lymphocytes apparently are derived from the bone marrow (cf. POLLIACK et al., 1973). The long finger-like projections (Mi) that emanate in large numbers from the surface of the leukocyte illustrated in Fig. 3 have the form of microvilli. This is probably a B-type lymphocyte derived from the bone marrow. In contrast, what probably represents a T-type lymphocyte derived from the thymus is illustrated in Fig. 4 and has an extremely smooth surface topography.

The scanning electron microscope images of platelets illustrated in the previous section were taken from blood fixed immediately with buffered glutaraldehyde and under these conditions the platelets usually take the form of thin flat discs or ovals. In contrast, the platelets (Pl) illustrated in Fig. 1 all display surface activity which is reflected in the presence of a variable number of slender extensions that have been called pseudopods (pods) (cf. WALSH and BARNHART, 1973). These morphological alterations in the platelets give them a dendritic form, which is an indication of platelet activation. The platelets illustrated in Fig. 1 apparently responded to contact with the substrate by rounding up and extending dendritic processes (pseudopods) that assist in anchoring or attaching the platelet. A number of different platelet aggregators such as adenosine diphosphate, collagen, serotonin, and epinephrine have been described by WALSH and BARNHART (1973) to cause aggregation, sphering, and extensive pseudopod formation by the platelets. Collagen activated platelets were observed to have as many as 16 to 20 pseudopods. Scanning electron microscopy is clearly able to demonstrate that blood platelets are dynamic elements of blood and its continued use in experimental studies of various blood diseases should help to clarify many aspects of platelet function.

Bl   Blebs
Le   Leukocytes
Mi   Microvilli
Pl   Platelets

Fig. 1, ×3750; Fig. 2, ×9370; Fig. 3, ×14400; Fig. 4, ×14250.

References

POLLIACK, A., LAMPEN, N., and DE HARVEN, E.: "Comparison of air drying and critical point drying procedures for the study of human blood cells by scanning electron microscopy." In Scanning Electron Microscopy/1973, O. JOHARI and I. CORVIN, eds. IITRI 6, 529 (1973).

WALSH, R., and BARNHART, M.: "Blood platelet surfaces in 3-dimension." In Scanning Electron Microscopy/1973, O. JOHARI and I. CORVIN, eds. ITRI 6, 481 (1973)

WETZEL, B., ERICKSON, B., and LEVIS, W.: "The need for positive identification of leukocytes examined by SEM." In Scanning Electron Microscopy/1973, O. JOHARI and I. CORVIN, eds. IITRI 6, 535 (1973).

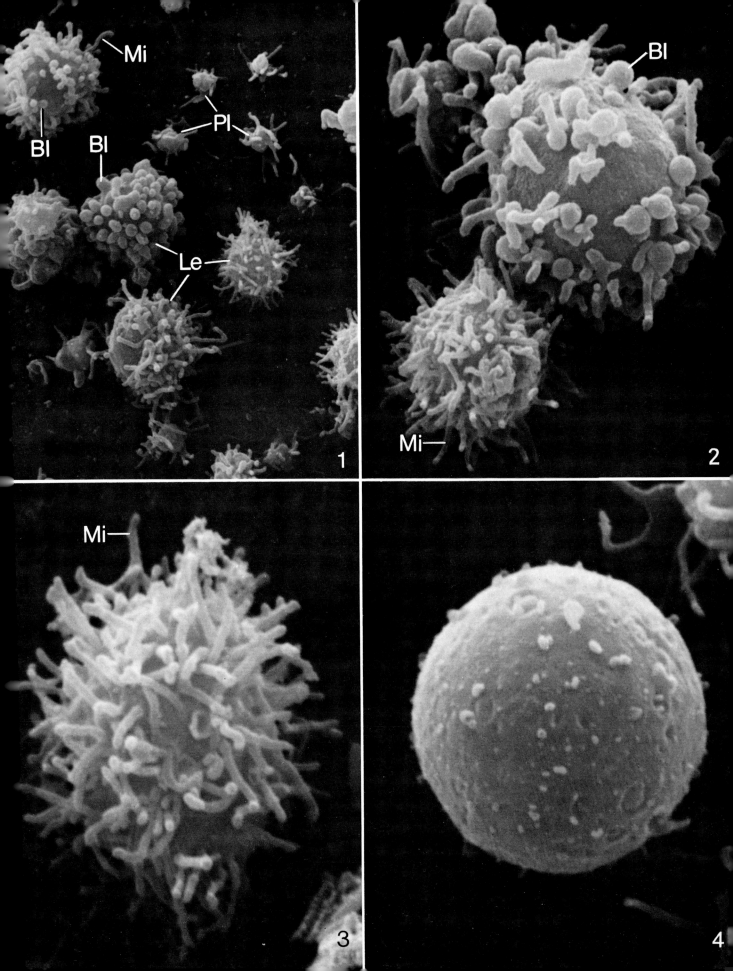

Surface Effects of Intraaortic Balloon Counterpulsation

Er Erythrocyte
Fi Fibrin

Arrow: Swollen dendritic platelets

Fig. 1, × 10500; Fig. 2, × 10500.

To assist patients in cardiogenic shock, intraaortic balloon counterpulsation (IABC) has been used to provide temporary mechanical circulatory aid. However, it has been suggested, for example, that balloon overinflation may produce thrombus formation. The SEM has been used to study the interaction between IABC, the blood tissue, and the aortic intima adjacent to the pumping chamber (SCHNEIDER and ECKNER, 1973). SCHNEIDER and ECKNER (1973) have described several morbid lesions associated with the luminal endothelium of the calf thoracic aorta after three days of continuous IABC. A single layer of simple squamous endothelium normally forms a continuous layer adjacent to the lumen of all blood vessels and represents the only layer present in capillaries. Mechanical aortic balloon pumping for 3 days results in (1) deformation of the luminal endothelium, (2) patchy losses of endothelium, and (3) mural thrombus formation in the experimental animals.

Mural thrombus formation induced by the IABC in a calf thoracic aorta is illustrated in Figs. 1 and 2. The background in each figure is the luminal intima of the blood vessel. The microthrombus illustrated in Fig. 1 consists of clusters of swollen, dendritic platelets (arrow) together with fibrin filaments (Fi). Deformed red blood cells (Er) are also associated with the thrombus.

In Fig. 2 the fibrin filaments appear to be emerging from surfaces of disfigured platelets to link with other platelets and to entrap deformed erythrocytes. Constituents of the microclot adhere to the subendothelial layer of the aorta. The results of this study indicate that the device used in IABC can affect the integrity and permeability of the luminal lining of the calf thoracic aorta. This impairment in the structure and function of the aortic intima probably results in the release of clot-promoting substances so that as a result a platelet microthrombus forms which adheres to the subendothelial layer of the vessel. The subsequent release of these microclots can apparently result in embolization downstream in the blood vascular system resulting in eschemia leading to observed renal infarction (SCHNEIDER and ECKER, 1973). (Micrographs kindly provided by Dr. M. D. SCHNEIDER.)

References

SCHNEIDER, M.D., and ECKNER, F.A.O.: "Surface effects of aortic balloon counterpulsation." In Scanning Electron Microscopy/1973. O. JOHARI and I. CORVIN, eds. IITRI **6**, 451—458 (1973).

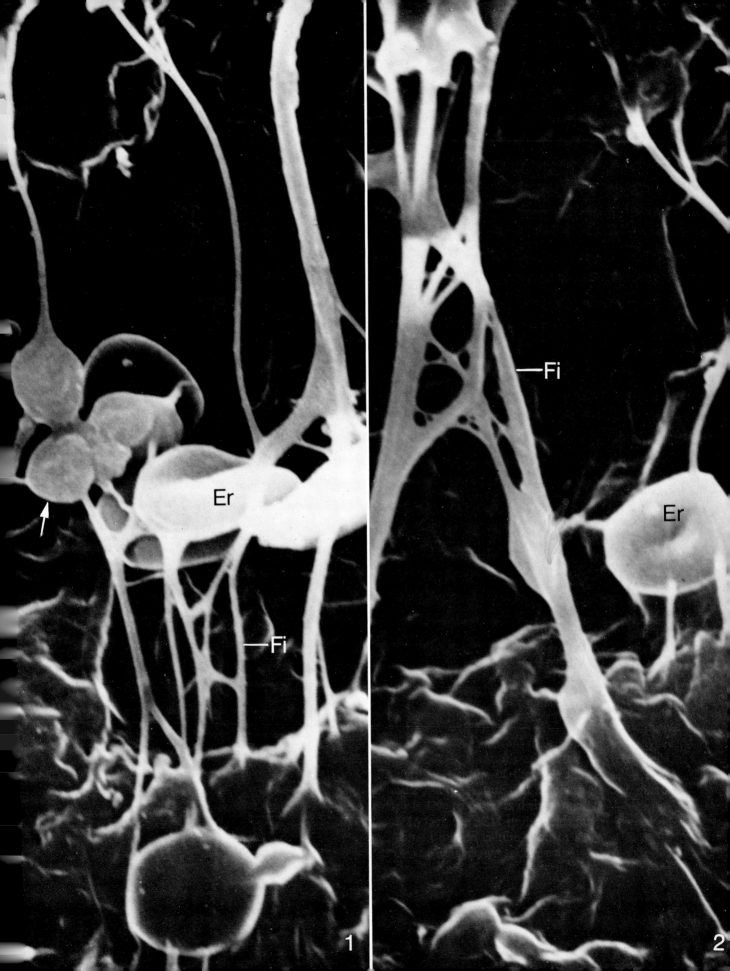

Endothelial Damage to Veins by Leukocytes

BM   Basement membrane
EL   Endothelial cell layer
Le   Leukocyte (neutrophil)
Su   Subendothelial region (of vessel wall)

Arrow: Discontinuities in endothelial cell layer

Fig. 1, ×330; Fig. 2, ×3900; Fig. 3, ×1090; Fig. 4, ×3850; Fig. 5, ×3750; Fig. 6, ×12000.

The luminal surface (endothelial layer of the tunica intima) of canine jugular veins under normal and abnormal (inflammatory process) conditions has been studied by both scanning and transmission electron microscopy by STEWART *et al.* (1973). This technique makes possible the examination of large areas of the lumen surface of blood vessels at a magnification and resolution which clarifies the nature of blood vessel diseases.

The endothelial lining of normal veins is a continuous sheet of cells. Under spontaneous pathological conditions large numbers of polymorphonuclear (PMN) leukocytes were observed to adhere to the intima layer (Fig. 1). This phenomena was associated with damage to the endothelial sheet of the intima so that interendothelial gaps (arrows) appeared within which PMN leukocytes were observed (Figs. 2 and 3). During this pathological process white blood cells pass through the endothelium by forcing apart endothelial cells at their junctions. The transmission electron micrograph (Fig. 6) illustrates a portion of a PMN leukocyte (Le) partially in the subendothelial (Su) region. Note that the thin endothelial cell layer (EL) is interrupted above the leukocyte. This image can be compared with the SEM micrograph of Fig. 5, which illustrates a leukocyte (Le) within a discontinuity (arrow) of the endothelial lining (EL).

The massive invasion of PMN leukocytes across the endothelium and their subsequent accumulation in packets between the endothelial sheet and the underlying basement membrane (BM in Fig. 6) appears to result in permanent separations between cells (Fig. 3). The separation of the endothelial cells, which is associated with the localized accumulation of PMN leukocytes in the subendothelial layer, appears to result in localized removal of this region of the tunica intima. These areas then tend to serve as foci for fibrin formation and the entrapment of platelets and red blood cells. In addition, packets of PMN leukocytes in the subendothelial layer apparently can rupture through the endothelial cell layer (Figs. 3 and 4). The study by STEWART and colleagues (1973) clearly indicates that leukocyte invasion is a physical means of producing endothelial injury. (Micrographs kindly provided by Dr. G. J. STEWART.)

Reference

STEWART, G.J., RITCHIE, W.G.M., and LYNCH, P.R.: "A scanning and transmission electron microscopic study of canine jugular veins." In Scanning Electron Microscopy/ 1973. O. JOHARI and I. CORVIN, eds. IITRI **6**, 473—480 (1973).

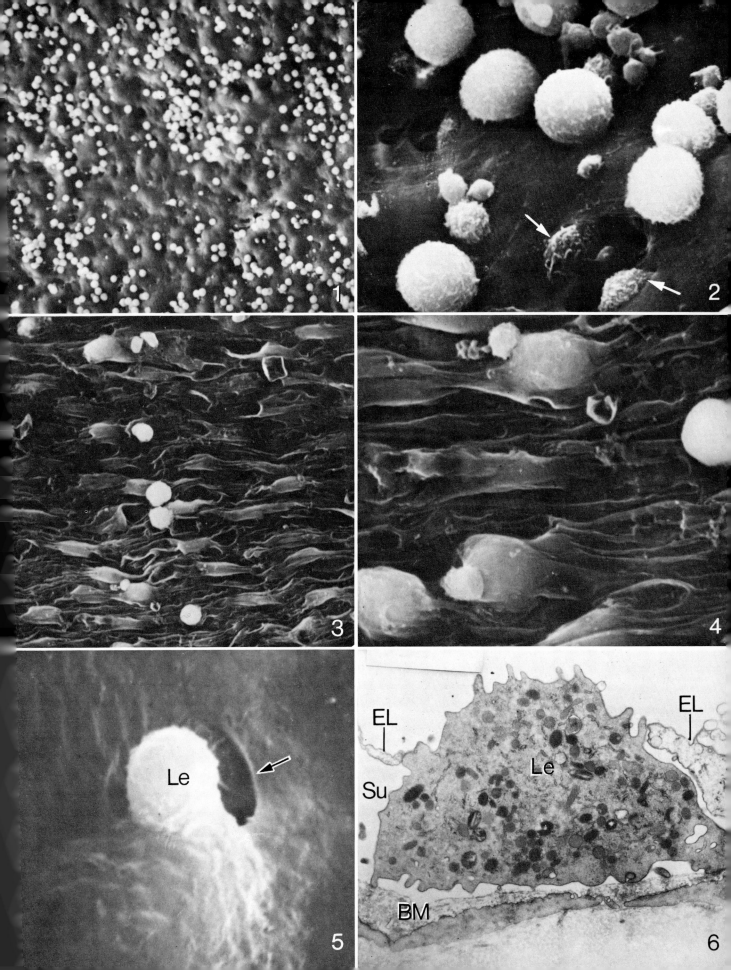

Muscle Tissue

Ca   Capillary
ID   Intercalated disc
Mi   Mitochondria
Nu   Nucleus
Po   Pores
RBC   Red blood cell
Sa   Sarcomere
Sar   Sarcolemma
TT   Transverse tubules

Arrows: cut edges of myofibrils

Fig. 1, ×4300; Fig. 2, ×8000; Fig. 3, ×4400; Fig. 4, ×30000.

The use of the SEM in the study of intracellular details of tissues and organs has not progressed to the extent that it has in the study of external surface structures. However, SYBERS and ASHROF (1973) have recently studied the dog heart muscle under a variety of preparative conditions. The results of their study indicate that the SEM can be usefully employed to study muscle tissue since intracellular organelles can be identified and studied at considerably higher resolution than with the light microscope. The heart muscle was fixed by coronary artery perfusion with 5% buffered glutaraldehyde. The trimmed tissue blocks were then dehydrated in ethanol and dried by the critical point method from Freon 113 (standard method). In some cases the dehydrated tissue was frozen in Freon 113 with liquid nitrogen and fractured. The fractured tissue was then thawed, rinsed in several changes of fresh Freon before critical point drying, coating with gold, and examination in the SEM.

Cardiac muscle prepared by the standard method just described is illustrated in Fig. 1. The cut edges of myofibrils are indicated at the arrows. The transverse tubules (TT) take the form of elevated ridges which traverse the myofibrils at the level of the Z-band. A nucleus (Nu) can be identified. A portion of a capillary (Ca) has been sectioned, thus exposing a red blood cell (RBC). A piece of cardiac muscle prepared by the cryofracture method is illustrated in Fig. 2. The sarcomeres (Sa) of the myofibrils are apparent due to their demarcation by the overlying transverse tubules (TT). A nucleus (Nu) and mitochondria (Mi) are easily identified. The cardiac muscle illustrated in Fig. 3 was also prepared by the cryofracture method with the plane of fracture apparently following the intercalated disc (ID). In Fig. 4 the surface of the sarcolemma (Sar) is illustrated and appears to contain pores (Po). A transverse tubule (TT) at the level of the Z-band is also illustrated. The pores in the sarcolemma may represent openings into the T-tubules. (Photographs kindly provided by Dr. H.D. SYBERS.)

Reference

SYBERS, H.D., and ASHRAF, M.: "Preparation of cardiac muscle for SEM. "In Scanning Electron Microscopy/1973, O. JOHARI and I. CORWIN, eds. IITRI **6**, 342—348 (1973).

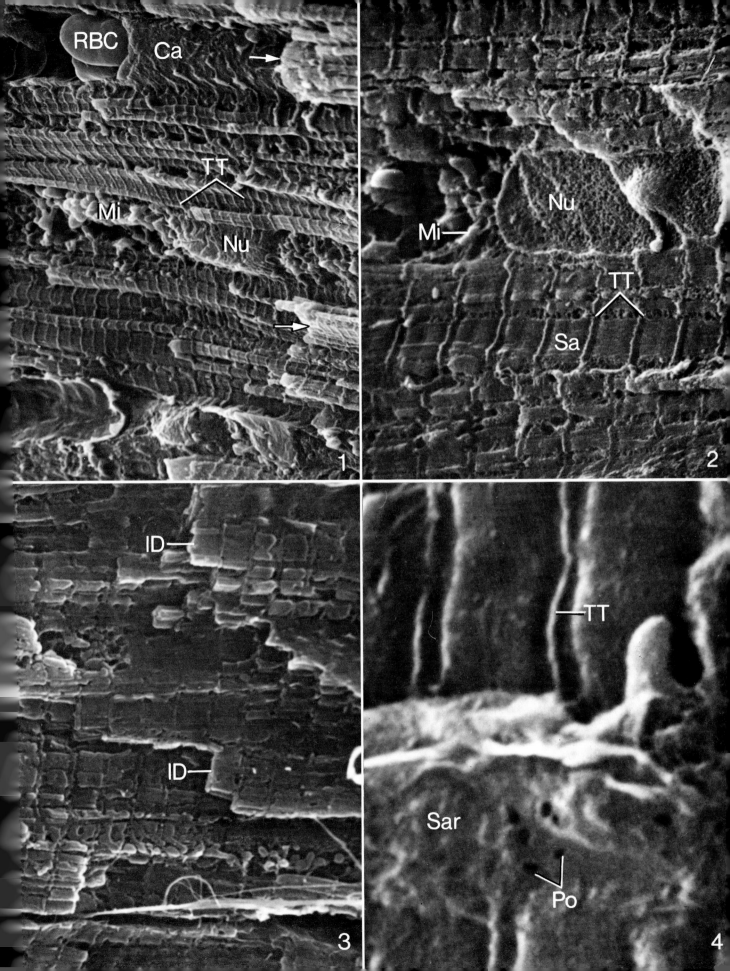

Digestive System

Tongue

Ba   Bacteria
CT   Cornified tips (of filiform papillae)
FP   Filiform papillae
FuP Fungiform papillae
SC   Squamous cells (on surface of fungiform papilla)

Fig. 1, × 500; Fig. 2, × 2975.

The tongue consists of an internal mass of striated muscle fibers surrounded by connective tissue and an epithelial covering (a mucous membrane). In man the tongue is used in feeding, swallowing, and speaking. On its dorsal surface the epithelial layer of the tongue is modified into a number of projections called papillae. Two types of papillae from the dorsal surface of the rat tongue are illustrated in Fig. 1. The most numerous type of extension is the filiform papilla (FP in Fig. 1). These have pointed tips, which are cornified (CT in Fig. 1), and they are arranged in rows. The epithelium of these papillae is provided with nerve endings which serve a tactile function. Fungiform papillae (FuP in Fig. 1) are less numerous and are scattered among the filiform papillae. They are short and have a flattened surface. Taste buds may be present in the epithelial lining at the sides of these papillae. The cell boundaries of flattened squamous cells (SC) comprising the outer layer of the epithelium covering the fungiform papillae are apparent in Fig. 2. Numerous chains of bacteria (Ba) adhere to the surface of the papillae (Fig. 2).

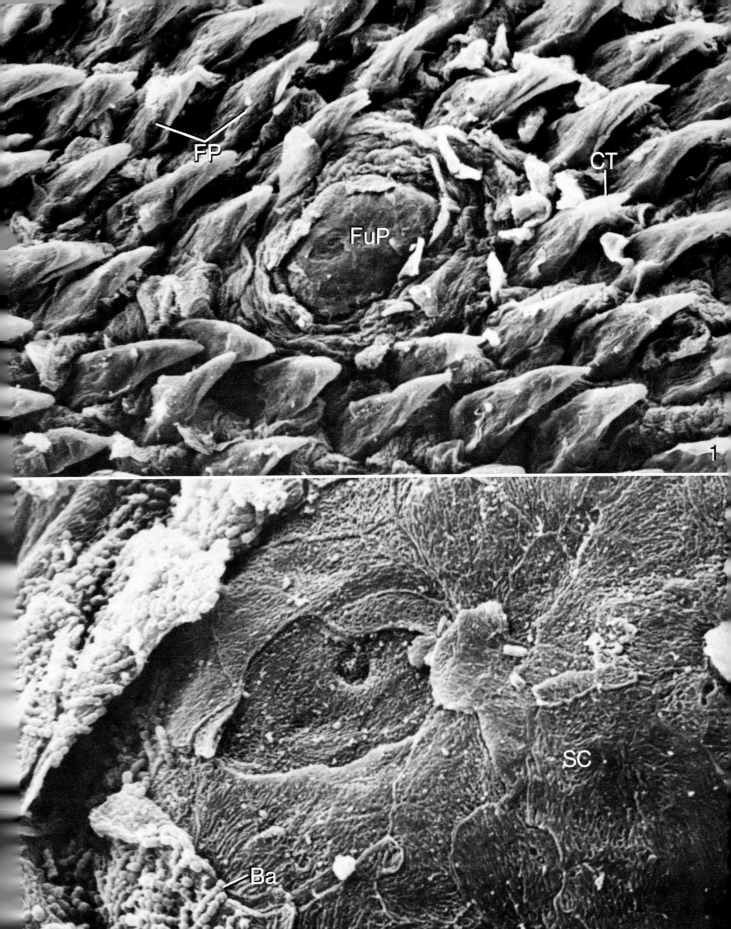

Tooth

DT   Dentinal tubules
EP   Enamel prisms

Fig. 1, ×755; Fig. 2, ×755; Fig. 3, ×1515; Fig. 4, ×1485.

The bulk of each adult tooth consists of enamel and dentin. Enamel, the hardest tissue found in the body, is a calcified substance covering the exposed portion (crown) of the tooth. The initial enamel matrix during its formation is secreted by epithelial-derived cells termed ameloblasts and this matrix subsequently becomes calcified. The matrix consists of protein and carbohydrate with calcium phosphate in the form of apatite. Mature enamel has a mineral content of about 95%. Because the ameloblasts have six sides, the enamel which each cell secretes tends to have the form of a hexagonal rod, and each of these is termed an enamel prism (EP in Figs. 1 and 3). In the honeycomb portion of the enamel surface (Fig. 1) the grooves represent the space filled in the living tooth by a Tomes' process (a slender cytoplasmic extension from the secretory end of an ameloblast) and the bases correspond to the ends of enamel rods. The enamel rods are also illustrated in longitudinal view in a fractured tooth in Fig. 3. The enamel surface of the human tooth is not smooth when viewed in the SEM (Fig. 1). The grooves in Fig. 1 correspond to the ends of individual enamel rods. The alternate elevations and furrows extending transversely as illustrated in Fig. 1 are called perikymata and result from the meeting of the enamel surface and the growth lines (Retzius' stripes). One Retzius' stripe (Fig. 2) is formed in about 10 days. When a tooth erupts, the ameloblasts which form the enamel are lost. Therefore, enamel is incapable of repair if injured by bacterial decay or other means.

Forming the major inner portion of an adult tooth is dentin, which also consists of a calcified matrix, but is softer than enamel. The specialized cells which secrete dentin are called odontoblasts. The dentin matrix consists of glycoproteins and collagenic fibers. It is thought that the glycoproteins secreted by the odontoblasts may play a role in binding calcium salts to the fine collagenic fibers as the dentin becomes calcified after its formation. As the odontoblasts secrete dentin matrix from one surface, the layer of cells withdraw as additional matrix is secreted. As this occurs, a very thin cytoplasmic process from the odontoblast remains extended into the formed dentin. This long, slender extension of the secretory surface of the odontoblast is called a Tomes' dentinal or odontoblastic process. The tiny canals in the dentin in which the odontoblastic processes reside are called dentinal tubules (DT in Fig. 4) and can be seen in a fractured tooth in Fig. 4. Unlike ameloblasts, odontoblasts do persist in the adult tooth and form a layer along the inner margin of the dentin and the pulp cavity which contains blood vessels and nerves. Dentin is, therefore, capable of repair and is more sensitive to stimuli than is enamel.

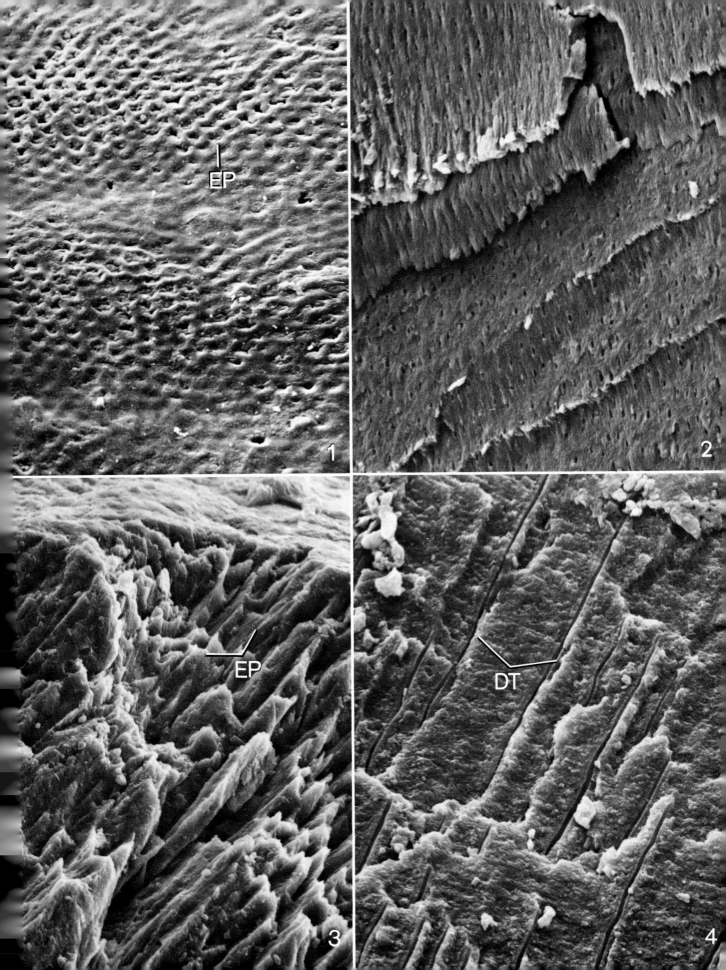

Stomach (Mouse)

The vertebrate stomach, extending between the esophagus and the small intestine, is both a reservoir and a digestive organ. A well-developed pyloric sphincter at the lower end retains the food while three layers of smooth muscle (muscularis externa) mix the food with the digestive enzymes of the stomach by peristaltic contractions. Three other layers of muscle interior to these (muscularis mucosa) mark the outer border of the mucous membrane lining the stomach cavity (Fig. 2) and provide some measure of independent movement for the mucosa, which is thrown into longitudinal folds (rugae) by contraction of these muscles.

The major inner layer of the stomach is called a mucosa or mucous membrane and contains a muscularis mucosa (MM in Fig. 2) and millions of simple tubular glands consisting of gastric glands (GG) and gastrics pits (GP). The cells lining the stomach surface and gastric pits consist of a single layer of tall columnar epithelial cells (EC in Figs. 1 and 2). The openings into the gastric pits are clearly observed in the SEM (GP in Fig. 1). Thus, the gastric pits are invaginations of the surface epithelium and communicate directly with the lumen of the organ.

The gastric pits are continuous below with the small lumen in the tubular gastric glands. The gastric glands are lined by cells that secrete digestive enzymes and include parietal or oxyntic cells which secrete hydrochloric acid in a concentration sufficient to dissolve zinc as well as chief or zymogenic cells which secrete pepsinogen, rennin, and lipase. The gastric juices (digestive enzymes) enter the lumen of the stomach via the gastric pits in response to stimulation of the vagus nerve or by reflex action within the mucosa when food enters the stomach.

The surface epithelial cells of the stomach secrete mucus, as do the cells lining the gastric pits. The free border of these cells has generally been considered to be free of microvilli (no striated border). However, short microvilli extend from the margins of the free surface of these cells (Mi in Fig. 4). These epithelial cells secrete considerable quantities of mucus (Mu in Fig. 3), which form an extensive sheet or covering (MS in Fig. 1) of the stomach. This mucus sheet provides some protection from digestive juices as well as protection from direct mechanical injury from bulky or sharp foods.

Under normal conditions the stomach does not digest itself although the powerful enzymes produced here are capable of such activity. The mucus sheet provides the first measure of protection against self-digestion. In addition, the structurally important junctional complexes between epithelial cells are ionic barriers and prevent hydrogen ions from penetrating and damaging the mucosa. The third line of defense is desquamation, for epithelial cells are continually flaking off the mucosal surface and are replaced by new cells at such a rate that the lining is completely replaced in three days. Even though a few superficial hemorrhages are produced during digestion of a meal by the stomach's own enzymes, the damaged cells may be replaced in a matter of hours.

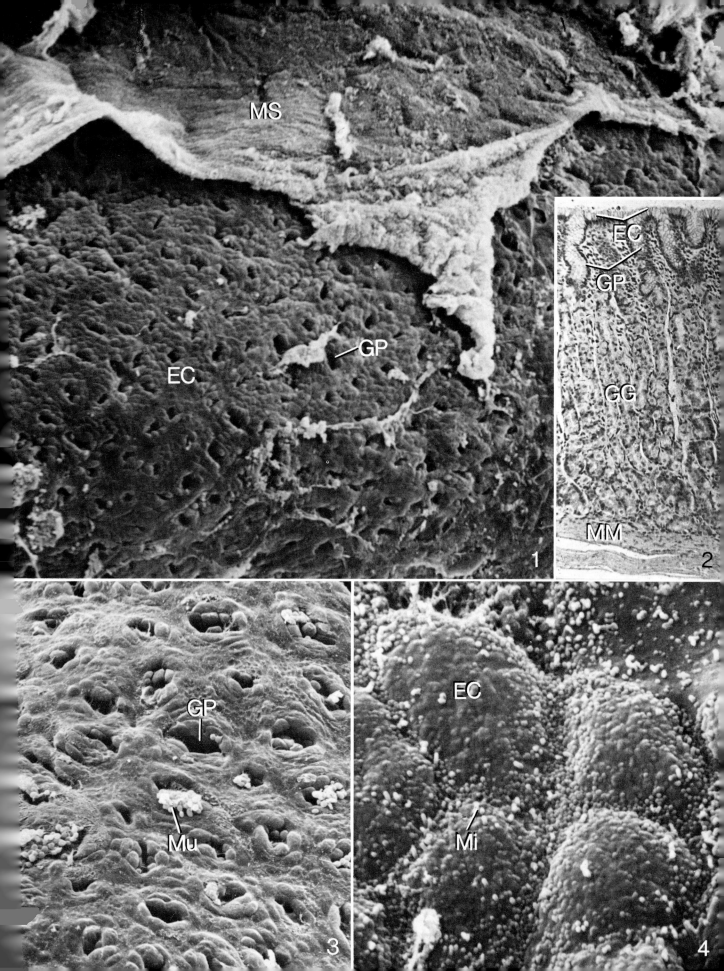

Small Intestine (Rat and Mouse)

The digestion of food and the subsequent absorption of the products of digestion occur primarily in the small intestine. To facilitate and accomodate the extensive absorption process that must occur, the inner layer of the small intestine is structurally modified to produce an extensive surface area. The wall of the small intestine, from inside to outside, consists of a mucosa, a submucosa, a muscularis, and, serosa layers. The serosa consists of a single layer of simple squamous epithelial cells and a thin underlying connective tissue layer. The muscularis, as in other regions of the digestive tract, consists of circular and longitudinal smooth muscle layers, which serve to propel the contents through the small intestine. The submucosa consists of connective tissue with blood vessels, nerve, and lymphatics. Depending upon the particular region of the small intestine, glands and lymphatic nodules may also be located in this layer. It is in the structure of the innermost mucosa layer that the small intestine significantly differs from the remainder of the digestive tube.

The mucosa is modified into numerous finger-like extensions called villi (Vi in Figs. 1 and 2). These evaginations of the mucosa consist of a single external cell layer containing both simple columnar epithelial cells (EC in Fig. 2) as well as goblet cells (GC in Fig. 2). The core (Co in Fig. 2) of the villus contains blood vessels, nerves, lymphatics, and smooth muscle. The tongue-shaped configuration of the villi in the mouse ileum is illustrated in Fig. 1. The villi are closely packed and arranged in a staggered fashion approximately in rows. A number of creases or folds (Fo in Figs. 1 and 3) are associated with the villi. These folds are oriented horizontally and not only increase the surface area of the villus but probably provide flexibility and freedom of movement to the villi. The small orifices associated with the villi, illustrated in Fig. 3, denote the position of goblet cells (GC) covering the villi. The surface of the goblet cell (GC) is illustrated at higher magnification in Fig. 5. Several microvilli and small secretion droplets are associated with the free surface of these cells. Goblet cells produce a mucus coat (Mu in Figs. 1 and 6) that protects the mucosa from physical damage and serves to entrap bacteria.

The mucosa of the small intestine also is differentiated into numerous tubular *invaginations* of its surface, and these structures are called crypts of Lieberkuhn (CL in Fig. 2). When the small intestine is fixed under conditions of some distension, the mucosal floor (MF in Fig. 3) surrounding the intestinal villi is more favorably demonstrated. In such preparations the openings into the crypts of Lieberkuhn (OCL in Fig. 3) are quite apparent. There are 15 to 20 crypt openings around each villus. The location of the goblet cells (GC in Fig. 3) in the epithelial lining of the mucosal floor is also favorably demonstrated.

The villus surface epithelium is composed of tall columnar absorptive (EC) cells that are continuous with those on the mucosal floor and those lining the crypts of Lieberkuhn. Removal of the lubricating mucus coat reveals the striated border of the mucosal epithelial cells, actually an apical region of tightly packed, highly absorptive microvilli (Mi in Fig. 6). A robust glycocalyx (observable with the TEM) completely fills the interstices between the microvilli and probably exerts some selectivity of substances to reach the plasmalemma. The microvilli on the mouse intestinal epithelial cell vary from 0.60 to 0.65 μm in length and 0.10 to 0.15 μm in width in this area of the intestine. There are approximately 75 microvilli per square micrometer. The size and density of microvilli may vary across the apex of an individual cell, suggesting differences in activity at different areas of the surface, and variations among species with respect to microvillar size and density are pronounced (ANDERSON and WITHERS, 1973).

As a result of normal wear in the intestine, the entire epithelial cell population is totally replaced in one to two days (ANDERSON and WITHERS, 1973). The worn-out cells are shed from an extrusion zone (EZ in Fig. 3) at the tips of the villi into the intestinal lumen. New cells are formed by mitosis which occurs deep within the crypts. These cells then migrate up the crypt walls, across the mucosal floor, and up the villus surface. Goblet cells proliferate in a similar manner. The morphological and functional integrity of the intestinal villi depends upon the continuous and synchronous processes of cell division and exfoliation. Ionizing radiation kills the proliferating cells in the crypts, while cell loss continues from the tips of the villi. This serious depletion of the cell population causes stunting, fusion, and denudation of the villi (Fig. 4), but repair of the damage may be accomplished fully in seven to eight days.

The surface of several serosal (mesothelial) cells is illustrated in Fig. 7. A number of microvilli (Mi in Fig. 1) project from the exposed surface of these cells. In addition, a single cilium (Ci in Fig. 7) projects from the central portion of the cell. Similar specializations have been noted on many mesothelial cell surfaces, and it has been suggested that the microvilli may serve to facilitate the movement of organs of the viscera over each other, as well as to transport of materials through the cells. The function of the single cilium is not clear. (Figures 3 and 4 kindly provided by Dr. J. H. ANDERSON.)

References

ANDERSON, J. H., and WITHERS, R. H.: "SEM studies of irradiated rat intestinal mucosa", Scanning Electron Microscopy/1973. O. JOHARI and I. CORVIN, eds. IITRI **6**, 565—572 (1973).

TAYLOR, A. B., and ANDERSON, J. H.: "SEM observations of mammalian intestinal villi, intervillous floor, and crypt tubules." Micron **3**, 430—453, (1972).

CE	Columnar epithelial cells
Ci	Cilium
CL	Crypt of Lieberkuhn
Co	Core (of villus)
EC	Epithelial cell surface
EZ	Extrusion zone
Fo	Folds (on villi)
GC	Goblet cells
MF	Mucosal floor
Mi	Microvilli
Mu	Mucus
OCL	Openings (into crypts of Lieberkuhn)
Vi	Villi

Fig. 1 (villi of mouse ileum), × 260; Fig. 2 (LM section of mucosa), × 110; Fig. 3 (rat ileum), × 480; Fig. 4 (rat ileal mucosa four days after irradiation), × 180; Fig. 5 (surface of goblet and epithelial cells), × 9375; Fig. 6 (microvilli of epithelial cells), × 44,065; Fig. 7 (surface of serosal cells), × 6610.

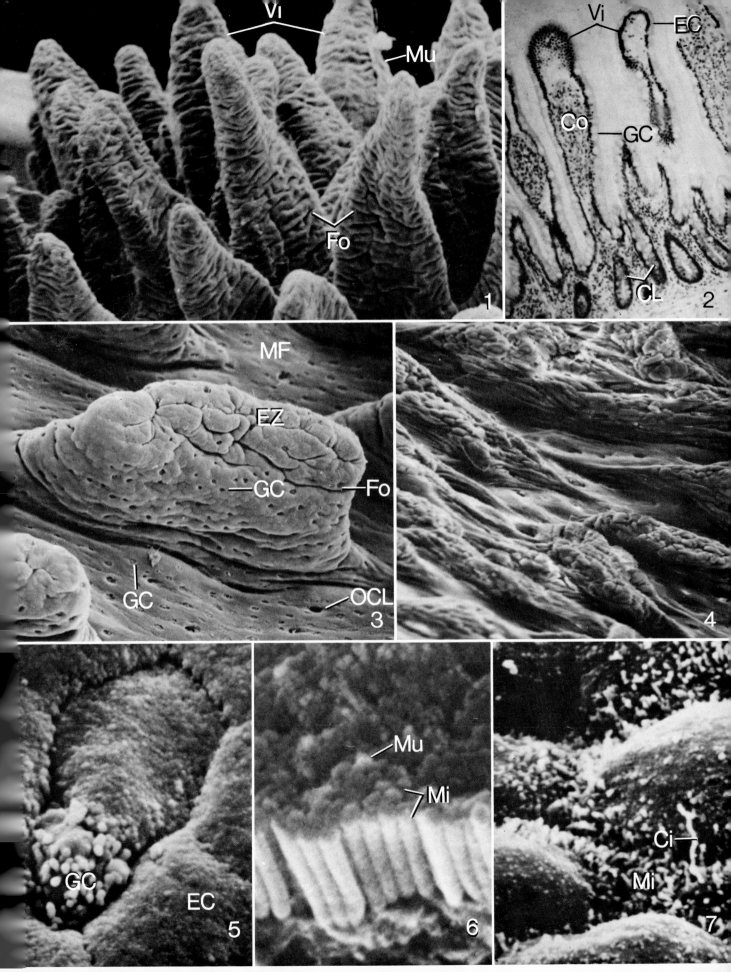

Large Intestine (Rat)

The large intestine, or colon, is an absorptive and propulsive organ. Water is recovered from the feces as it is propelled through the lumen (Lu in Fig. 1) by peristaltic contractions of the smooth muscle in its wall. A section through the wall of the colon as viewed with the SEM (Fig. 1) illustrates the principal layers of the wall; namely, a serosa (Se), muscularis externa (ME), the submucosa (Su), and the folded mucosa (Mu). At regions denoted by the arrows in Fig. 1 the tubular crypts of Lieberkuhn (CL) in the mucosa are apparent. A stained section of the colon is illustrated in the optical photomicrograph of Fig. 2, and all layers previously noted are illustrated, including the tubular crypts of Lieberkuhn (CL).

The epithelial lining (mucous membrane) of the organ consists of simple columnar epithelial cells and many unicellular mucous glands called goblet cells. The mucosal surface of the colon is illustrated in the scanning electron micrographs of Figs. 3 through 5.

This mucous membrane is interrupted by invaginations that represent openings, into the crypts of Lieberkuhn. The entrances to the crypts of Lieberkuhn (OC) are shown as are the surfaces of epithelial (EC) and goblet cells (GC). The crypts of Lieberkuhn greatly increase the absorptive and secretory surface of the colon. No major digestive enzymes are added in the colon, although there may be some digestion in process due to enzymes carried along with food from the ileum and to the putrefactive bacteria which thrive in the colon. The crypts and free surface of the mucosa are lined with tall columnar epithelial cells, each with an apical striated border of microvilli (Mi in Fig. 5) which serves to further increase the absorptive surface. Goblet cells (GC in Figs. 3 and 5) appear in the large intestine in great numbers and provide lubrication by their mucus secretion. Mucus may be released in the form of droplets (arrows in Fig. 3) from the recessed surfaces of the goblet cells which also possess microvilli.

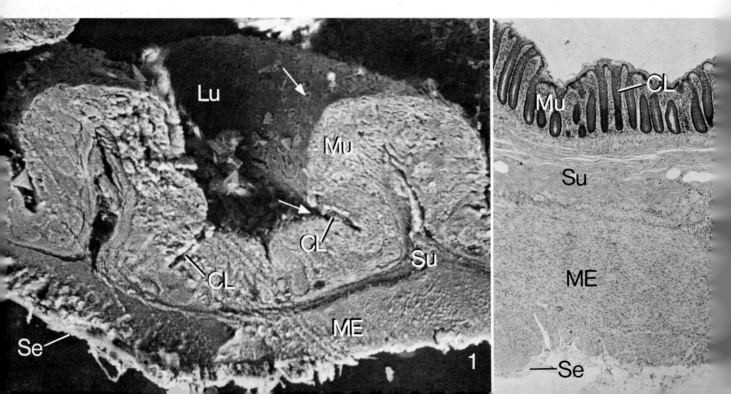

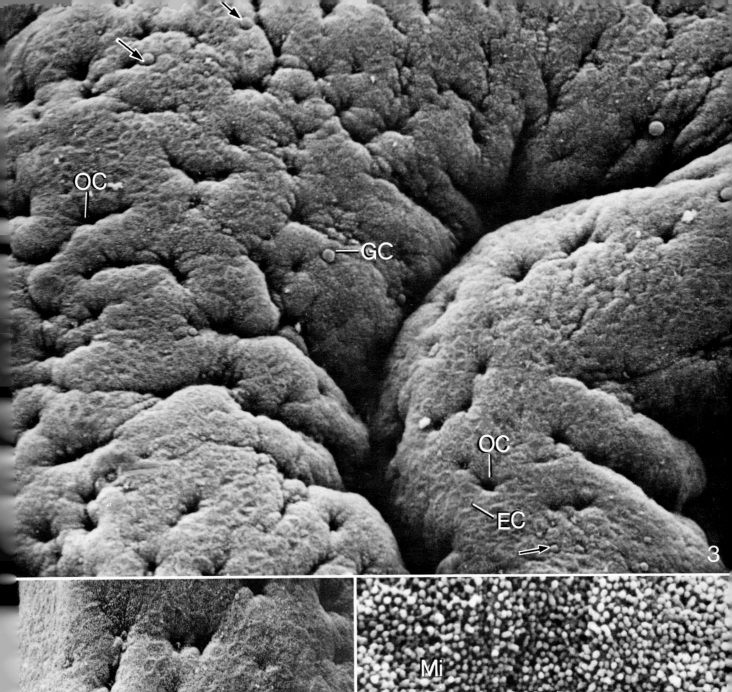

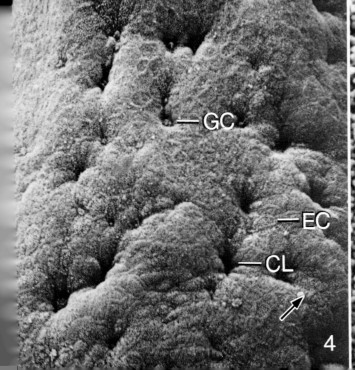

Liver (Rat)

BC	Bile canaliculi	HP	Hepatic plates
CV	Central venule (of hepatic lobule)	HS	Hepatic sinusoids
EC	Endothelial cells	PA	Portal area
Er	Erythrocytes in sinusoids	PV	Portal vein
Fe	Fenestrations in cells lining sinusoids	SD	Space of Disse
He	Hepatocytes		

Arrows: Openings of sinusoids into central venule

Fig. 1, ×705; Fig. 2, ×220; Fig. 3, ×285; Fig. 4, ×2170; Fig. 5, ×6470; Fig. 6, ×13725.

The major portion of the liver consists of cells (hepatocytes) which are arranged in the form of lobules. The hepatic lobule, the structural unit of the liver, has the form of a polyhedron which is approximately hexagonal in shape. There are approximately one million hepatic lobules in the liver of man. Each lobule measures about 2 mm long by about 0.75 mm wide. Within each hepatic lobule, the hepatocytes or hepatic parenchyma cells are arranged in the form of plates that are one cell thick, but many cells wide. The plates of cells branch and anastomose (HP in Figs. 1, 2). Many such plates or cords of hepatocytes (He in Figs. 2, 4, 5) are radially arranged about a central venule (CV in Figs. 1 through 3). The many spaces between the plates of hepatocytes in a lobule are called hepatic sinusoids (HS in Figs. 1, 4, 5). A portion of a hepatic lobule is illustrated in figures 1 and 2. In the case of figure 1, the liver was cut in thin slices while immersed in fixative. As a result, red blood cells (Er) are present in the sinusoids. For some of the illustrations, (Figs. 2 through 5), the liver was perfused for several minutes with an oxygenated physiological saline solution containing heparin and procaine to remove the blood from the vessels and sinusoids. This was followed by perfusion with glutaraldehyde fixative.

Many portal areas (PA in Fig. 2) or portal triads are present in the liver and they are located at the periphery of the lobules. This portal area contains a branch of the hepatic portal vein (PV in Fig. 2), hepatic artery, bile ductule, as well as branches of lymphatic vessels and nerves. The structures comprising the portal area are located in thin connective tissue septa at the lobule margin. The central venule (CV) has been sectioned longitudinally in figure 3 and the openings seen in the wall of this vein (arrows) represent points of continuity between the sinusoids and central venule.

The functional unit of the liver includes the portal area and adjacent portions of several lobules supplied by this portal area. End products of digestion in the intestine are carried via the hepatic portal vein to the liver where transformation and storage of the food products occur by the hepatocytes. Blood in the hepatic artery and hepatic portal vein drains into the sinusoids of the liver lobule and then flows into the central venule through openings in the venule wall (arrows in Fig. 3). The numerous sinusoids are usually 9—12 μm in width. Blood in the sinusoids is partially separated from the hepatocytes of the hepatic plates by endothelial cells and Kupffer's cells. The endothelial cells (EC in Figs. 5, 6) are thin and possess openings or fenestrations (Fe in Fig. 6). Kupffer's cells are phagocytic and are also fenestrated. The small space between the lining cells of the sinusoid and the hepatic parenchyma cells is termed the space of Disse (SD in Figs. 4, 5). The hepatic parenchyma cells perform a multitude of functions on the blood as it passes through the sinusoids. The fenestrated sinusoidal lining permits direct continuity between blood and hepatocytes. Blood passing through the sinusoid continues through the central venule to a sublobular vein, then to the hepatic vein and eventually into the inferior vena cava.

Bile canaliculi (BC in Fig. 4) are small intercellular channels or tubules between the hepatocytes in the hepatic plates. Bile is manufactured by the hepatocytes and is secreted into the tiny canaliculi between the hepatocytes. The bile moves in the canaliculi to bile ductules located in the portal area and eventually to the larger bile ducts which convey the bile from the liver.

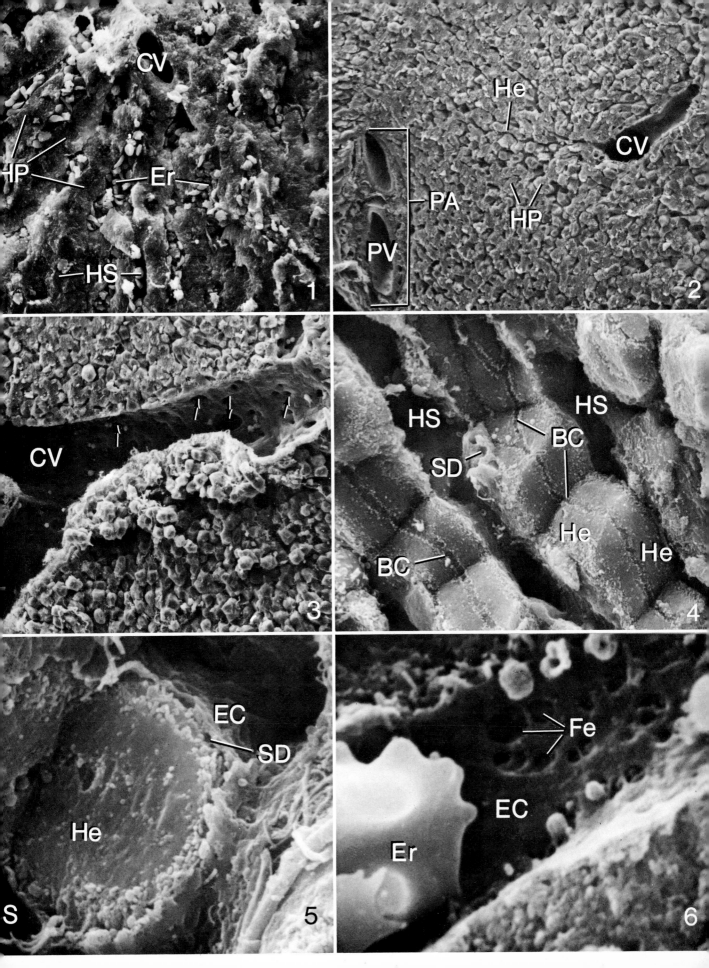

Respiratory System

Mucosa of Trachea

CE   Ciliated epithelial cells
GC   Goblet cells (surface)

Fig. 1, $\times 400$; Fig. 2, $\times 2370$; Fig. 3, $\times 5950$.

The respiratory system of mammals consists of a number of tubes and passageways which serve, in part, to conduct oxygen and carbon dioxide between the lungs and the exterior. These passageways, including the nasal cavities, trachea, and bronchi, are lined with a pseudostratified epithelium, the organization of which is illustrated in stained sections as viewed with the LM (Fig. 1). This respiratory epithelium is composed of two cell types; ciliated epithelial cells (CE in Figs. 1 through 3) and goblet cells (GC in Figs. 1 through 3).

The surface of the pseudostratified ciliated columnar epithelium with goblet cells is illustrated from the rat trachea in Figs. 2 and 3. The large amount of mucus secreted by the goblet cells onto the surface epithelium serves a number of functions including (1) entrapment of particles in the conducting tubes, (2) moistening or humidifying the air pulled into the system en route to the lungs, and (3) provision of bactericidal properties. The primary function of the cilia is to propel any particles entrapped in the mucus toward the oral pharynx, where they are removed from the system by the cough reflex.

GC CE

CE

GC

1

2

CE

GC

3

Lung

The preparations of the rat and mouse lung illustrated here were prepared for SEM in the following manner. The lungs were perfused through the bronchi with either 3% phosphate buffered glutaraldehyde (pH 7.3) or a dilute solution of Karnovsky's paraformaldehyde-glutaraldehyde fixative. Fixative was also pipetted onto the surface of the lung during perfusion. The lung was then excised and placed in fresh fixative for a period of several hours to overnight. After secondary fixation in 1% phosphate buffered osmium tetroxide (pH 7.3), the lungs were dehydrated in ethanol. During dehydration, the lung was either cut into thin slices or frozen in liquid nitrogen and fractured into smaller pieces. The dehydration was then completed and the specimens were dried by the critical point method.

The lungs are attached in the thorax only at the hilus and are covered by a layer of simple squamous epithelium called the visceral pleura. The extrapulmonary bronchus as well as the pulmonary artery and vein enter or leave the lung at the hilus. The visceral pleura is continuous with a structurally similar parietal pleura which forms the lining of the thoracic cavity. The visceral and parietal pleura, separated by only a thin lubricating pleural fluid, slide over each other during respiratory movements. The single layer of visceral pleura cells covering the lung are mound-shaped and are covered with numerous microvilli which probably provide an additional cushion between the two sliding surfaces.

Extrapulmonary bronchi, upon entering the lung, become intrapulmonary bronchi and repeatedly divide to supply all lobes of the lung. The smaller branches of this bronchial tree are called bronchioles (Br in Fig. 1). Branches of the pulmonary artery (PA in Fig. 1) and pulmonary vein (PV in Fig. 1) closely parallel the branches of the bronchial system. Smaller branches of the bronchioles are called terminal bronchioles (TB in Fig. 3) and respiratory bronchioles (RB in Fig. 3). The partial contraction of smooth muscle in the wall of the bronchi and bronchioles tends to cause the mucosa or inner lining of these structures to be folded. The folded mucosa (FM) of a bronchiole is illustrated in Fig. 4. The respiratory bronchioles open into the respiratory portion of the lung, a sponge-like lattice of terminal air spaces (Figs. 5, 6) of different size, shape, and arrangement. Bronchioles conduct air into the largest spaces called alveolar ducts (AD in Figs. 2 and 3) which, in turn, lead to alveolar sacs containing smaller terminal divisions or saccules called alveoli (Al in Figs. 4 through 6). The alveoli, usually polyhedral in shape and 50 to 100 µm in diameter, are formed by branching and anastomosing trabeculae called alveolar septa (AS in Figs. 5, 6). The lining of each air sac or alveolus consists of a complete layer of simple squamous epithelial cells covered by a basement lamina. Between adjacent saccules, in the alveolar or interalveolar septa, are located endothelial capillaries also surrounded by a basal lamina as well as fine reticular and collagenic fibers. The septa are about 10 µm thick and diffusion readily occurs between the air spaces and capillaries in the septa.

The investing epithelial membrane of the air sacs contains two cell types; one is a highly flattened cell (about 0.1 to 0.2 µm thick) which presents little resistance to diffusion, while the second cell type is called an alveolar secretory cell (ASC in Fig. 6) or an alveolar type II cell. These cells are more rounded and can be seen bulging from the septal surface in Fig. 6. In some species the secretory cells are crowned with microvilli, especially near the cell border, but in the rat not all secretory cells have such surface modifications. The secretory cells are thought to secrete a substance that acts as a surfactant. That is, the material reduces the surface tension in the thin film of fluid that covers the septal surfaces. This tension is great enough to collapse the fine delicate alveoli at expiration. Other cells, called alveolar phagocytes (AP in Fig. 6), may be seen in the alveolar spaces. They remove particles such as dust and carbon that may be inhaled into the lungs and can migrate between the interalveolar septa and alveolar spaces. When filled with foreign debris, they can migrate up the bronchial tree to the trachea where, by action of the cilia lining this tube, they are subsequently passed into the esophagus for removal.

Alveolar septa are often perforated by small pores (Po in Fig. 5) to permit communication between individual alveoli. Some authors believe these always to result from a previous disease process, but in view of the widespread occurrence of these pores within individual specimens and across vertebrate classes, it seems unlikely that these pores are merely defects in septal structure. Small communications also exist between adjacent respiratory bronchioles, and they are known as Lambert's sinuses (LS in Fig. 3). Together with the alveolar pores, they provide alternate routes of entry and escape to the terminal alveoli should localized regions of the lung become fibrotic or the normal passageways occluded in some manner.

AD	Alveolar duct
Al	Alveolus
AP	Alveolar phagocytes
AS	Alveolar (interalveolar) septa
ASC	Alveolar secretory cell
Br	Bronchiole
FM	Folded mucosa (of bronchiole)
LS	Lambert's sinus
PA	Pulmonary artery (branch)
PV	Pulmonary vein (branch)
Po	Pores (in alveoli)
RB	Respiratory bronchiole
TB	Terminal bronchiole

Fig. 1, ×67; Fig. 2, ×132; Fig. 3, ×276; Fig. 4, ×680; Fig. 5, ×1320; Fig. 6, ×1175.

References

NOWELL, J. A., GILLESPIE, J. R., and TYLER, W. S.: "Scanning electron microscopy of chronic pulmonary emphysema. A study of the equine model", In: Scanning Electron Microscopy/1971, O. JOHARI and I. CORVIN, eds. IITRI 4, 297—306 (1971).

NOWELL, J. A., PANGBORN, J., and TYLER, W. S.: "Stabilization and replication of soft tubular and alveolar systems. A scanning electron microscope study of the lung", In: Scanning Electron Microscopy/1972, O. JOHARI and I. CORVIN, eds. IITRI 5, 306—314 (1972).

TYLER, W. S., DELORIMIER, A. A., and MANUS, A. G.: "Surface morphology of hypoplastic and normal lungs from newborn lambs", In: Scanning Electron Microscopy/1972, O. JOHARI and I. CORVIN, eds. IITRI 4, 305—312 (1971).

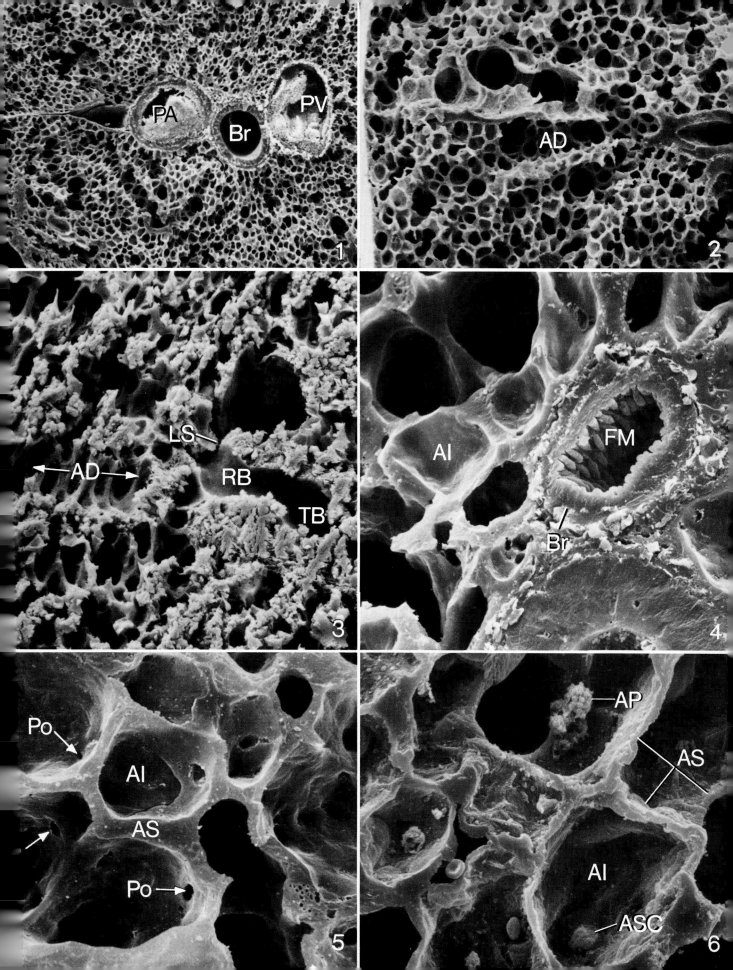

Excretory System

Kidney (Mouse and Rat)

Fig. 1 (kidney cortex), ×420; Fig. 2 (parietal layer of Bowman's capsule), ×4070; Fig. 3 (proximal tubule), ×4200; Fig. 4 (renal corpuscle), ×2400.

The kidney in land dwelling vertebrates is an organ for the production of urine containing the waste products of protein metabolism as well as an organ that regulates water and electrolyte balance in the blood and intercellular fluids. In addition, the kidney is an endocrine gland and secretes two known hormones; erythropoietin and renin, which stimulates red blood cell proliferation in bone marrow and elevates blood pressure, respectively. The organ is a compact mass of osmotically active tubules, microscopic in dimension, which receive large quantities of tissue fluid produced from special filtration units (called renal or Malpighian corpuscles) and convert it to urine. The tubules remove most of the water and electrolytes from this tissue fluid, and recapture essential proteins and glucose that escaped the filter. The waste products of protein metabolism including urea and certain organic acids and bases are added to the filtrate in the tubules. This hypertonic urine is then funneled into the ureter at the renal pelvis for removal via the bladder and urethra.

The beginning of each tubule is differentiated into a renal or Malpighian corpuscle (RC in Fig. 1) which is an oval appearing unit in the kidney cortex about 150—250 µm in diameter. The renal corpuscle consists of a Bowman's capsule, which is two layered, as well as a highly coiled capillary called a glomerulus. The renal corpuscle together with its differentiated tubule is termed a nephron. There are approximately 1,300,000 nephrons packed into each kidney in man. The remaining space is filled with blood vessels and loose, nonfibrous connective tissue which does not hinder the osmotic activity of the delicate tubules.

A low magnification SEM view of the cortex of the rat kidney is illustrated in figure 1. A portion of the connective tissue capsule (Ca) is identified as well as sections of the uriniferous tubules (Tu). Several renal corpuscles (RC) are illustrated, one of which contains the glomerular capillary and associated cells of the visceral layer of Bowman's capsule. The cells comprising the visceral layer of Bowman's capsule are called podocytes (Po) and are illustrated at a higher magnification in figure 4. The interior of the renal corpuscle was removed during cutting in two cases so that only the cup-shaped parietal layer (PL) of Bowman's capsule is revealed. The urinary pole orifice (UP) is apparent in figure 1 and marks the point where the parietal layer of Bowman's capsule becomes continuous with the proximal convoluted tubule.

The outer layer of Bowman's capsule (called the parietal layer) consists of a single layer of flattened cells. The space adjacent to this layer is the lumen of the glomerular capsule. The inner surface of the flattened cells comprising the parietal epithelium is illustrated at higher magnification in figure 2. One or two cilia (Ci in Fig. 2) extend from the central region of the lumenal surface of the parietal cells. The cilia are approximately 10 µm in length and 0.2 µm wide. They appear to be curved and frequently the distal ends appear to attach to the cell surface. There are also a few small microvilli associated with the lumen surface of the parietal cells.

The brush border associated with the luminal surface of the proximal convoluted tubule (PT in Fig. 1) consists of many microvilli (Mi in Fig. 3). One or more cilia (Ci in Fig. 3) are associated with the surface of the proximal tubule cells. They are approximately 2.5 µm in length and frequently appear bent (ANDREWS and PORTER, 1974). The functional significance of the presence of cilia throughout the renal corpuscle and uriniferous tubules has not been established. It has been suggested that they may assist with propulsion of fluid through the tubules, or that they may have some sort of sensory function.

Reference

ANDREWS, P.M., and PORTER, K.A.: A scanning electron microscopic study of the nephron. Am. J. Anat. **140**, 81—116 (1974).

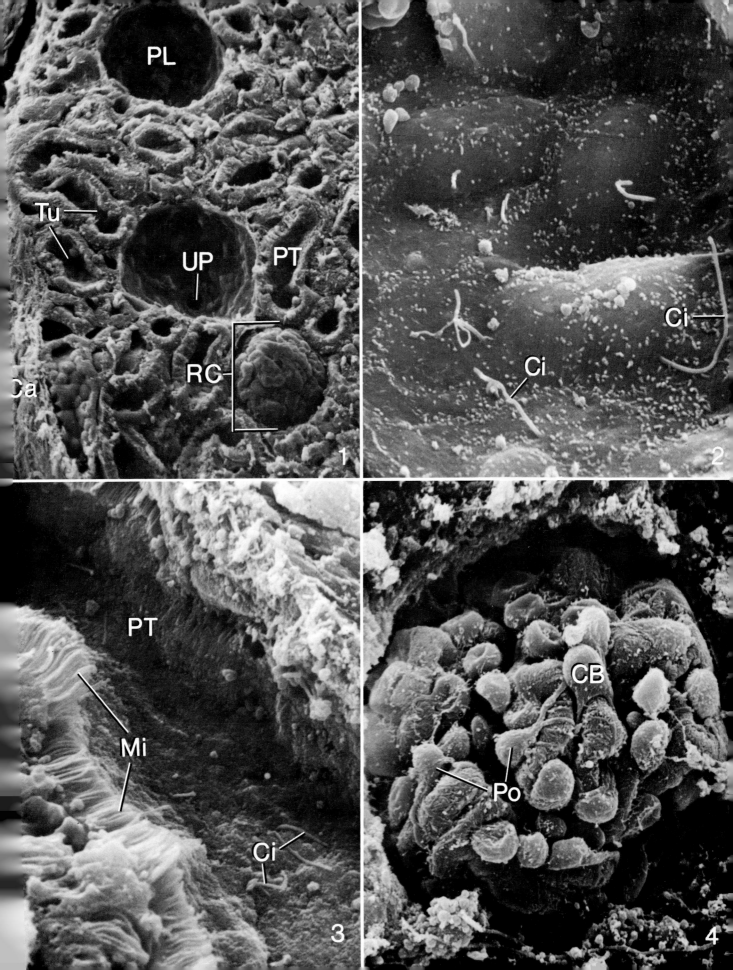

Kidney (Mouse and Rat)

(continued)

BL   Basal lamina
CB   Cell body (of podocyte)
Er   Erythrocyte
Fe   Feet (of podocyte)
FP   First-order podocyte branch
Fo   Folds in capillary endothelium
FS   Filtration slits
GC   Glomerular capillary
Mi   Microvilli
PE   Pores in endothelium
Po   Podocytes
SB   Stalked bud (on podocyte processes)
SP   Second-order podocyte branch
TP   Third-order podocyte branch

Arrows: Blebs on podocyte branches

Fig. 5, ×6,500; Fig. 6, ×14,250; Fig. 7, ×19,950; Fig. 8, ×21,800.

The inner layer of Bowman's capsule is composed of highly branched cells called podocytes that form a complete covering over the glomerular capillary and basement lamina. The cell body of the podocyte (CB in Figs. 4, 5) is dome-shaped and contains the nucleus. Its cytoplasm is highly branched so that many processes of different size extend from the cell and closely surround the underlying glomerular capillary and basal lamina (cf. ARAKAWA and TOKUNAGA, 1972; BUSS *et al.*, 1973). The cytoplasmic branches of the podocyte end in 'feet' (Fe in Figs. 5, 6); hence, the name podocytes. The podocytes branch into first-order (FP in Figs. 5, 6), second-order (SP in Figs. 5, 6), and finally third-order processes (TP in Figs. 5, 6). The terminal or third order processes interdigitate like fingers in a pair of folded hands. The clefts between these terminal branches are called filtration slits (FS in Figs. 5, 6). Note that small rounded blebs or buds of different size are associated with the surface of the podocytes (arrows in Figs. 5 and 6). Their functional significance is not yet clear. In some cases blebs or buds appear to be connected to the podocyte processes by a stalk (SB in Fig. 6). Short microvilli (Mi in Figs. 5, 6) are also associated with the podocyte cell body as well as its cytoplasmic branches.

When sections are made of the kidney fixed by perfusion fixation, the coiled glomerular capillary is exposed (Fig. 7). The relationship between the glomerular capillary (GC), basement lamina (BL), and foot processes of the podocytes (Po) is illustrated in Figure 7. An erythrocyte (Er) is present in the capillary lumen. At high magnification of the inner surface of the glomerular capillary endothelium, the rounded or polygonal pores (PE in Fig. 8) are apparent. These pores have been described as ranging from about 500 Å to 1000 Å in diameter. The surface of the glomerular capillary also exhibits thickenings or branched folds (Fo in Fig. 8). Such regions contain microtubules when viewed with the TEM and it has been suggested that they may provide a cytoskeletal-like support to the thin fenestrated endothelium (ANDREWS and PORTER, 1974).

Renal corpuscles in the human kidney may produce as many as 200 liters of ultrafiltrate in a 24-hour period, but about 99% is resorbed in the tubular portion of the nephrons. The basis for tissue fluid production in the renal corpuscle is a capillary tuft or glomerulus which is surrounded by the podocytes. The capillary network is interposed between two arterioles (afferent and efferent arterioles) of similar diameter. This unique circulatory arrangement results in no appreciable hydrostatic pressure loss across the length of the glomerular capillary. Therefore, large quantities of blood serum (tissue fluid) are forced out of the fenestrated capillary endothelium by the unusually high hydrostatic pressure (about 70 mm Hg); hence the term ultrafiltrate. The pores in the capillary endothelium are sufficient to retain the cellular portions of the blood, while most of the proteins are held back by either the basal lamina between capillary and podocyte, or the filtration slits themselves. Once the ultrafiltrate has passed through the capillary endothelium, the basal lamina, and the filtration slits, it is directed into the lumen of the proximal tubule. The rate of ultrafiltrate production can be varied by changes in filtration slit width.

References

ANDREWS, P.M., and PORTER, K.R.: A scanning electron microscopic study of the nephron, Am. J. Anat. **140**, 81—116 (1974).

ARAKAWA, M., and TOKUNAGA, J.: A scanning electron microscope study of the glomerulus. Further consideration of the mechanisms of the fusion of podocyte terminal process in nephrotic rats, Lab. Invest. **27**, 366—371 (1972).

BUSS, H., LAMBERT, B., and BRASS, H.: Orthology and pathology of the renal podocytes, *in,* Scanning Electron Microscopy/1973, O. JOHARI and I. CORVIN, eds., IITRI, **6**, 573—580 (1973).

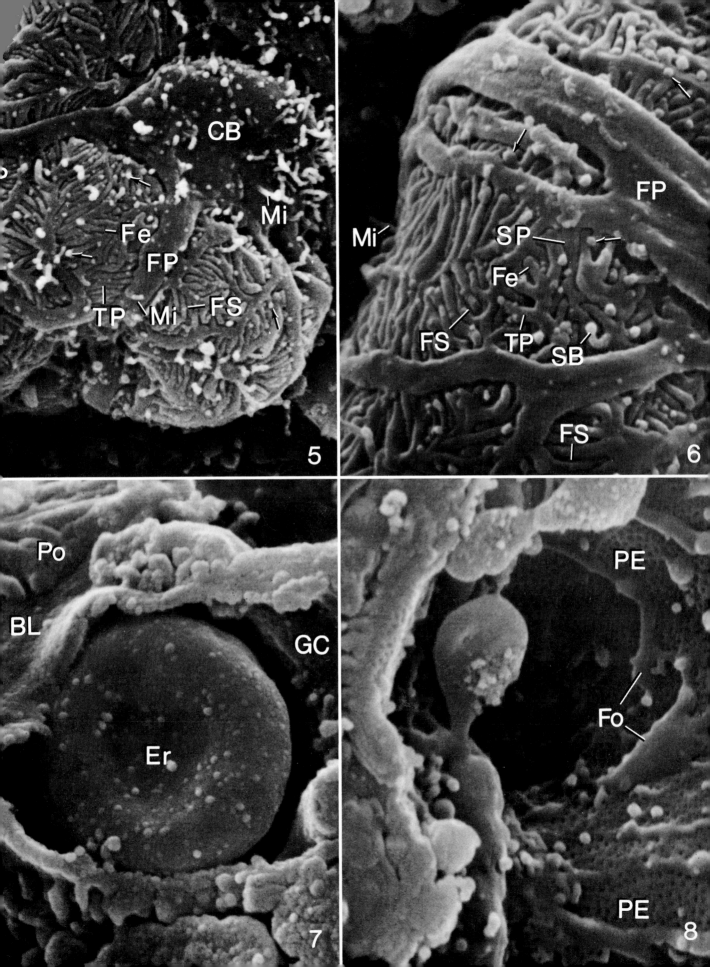

Male Reproductive System

Rat Testis and Epididymis

The reproductive organs in the male consist of paired testes which produce the male gametes (spermatozoa) and associated genital ducts whose function is to store, nourish, and transport the spermatozoa to the outside during ejaculation. In addition, genital ducts possess epithelial outpouchings that are differentiated into exocrine units called accessory or genital glands. Secretion from these glands provides a vehicle for the sperm (semen) as well as additional nutrient material. Embryologically, the testes develop in the abdomen and descend before birth into the scrotum, which lies outside the body cavity and provides a cooler environment necessary for the production of normal sperm. The testis serves two distinct but related functions in the reproductive life of the adult male. In addition to the function of providing germ cells, male sex hormones are secreted by the interstitial cells of the testis. These hormones are essential for the maintenance of secondary sexual characters (genitals) and sexual libido. Male sex hormones are known as androgens, the major one being testosterone. Normal function in the testis cannot be achieved or maintained without the continued release of two anterior pituitary gonadotropins: follicle stimulating hormone (FSH), which stimulates spermatogenesis, and leuteinizing hormone (LH), which stimulates the activity of the interstitial cells.

The testis consists of a tightly packed mass of coiled tubules surrounded by connective tissue. Within these tubules germ cell differentiation occurs. The wall of the seminiferous tubule consists of an outer layer of connective tissue cells (CT in Fig. 1), a thin basement membrane, and a seminiferous epithelium (SE in Fig. 1). The epithelial lining of the tubules contains two kinds of cells: the spermatogenic cells and the sustentacular cells. Among the spermatogenic cells, several distinct types can be seen that represent different stages in the development of mature sperm cells. The rounded spermatogonial cells (Sg in Fig. 2) lie against the basement membrane and divide to give rise to either reserve spermatogonia or primary spermatocytes. The primary spermatocytes divide meiotically to become secondary spermatocytes with a haploid chromosome number, and the secondary spermatocytes then divide by mitosis to give rise to spermatids. The final differentiation step in which the spermatid is transformed into a high-attenuated, motile spermatozoan is called spermiogenesis. As cell division occurs in the seminiferous tubules, the daughter cells become displaced toward the center of the lumen and the spermatids become surrounded by the cytoplasm of the sustentacular cells. The sustentacular cells are tall, irregular columnar cells that rest upon the basement membrane of the seminiferous tubule and send long processes toward the lumen to enfold the transforming spermatids.

Spermiogenesis consists of several simultaneous processes of morphological change and rearrangement of cellular organelles. The nucleus is condensed and elongated; the Golgi apparatus produces the acrosome which contains a lytic acrosomal granule at the anterior end; the centrioles are located near the posterior or caudal end of the spermatid and differentiate a flagellum or tail (ST in Fig. 1); the mitochondria align along the proximal end of the flagellum in the mid-piece; and the remaining cytoplasm is shed as the residual body. The head of a mature rat spermatozoan (Figs. 3 and 4) is hook-shaped. The nucleus is condensed in the form of a short rod, and the pointed acrosome ensheathes the nucleus over most of its length. The plasmalemma is closely applied over these structures, and the posterior limit of the acrosome can be distinguished by the slight surface elevations (arrows) above the caudal pole (CP) of the nucleus (Figs. 3 and 4). The posterior extent of the acrosome on the dorsal and ventral sides at the sperm head is denoted by arrows in Fig. 4. Anteriorly, lateral fin-like extensions project from the acrosome (AE). The acrosome does not cap the caudal pole of the nucleus, which extends past the junction of the mid-piece (MP) and the head. The surface of the mid-piece shows transverse or annular ridges, probably corresponding to mitochondrial positions along the flagellum and just beneath the plasmalemma.

The sperm move from the seminiferous tubules to the epididymis, by way of tubules of the rete testis and efferent ductules. The epididymis is continuous with the vas deferens at its distal end and is attached to the posterior aspect of the testis. It is an epithelial tube, highly folded upon itself and covered by the protective tunica albuginea of the testis. The epididymis (Ep in Fig. 5) is a reservoir for the storage of spermatozoa (SZ in Fig. 5), and in many species the final maturation of the sperm occurs here, including loss of the residual body. By virtue of a circular smooth muscle layer around the tube, sperm are released actively from the epididymis at ejaculation. The cells lining the epididymis are pseudostratified columnar cells of a secretory nature that provide nutrients to the spermatozoa stored in the lumen. Some of these secretory cells, especially the taller ones, possess tufts of long, branching cytoplasmic processes at their free margins. These processes are called stereocilia (Sc in Fig. 6) but are actually microvilli since they are non-motile and do not contain the tubular filaments characteristic of true cilia. The function of the stereocilia in the epididymis is not precisely known, but they may play a role in maintaining the proper environment for support of the spermatozoa.

AE   Acrosomal extension
CP   Caudal pole of nucleus
CT   Connective tissue of seminiferous tubule
Ep   Epididymis
MP Mid-piece
Sc   Stereocilia
SE   Seminiferous epithelium
Sg   Spermatogonium
ST   Sperm tails
SZ   Spermatozoan

Arrows: Posterior limit of acrosome

Fig. 1 (a seminiferous tubule of rat), ×450; Fig. 2 (spermatogonium in seminiferous tubule), 8100; Fig. 3 (mature sperm), ×10105; Fig. 4 (mature sperm), ×9375; Fig. 5 (epididymis), ×220; Fig. 6 (lumen surface of epididymis), ×4200.

References

FUJITA, T., MIYOSHI, M., and TOKUNAGA, J.: "Scanning and transmission electron microscopy of human ejaculate spermatozoa with special reference to their abnormal forms." Zeit. Zellforsch. **105**, 483—497 (1970).
GOULD, K.G., ZANEVELD, L.J.D., and WILLIAMS, W.L.: "Mammalian gametes: A study with the scanning electron microscope." In Scanning Electron Microscopy/1971. O. JOHARI and I. CORVIN, eds. IITRI **4**, 289—296 (1971).

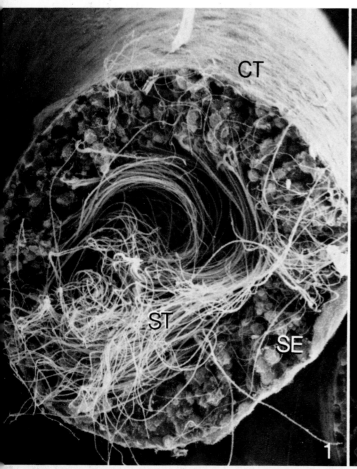

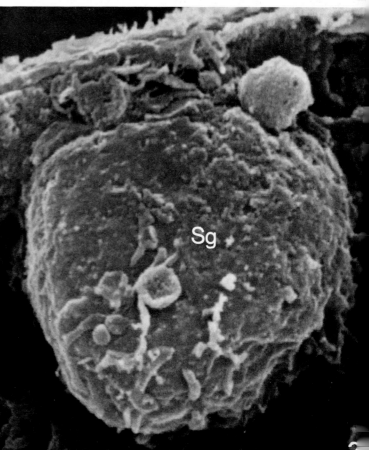

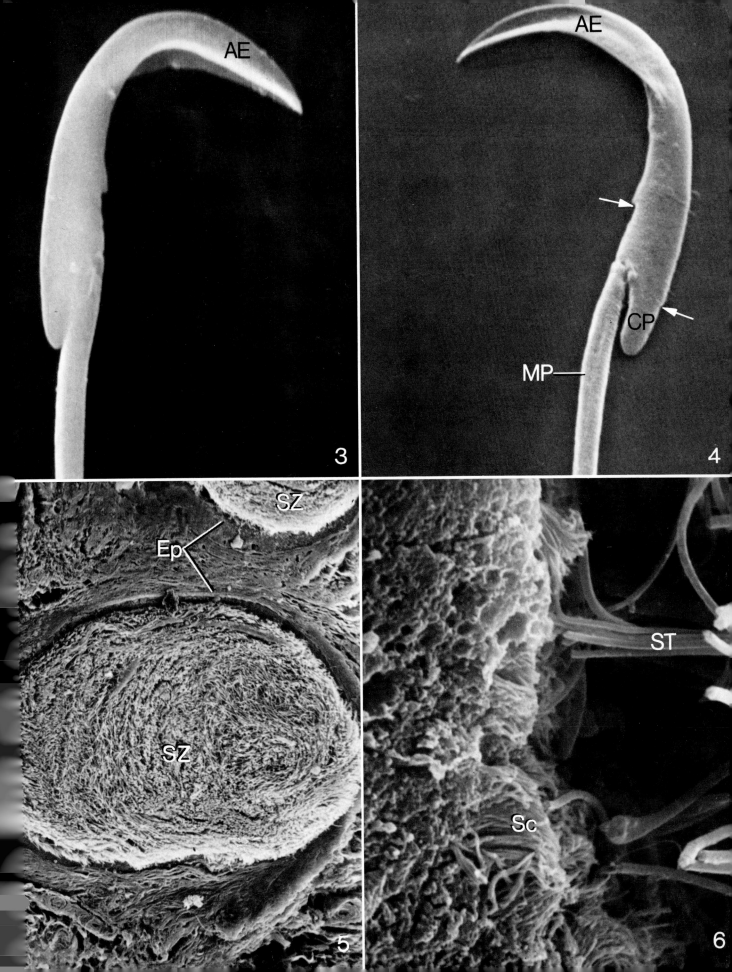

Female Reproductive System

Oviduct (Mouse)

BB   Basal bodies (of cilia)
CC   Central cilium
Ci   Cilia
CiC   Ciliated cell
Fi   Fimbriae (of infundibulum)
Mi   Microvilli
Os   Ostium
SC   Secretory cell

Fig. 1 (infundibulum), ×280; Fig. 2 (infundibulum), ×51000; Fig. 3 (infundibulum), ×6600; Fig. 4 (ampulla), ×8840; Fig. 5 (isthmus), ×2600.

The major regions of the mammalian oviduct consist of a long ampulla and a shorter, smaller isthmus. Near the ovary the ampulla terminates in the form of a fimbriated infundibulum. That portion of the oviduct which penetrates the uterine wall is frequently called the intramural portion. The epithelium of the oviduct mucosa consists of ciliated columnar cells and nonciliated (secretory) cells. Cilia (Ci) are especially numerous on cells comprising the fibriated lips, or fimbriae (Fi in Fig. 1), but few cilia are present in the isthmus.

The fimbriae or folds of the infundibulum are extensively ciliated in the mouse. The fimbriae and the cilia are illustrated in a region close to the opening, or ostium (Os in Fig. 1), into the oviduct. The transmission electron micrograph (Fig. 2) illustrates the apical ends of three ciliated cells from the fimbriae. In the figure the cilia (Ci) are sectioned transversely and longitudinally, and the basal bodies (BB) are aligned beneath the apical portion of the plasma membrane. The cilia are about 5 µm long and have blunt tips. The direction of the effective beat of these cilia is toward the ostium (Dirksen and Satir, 1972). Some of the cells of the fimbriae are not ciliated, but their surfaces are flat and have a peripheral ring of short microvilli (Mi in Fig. 3). Some of the nonciliated cells of the fimbriae appear to have a single, central cilium (CC in Fig. 3).

In the ampulla the mucosa is highly folded so that the lumen is quite narrow. There are fewer ciliated cells (CiC in Fig. 4) here than in the infundibulum, and the cilia beat toward the uterus. Also, the nonciliated cells in the ampulla have more and longer microvilli associated with their free surface (Mi in Fig. 4). The number of ciliated mucosal cells is further reduced in the isthmus (Fig. 5). The nonciliated (secretory) cells (SC in Fig. 5) are quite numerous and are also covered with many microvilli (Mi in Fig. 5).

In the mouse, eggs are released from the ovary into a periovarial space which is some distance from the ostium of the oviduct. But because of the mesovarium the ova are prevented from entering the peritoneal cavity. When the eggs in the fluid of the periovarial space come in close proximity to the oviduct, they are directed by the beat of cilia on the fimbriated lips into the ostium. In the lower portion of the ampulla, because of the relative scarcity of ciliated cells, it appears that muscle activity in the oviduct wall (peristalsis) serves to transmit the ovum toward the uterus. (Micrographs kindly provided by Drs. E. R. Dirksen and P. Satir.)

Reference

Dirksen, E. R., and Satir, P.: "Ciliary activity in the mouse oviduct as studied by transmission and scanning electron microscopy." Tissue and Cell **4**, 389—404 (1972).

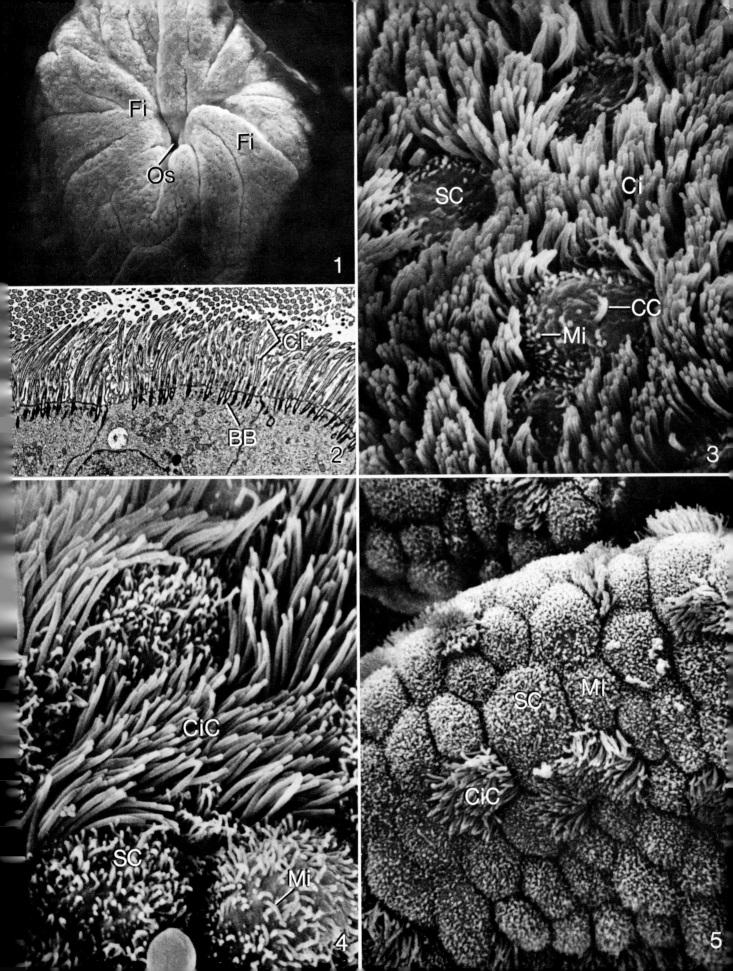

Normal Uterine Surface (Human)

CC   Ciliated epithelial cells
Mv   Microvilli
Os   Ostia of uterine glands
RBC   Red blood cell
Se   Secretory epithelial cells

Fig. 1, ×60; Fig. 2, ×1150; Fig. 3, ×5600; Fig. 4, ×122; Fig. 5, ×5800; Fig. 6, ×2300.

The surface epithelial cells of the reproductive tract commonly undergo cyclic changes, especially in response to changes in endocrine function. For example, the extensive development of cilia and microvilli on endometrial cells in the uterus during the proliferative phase of the menstrual cycle suggests that estrogens are responsible for the development of these surface specializations. The endometrium is the inner layer of the uterus and during the proliferative phase it is characterized by clusters of surface epithelial cells having cilia (CC in Fig. 3) interspersed with nonciliated cells having numerous microvilli (Mv in Fig. 3) on their free surfaces (cf. WHITE and BUCHSBAUM, 1973). There are approximately 30 nonciliated cells to 1 ciliated cell (FERENCZY and RICHART, 1973). In the proliferative phase the endometrial surface is relatively flat, and the openings or ostia (Os in Fig. 1) to many uterine glands are located about 125 μm apart. The ostia may be circular or oval and the endometrial surface is slightly elevated between these ostia. Ciliated (CC) and secretory (Se) cells are illustrated in Fig. 2. Red blood cells (RBC) are often located in the ostia of the uterine gland (Fig. 2). Both the secretory and ciliated cells measure 5 to 6 μm in diameter (Fig. 3). The cilia number about 75 per cell and they are approximately 3 μm long and 2500 Å in diameter. Secretory droplets may be seen on the surface. Secretory cells are covered with microvilli measuring 1500 Å in width and 3000 Å in length. There may be as many as 200 microvilli per secretory cell, but fewer microvilli are associated with the surface of ciliated cells. During the proliferative phase there is a general uniform appearance to the surface cells and they are arranged in a regular fashion.

During the secretory phase the general surface appears more convoluted (Fig. 4), and individual gland ostia are rarely observed. The wrinkled appearance is due to cellular proliferation on the surface. Ciliated cells are numerous on the apices of the ridges. Micro-villi are prominent and there are more secretory droplets than during the proliferative phase (Fig. 5). On day 24 of the menstrual cycle protuberances are seen on the secretory cells which bulge into the lumen of the endometrial cavity. Late in the secretory phase the surfaces of the secretory cells are convex and secretory droplets appear to emanate from the microvillous border (Fig. 6). During the peak of secretory activity (day 20 to day 24 of the menstrual cycle) the number of ciliated cells decreases, as does the length of the microvilli on the nonciliated cells. Atrophic endometria have epithelial cells with flattened surfaces. They are sparsely covered with short microvilli and virtually no ciliated cells are present. The decrease in, and eventual loss of some surface structures may be correlated with the lack of ovarian hormone production (estrogen). (All micrographs and a portion of the text material were kindly provided by Drs. A.J. WHITE and H.J. BUCHSBAUM.)

References

FERENCZY, A., and RICHART, R.M.: "Scanning electron microscopy of hyperplastic and neoplastic endometria", in Scanning Electron Microscopy/1973. O. JOHARI and I. CORVIN, eds. IITRI **6**, 613—619 (1973).

WHITE, A.J., and BUCHSBAUM, H.J.: "Scanning electron microscopy of the human endometrium. I. Normal". Gynecologic Oncology **1**, 330—339 (1973).

WILLIAMS, A.E., JORDAN, J.A., MURPHY, J.F., and ALLEN, J.M.: "The surface ultrastructure of normal and abnormal cervical epithelia", in Scanning Electron Microscopy/1973, O. JOHARI and I. CORVIN, eds. IITRI **6**, 597—604 (1973).

Human Endometrial Hyperplasia and Adenocarcinoma (Cancer)

CC   Ciliated cells
Fo   Folds on cell surface
IS   Intercellular strands
Os   Ostia of uterine glands

Fig. 1, ×58; Fig. 2, ×2300; Fig. 3, ×2250; Fig. 4, ×13500; Fig. 5, ×2650; Fig. 6, ×9750.

Under conditions of adematous hyperplasia (and other conditions of high estrogen production) an increase in number of ciliated cells compared to normal proliferative endometria has been observed by FERENCZY and RICHARTT (1973). The ciliated to nonciliated cell ratio is about 10 to 1. Further, nonciliated cells have more extensive microvillous projections, there being an increase in both number and length of microvilli compared to normal endometrial cells. These changes, at least in part, appear to be under estrogenic influence. However, in cases of hyperplastic endometria treated with progestins, there is a marked decrease in cilia and microvilli. This suggests that progesterone and progestational agents have an inhibitory effect on the surface specializations of the endometrial epithelial cells. In well differentiated states of endometrial cancer the endometrial cells are large and pleomorphic, lack cellular adhesiveness, contain decreased numbers of microvilli, and lack cilia (WHITE and BUCHSBAUM, 1974).

In endometrial hyperplasia the surface of the endometrium is generally flat, and at low magnification multiple gland openings are observed (Os in Fig. 1). At higher magnifications the two basic cell types are still noted, but there are more ciliated cells (CC) than in either phase of the normal ovarian cycle (Fig. 2). The cells on the endometrial surface may be larger than normal; some cells measure 50 μm in diameter (cf. WHITE and BUCHSBAUM, 1974).

In adenocarcinoma of the endometrium there is great variation in cell size and a general pattern of disorganization (Fig. 3). There are many large cells and there is a loss of the normal mosaic pattern. Ciliated cells are rare and they are absent in the poorly differentiated malignancies (cf. WHITE and BUCHSBAUM, 1974). At higher magnification, changes in the microvilli and cell membranes are found. On any individual cell, microvilli may vary greatly in length (arrow in Fig. 4). They may be sparse in other areas or large

and knobby (arrow in Fig. 5). Microvilli may exhibit a tenfold increase in length and a fourfold increase in width. In many instances the microvilli may be flattened as if fused to the cell surface (Fig. 6). The cell surface may possess wrinkles or folds (Fo), and in some cases the surface may be thrown into flap like appendages. In many cases fine intercellular strands (IS) are observed, as if the cells were slightly separated. (All micrographs and a portion of the text were kindly provided by Drs. A.J. WHITE and H.J. BUCHSBAUM.)

References

FERENCZY, A., and RICHART, R,M.: "Scanning electron microscopy of hyperplastic and neoplastic endometrium", in Scanning Electron Microscopy/1973, O. JOHARI and I. CORVIN, eds. IITRI 6, 613—619 (1973).

WHITE, A.J., and BUCHSBAUM, H.J.: "Scanning electron microscopy of the human endometrium. II. Hyperplasia and adenocarcinoma". Gynecologic Oncology, in press (1974).

WILLIAMS, A.E., JORDAN, J.A., MURPHY, J.F., and ALLEN, J.M.: "The surface ultrastructure of normal and abnormal cervical epithelia", in Scanning Electron Microscopy/1973, O. JOHARI and I. CORVIN, eds. IITRI 6, 597—604 (1973).

Sense Organs

Eye (The Mouse Iris)

AS   Anterior surface of iris
CF   Contraction furrows
CM   Circular muscle
Mi   Microvilli
Pr   Protrusions (on surface of pigmented epithelial cells)
Pu   Pupil
PS   Posterior surface of iris

Fig. 1 (anterior surface of iris), × 195; Fig. 2 (rim of anterior surface of iris), × 1 500; Fig. 3 (posterior surface of iris), × 195; Fig. 4 (rim of posterior surface of iris), × 690; Fig. 5 (epithelial cells on posterior surface of iris), × 14970.

The iris is a pigmented, opaque disc with a central aperture or pupil (Pu in Figs. 1 and 2), which can regulate the amount of light admitted to the lens and posterior parts of the eye. The anterior surface (AS in Fig. 1) of the iris is bathed by aqueous humor in the anterior chamber and the posterior surface (PS in Fig. 3) is closely applied to the anterior surface of the lens. In addition to its mydriatic and myotic function (dilation and contraction), the iris also serves as a one way valve for aqueous humor. Aqueous humor, secreted into the posterior chamber behind the iris by the ciliary body, flows through the pupil into the anterior chamber, but is prevented from returning by the flap-like action of the iris pressing against the anterior surface of the lens. The aqueous humor is removed or drained from the anterior chamber via the canal of Schlemm.

The inner edge of the iris is provided with smooth muscle fibers that are disposed circularly around the pupillary opening. The internal position of the muscle is denoted in Figs. 3 and 4 (CM). Contraction of this spincter muscle constricts the pupil and causes extensive folding of the surface epithelium (Figs. 1 and 4). These contraction furrows (CF in Fig. 1 and 4) are oriented radially and are readily demonstrated on both surfaces of the rat iris. The furrows on the anterior surface (Fig. 2) are deeper and more regular than those on the posterior surface (Fig. 4), which are fewer and more shallow. The entire posterior surface of the iris is flattened because it closely adheres to the adjacent lens capsule (Fig. 3 and 4). The interior stroma of the iris also contains radially oriented, fusiform myo-epithelial cells which contract and cause dilation of the pupil. Upon dilation, several shallow furrows appear which are oriented perpendicular to the radial furrows caused by contraction of the sphincter muscle.

The surface of the epithelial cells covering the sphincter muscle near the pupil possess small and irregular processes, or microvilli (Mi in Fig. 2), and the concentration furrows in this area are extensive enough to form crypts in the epithelial surface (Fig. 2). The remainder of the posterior surface of the iris is pigmented and continuous at its margin with the pigmented epithelium of the retina. The numerous rounded protrusions (Pr in Fig. 5) observed on the surface of these cells mark the location of many intracellular pigment granules.

Reference

NISHIDA, S., and MIZUNO, K.: "Observations of the posterior surface of the cat iris." JEOL 8 B **2**, 10—11 (1970).

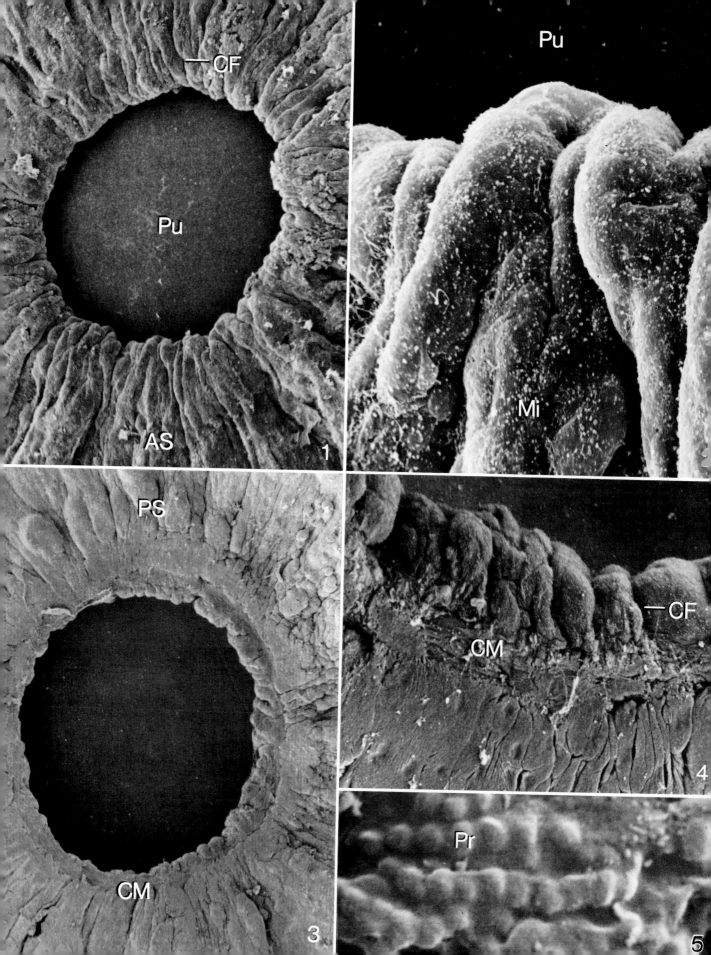

Eye (The Rat Lens)

Visual accomodation is achieved by shape changes in the transparent lens of they eye. These shape changes are produced by the action of ciliary muscles. The lens is a thick biconvex disc composed of attenuated epithelial cells layered upon each other like the scales of an onion (Fig. 1). In their mature form these cells are devoid of nuclei and most cytoplasmic organelles. In such specialized form they are called lens fibers and are arranged approximately parallel to the axis of vision. Therefore, a section through the equator (the area of attachment of the ciliary zonule) produces transverse sections of fibers arching from the anterior to the posterior pole of the lens. This arrangement provides the lens with its unique properties of transparency and flexibility.

In mammals the lens fibers (LF in Fig. 2) are flattened, six-sided prisms in transverse section. Small cytoplasmic processes (CP in Fig. 2) project from each ridge or crest of the prism and interdigitate with similar extensions of adjacent fibers. Since visual accomodation in mammals and other vertebrates is based upon changes in shape of the lens, these interlocking processes are thought to prevent excess aberration of individual fibers or at least somewhat confine their position with respect to each other. Marked differences in the shape and arrangement of the lens processes characterize different vertebrate classes and may prove to be of interest from an evolutionary standpoint.

Different areas of the biconvex lens are subjected to different magnitudes of stress and movement during accomodation. Therefore, it is not surprising that lens fibers and processes differ from area to area within a single lens. In the cortical region of an adult rat lens (near the surface) the fibers are regular and hexagonal in cross fracture, having relatively straight edges with numerous short cellular processes (Figs. 1 and 2). Deeper in the subcortical region of the lens the lateral surfaces (LS in Figs. 3 and 4) of the lens fibers are thrown into regular waves rather than being straight, and the cell processes (CP in Fig. 4) are fewer in number but longer since the lens fibers appear to be less densely packed. Near the lens equator, where new fibers are formed, the cells become much less regular in appearance and bear numerous club-shaped (CsP in Fig. 5) processes which fit into deep indentations (IN in Fig. 5) in neighboring cell surfaces. The extracellular material is more abundant in this region (Fig. 5). The nuclear portion of the lens is difficult to study since it is nearly impossible for fixatives to penetrate to this depth. However, it appears that the fibers in this region possess almost no processes and indentations, perhaps due to lack of movement during accomodation in this central area of the lens.

References

Fujita, T., Tokunaga, J., and Inoue, H.: Atlas of SEM in Medicine. New York: Elsevier Publishing Company, pp. 66—69 (1971).

Hansson, H.-A.: "SEM of the lens of the adult rat." Zeit. Zellforsch. **107**, 187—198 (1971).

Eye (The Rat Lens)

(continued)

CP   Cytoplasmic processes (of lens fibers)
CsP   Club-shaped processes
IN   Indentation (of lens fibers)
LF   Lens fibers
LS   Lateral surface (of lens fibers)

Fig. 1 (anterior and cortical view of lens), ×2375; Fig. 2 (anterior and cortical area of lens), ×2750; Fig. 3 (lens fibers in subcortical region of lens), ×1430; Fig. 4 (subcortical lens fibers), ×2990; Fig. 5 (fibers in equatorial region of lens, ×2890.

Eye (The Retina)

The retina is a sheet of nervous tissue applied to the inner surface of the posterior half of the eyeball. Its cellular population is divided into those cells that are specialized to receive the light and convert the energy of photons into electrical impulses and those cells specialized to process and transmit these electrical signals to the brain. Cells that perform a supportive or maintenance function are also present. Ten layers are noted by histologists and are listed in Fig. 1. The visual cells lie on the outer aspect of the retina, and the cells of subordinate nervous function and support comprise the other layers (Fig. 1). Incident light, after passing through the refractive media of the eye, must pass through the entire thickness of the retina before striking the visual cells to become a nervous impulse.

The visual cells, the rods (RC) and cones (CC), are highly elongate cells, consisting of an outer segment (OS) specialized for photoreception, and an inner segment (IS), or ellipsoid, containing the nucleus and synthetic organelles of the cell, and a synaptic portion (Sy) that communicates with neurons in another layer of the retina (Figs. 1, 2). Rods and cones in the vertebrate retina are given their names because of the shape of the photoreceptive segment, which in the rods measures 7.5 to 12 μm in diameter and 20 to 30 μm in length; the dimension of the cones is 5.5 to 9 μm by 1.6 to 29 μm. Rods are discriminatory of black and white; that is, they respond to all wavelengths of light in the visible spectrum, while the cones detect color, being responsive to only certain wavelengths. However, rods are much more effective in dim light than cones, since they have a lower threshold of response. The outer segment in both rods and cones develops from a single cilium and in mature form contains a tall stack of flattened membranous saccules that resembles a stack of coins. The stack of membranous saccules is ensheathed in a plasmalemma to give the characteristic rod or cone shape to the outer segment. It is upon these membranous discs that the visual reaction occurs. Visual pigment molecules, disposed along the saccular membranes, undergo a configurational change when struck by light energy, and an electrical impulse is generated.

Externally, the rod outer segments have 20 to 30 longitudinal grooves (arrows) that give a scalloped appearance in cross section (Fig. 5), while the cones are smaller and smooth-surfaced (Figs. 3 and 5). Heavy dendrites (De) whose cytoplasm is continuous with the inner segment support the outer segment near the base (Fig. 7). The membranous discs within the outer segments are dynamic structures and are continually displaced apically to be shed, a few at a time, at the tip. New saccules are formed by invaginations of the plasmalemma near the base, which pinch off internally, and some of these saccules retain communication with the surface in both rods and cones. The outer segments, enmeshed in a web of supporting fibrillar material (Fig. 6), are arranged like palisades, with bases toward the center of the eyeball and tips pointing outward (Fig. 2). The tips are embedded in the outer surface of the retina; the pigmented epithelium (PE) is composed of simple cuboidal cells (Fig. 1). This layer of cells is active in phagocytosing the discs that are continually shed from the apexes of the outer segment.

The ends of the rod and cone cells opposite the outer segment are differentiated into synapses with bipolar neurons whose nuclei comprise the inner nuclear layer (IN in Fig. 1). These, in turn, synapse with the neurons of another layer, the ganglion cell layer (GC in Fig. 1). There is not a one-to-one relationship between the visual cells and the bipolar neurons, and the ganglion cells also appear to intercommunicate. This arrangement allows the information from many visual cells to be somehow processed within the retina and sent in electrical code to the brain for interpretation.

Supporting the layers of the retina are neuroglial cells, commonly associated with nervous structures, and Mueller's cells, a special type of glial cell whose nuclei are found in the layer of bipolar neurons. These cells have elongate bodies and processes that span the width of the retina, and the processes are fused together at the very ends by desmosomes. Thus anchored, these cellular processes provide a supporting framework for the entire retina, extending from the inner surface across the layers of cells to the bases of the outer segments, where they terminate in microvilli that surround the visual cells (Mi in Fig. 7).

References

HANSSON, H.-A.: "Scanning electron microscopy of the rat retina." Zeit. Zellforsch. **107**, 23—44 (1970).

LEWIS, E.R., ZEEVI, Y.Y., and WERBLIN, F.S.: "Scanning electron microscopy of vertebrate visual receptors." Brain Res. **15**, 550—562 (1969).

BL	Bacillary layer (rod and cone outer segments)
CC	Cone cell
De	Dendrite supporting outer segment
GC	Ganglion cell layer
IN	Inner nuclear layer
IP	Inner plexiform layer
IS	Inner segment
LO	Layer of optic nerve fibers
Mi	Microvilli
OL	Outer limiting membrane
ON	Outer nuclear layer
OP	Outer plexiform layer
OS	Outer segment
PE	Pigmented epithelium of the retina
RC	Rod cell
Sy	Synapse

Arrow: Longitudinal furrows in outer segment of rod

Fig. 1 (LM of human retina), ×900; Fig. 2 (frog retina), ×630; Fig. 3 (mouse cone cell), ×10710; Fig. 4 (frog rod cell), ×2065; Fig. 5 (frog rod and cone), ×3075; Fig. 6 (frog rod cells), ×1575; Fig. 7 (frog retina), ×2065.

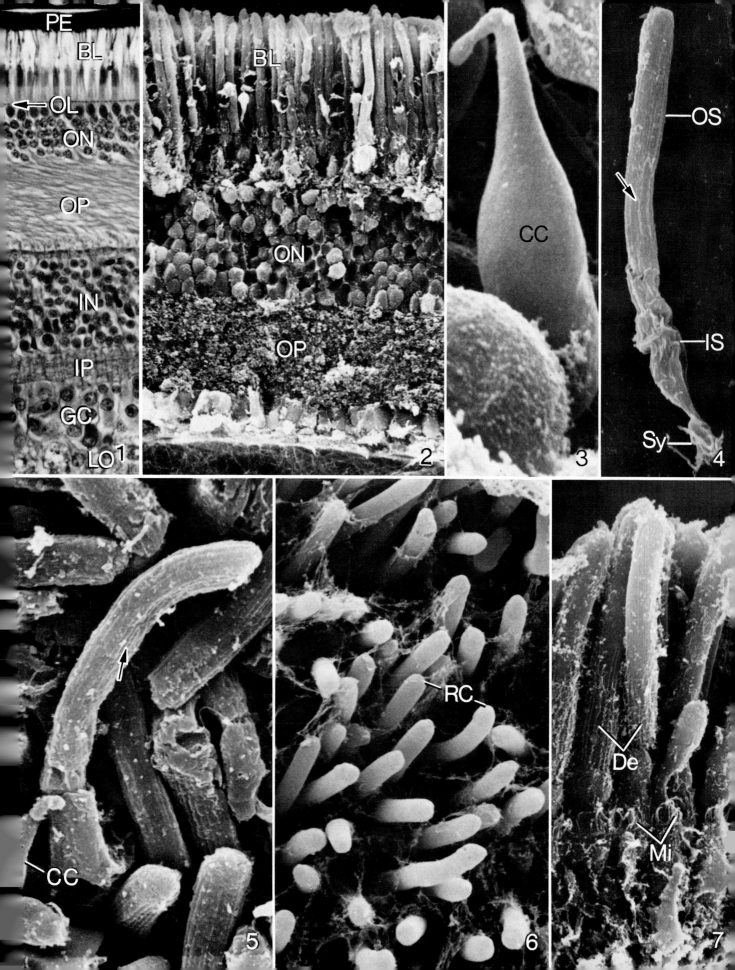

Ear (The Organ of Corti)

The end organ of hearing, the organ of Corti, is a delicate ribbon of neuroepithelial tissue which is suspended in a tube of fluid by a resonant membrane and is specialized to transform vibrations within the fluid into nervous impulses. The cochlea of the inner ear, of which the organ of Corti is a part, is composed of three membranous tubes filled with a clear fluid. These fluid-filled tubes are the resonating structures that transmit vibrations from the bones of the middle ear to the organ of Corti during audition. The organ of Corti (OC) is disposed along the entire length of the middle tube; the cochlear duct (CD), or scala media, and is bathed by endolymph fluid contained therein (Fig. 1). The other two tubes, the scala vestibuli (SV) and the scala tympani (ST), are both filled with perilymph fluid (Fig. 1). The three tubes, lying parallel to one another and separated by only thin membranes, are wound in a tight spiral and encased in the temporal bone (TB in Fig. 1) of the skull. The portion of the temporal bone that extends down the middle of the spiral is called the modiolus. Acoustical vibrations pass from the stapes of the middle ear to the oval window and into the perilymph of the scala vestibuli. The vibrations cross a thin membrane, Reissner's membrane (RM in Fig. 1), into the endolymph of the cochlear duct, causing the basilar membrane (BM in Figs. 1 and 2) and the organ of Corti to oscillate. The oscillations result in the nervous sensation of audition. From the basilar membrane the vibrations are transmitted from the inner ear through the perilymph of the scala tympani to the round window and into the air space of the tympanic cavity (middle ear).

The organ of Corti is suspended across the floor of the cochlear duct along its entire length by the basilar membrane and consists of four rows of specialized sensory cells called hair cells. The hair cells are surrounded and supported by accessory cells called phalangeal cells (PC in Figs. 2 and 3). Other support cells, called pillar cells (Pl in Figs. 2 and 3), line the tunnel of Corti (CT in Fig. 2), which divides the three rows of outer hair cells (OHC in Figs. 2 and 3) from the single row of inner hair cells (IHC in Figs. 2 and 3). At the apices of the hair cells is a protective cuticle through which large stereocilia project (St in Fig. 4). The stereocilia of the sensory hair cells, when bent or displaced, cause nervous impulses to be generated in the hair cells, which are then passed to fibers of the acoustic nerve with which the hair cells synapse. The stereocilia are overlain by the tectorial membrane (TM in Figs. 1 and 2), which projects from a ridge of tissue (vestibular lip) at one side of the cochlear duct. When the basilar membrane and the organ of Corti oscillate in response to acoustic vibrations, the stereocilia are pressed against the underside of the tectorial membrane, and the nervous sensation of hearing results.

The SEM provides three-dimensional information about the organ of Corti not easily gleaned from serial sections in the light and transmission electron microscopes.

In Figs. 5 and 6 the cuticular plates (CP) at the apices of the hair cells through which the stereocilia project are roughly octagonal. The plates are smooth-surfaced, except for an occasional microvillus (Mi). The phalangeal cells, positioned between the rows of hair cells, also possess microvilli associated with their free surfaces. These microvilli appear uniformly distributed on the cell apices near the row of inner hair cells (Fig. 3), but the phalangeal cells supporting the outer hair cells show more variation in the pattern of microvilli (Figs. 3—6). On some they are arranged only in a circle at the cell periphery (Figs. 3, 4, and 6). On others the cell surface displays a more even distribution of microvilli of variable height (Figs. 3, 4, and 5). The apical surface of the pillar cells enclosing the tunnel of Corti can also be seen, and the microvilli on these cells appear more dense near the inner hair cells (Fig. 3).

Ear (The Organ of Corti)

(continued)

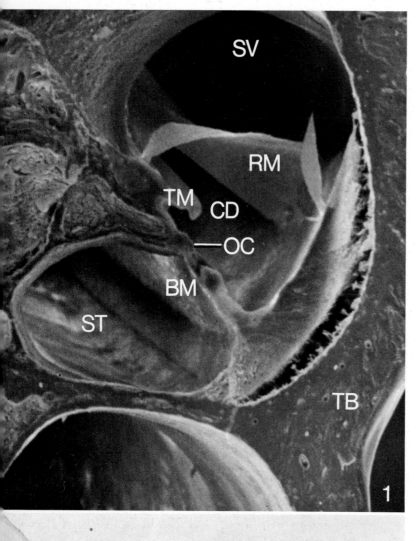

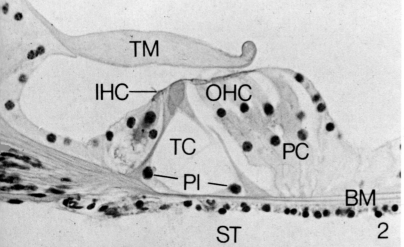

BM   Basilar membrane
CD   Cochlear duct (or scala media)
CP   Cuticular plates (of hair cells)
IHC Inner hair cells
Mi   Microvilli
OC   Organ of Corti
OHC Outer hair cells
PC   Phalangeal cells
PI   Pillar cells
RM Reissner's membrane
ST   Scala tympani
St   Stereocilia
SV   Scala vestibuli
TB   Temporal bone
TC   Tunnel of Corti
TM Tectorial membrane

Arrows: bulbous nodules at tip of stereocilia

Fig. 1, ×130; Fig. 2, ×540; Fig. 3, ×1955; Fig. 4, ×3375; Fig. 5, ×7650.

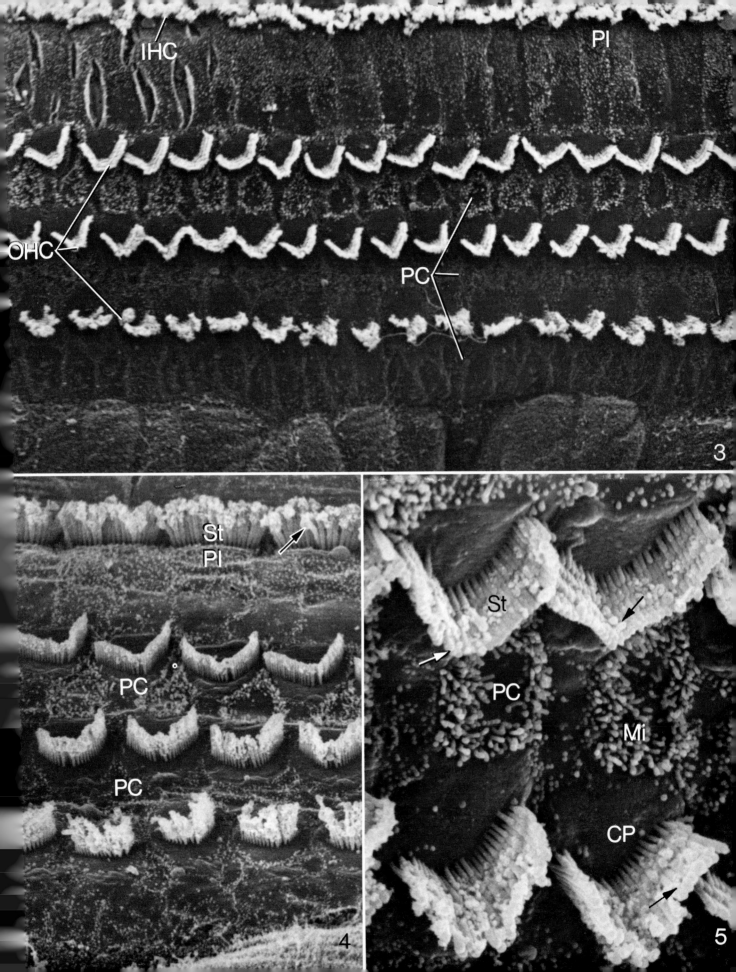

Ear (The Organ of Corti)

(continued)

CP   Cuticular plates (of hair cells)
FL   Fibrous layer (of tectorial membrane)
GL   Homogeneous granular layer (of tectorial membrane)
IHC   Inner hair cells
Mi   Microvilli
OHC   Outer hair cells
St   Stereocilia

Fig. 6, ×7875; Fig. 7, ×1495; Fig. 8, ×6800; Fig. 9, ×13600.

The stereocilia (St) on the three rows of outer hair cells are arranged in a "V" pattern, with the point of the V facing outward or away from the inner hair cell (Figs. 3, 4, 5, and 6). The stereocilia show gradations of height on individual hair cells; the tallest stereocilia are at the point of the V, with the shorter members arranged in a pipe-organ fashion toward the tunnel of Corti and the inner hair cells (Fig. 6). The stereocilia on the single row of inner hair cells are oriented linearally and are larger than those of the outer hair cells (Figs. 3 and 4). The line of stereocilia is 3 to 5 deep, with the taller members facing toward the outer hair cells and the shorter ones behind (Fig. 4). These stereocilia are thicker than those of the outer hair cells, and some stereocilia upon both kinds of hair cells can be seen to end in small bulbous nodules at their tips (arrows in Figs. 4, 5, and 6). The difference in size and arrangement of the stereocilia suggests a difference in function, and indeed, the outer hair cells are known to be able to detect feeble vibrations, while the less sensitive inner hair cells enable sounds to be heard with precision and clarity. The nodules on the stereocilia may have some unknown functional significance, or they may result from rubbing against the tectorial membrane. In fact, the outermost row of hair cells contains the greatest number of stereocilia-bearing nodules, and these are displaced the greatest amount during audition.

The tectorial membrane is a semi-gelatinous structure consisting of two main layers and an apical network of thick covering fibers (Figs. 7 and 8). The upper layer is fibrous in nature (FL in Figs. 7 and 8), and the homogenous granular layer (GL in Figs. 7 and 8) on the underside is in contact with the stereocilia. The tectorial membrane is attached firmly to the stereocilia on the outermost row of hair cells, and the area of attachment can be seen as a series of indentations arranged in a characteristic V pattern on the underside into which the stereocilia fit (Fig. 9).

Evidence suggests further attachment of the tectorial membrane to the supporting cells of this area for stability, but little is known about the nature of the attachment to the inner hair cells.

The basilar membrane (BM in Figs. 1 and 2), upon which the organ of Corti rests, is made highly resonant by the large number of tightly stretched fibers within it called auditory strings. The basilar membrane is the basis for pitch discrimination. It varies in thickness throughout the cochlea such that different areas of the basilar membrane resonate sympathetically with different frequencies of sound vibrations. High-pitched sounds are heard near the base of the cochlea, where the membrane is the thinnest, and low pitch sounds are heard at the apex of the spiral, where the basilar membrane is thickest. (Figs. 1, 7, 8 and 9 kindly provided by Dr. T. TANAKA.)

References

BREDBERG, G., LINDEMAN, H., ADES, H., WEST, R., and ENGSTROM, H.: "SEM of the organ of Corti." Science 170, 861—863 (1970).

KIMURA, R.: "Hairs of the cochlear sensory cells and their attachment to the tectorial membrane." Acta Otolaryng. 61, 55—72 (1966).

KOSAKA, N., TANAKA, T., TAKIGUSHI, T., and OZEKI, Y.: "Observations on the organ of Corti with the SEM." Acta Otolaryng. 72, 337—384 (1971).

LIM, D.: "The fine morphology of the tectorial membrane." Arch. Otolaryng. 96, 199—215 (1972).

LINDEMAN, H., ADES, H., BREDBERG, G., and ENGSTROM, H.: "The sensory hairs and the tectorial membrane in the development of the cat's organ of Corti." Acta Otolaryng. 72, 229—242 (1971).

MAROVITZ, W., THALMANN, R., and ARENBERG, I.: "SEM of freeze-dried guinea pig organ of Corti." In Scanning Electron Microscopy/1970. O. JOHARI and I. CORVIN, eds. IITRI 3, 273—280 (1970).

TANAKA, T., KOSAKA, N., TAKIGUSHI, T., AOKI, T., and TAKAHARA, S.: "Observations on the cochlea with the SEM." In Scanning Electron/Microscopy/1973. O. JOHARI and I. CORVIN, eds. IITRI 6, 427—434 (1973).

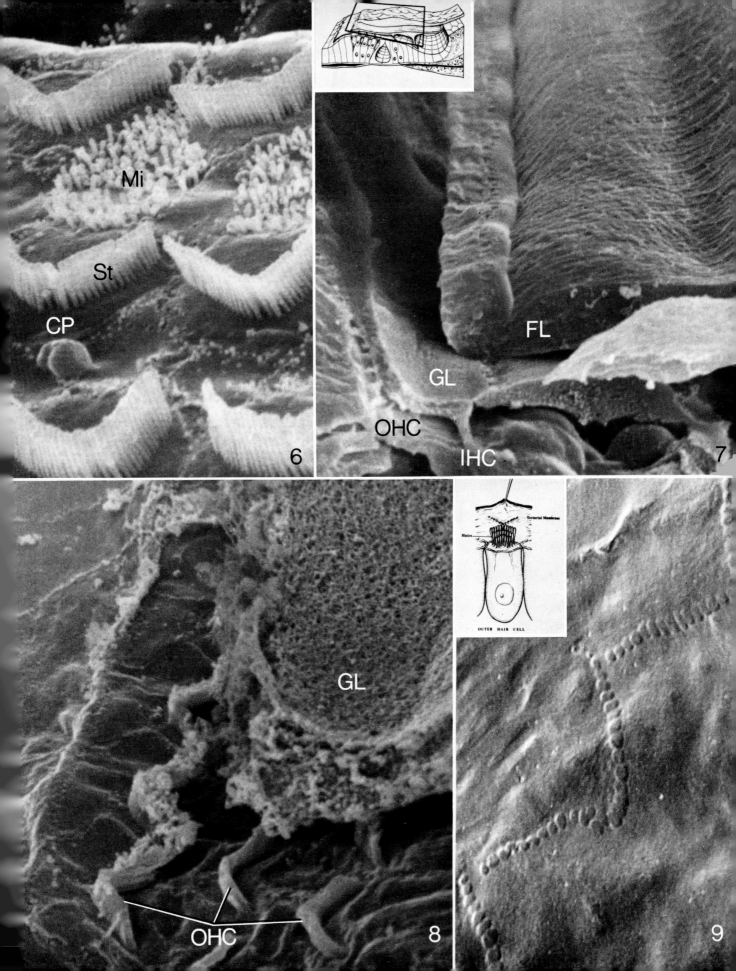

Ear (Vestibular Sensory Receptors)

Vestibular hair cells are present in localized sensory regions of the vertebrate internal ear. Such sense organs are present, for example, in the sacculus (macula saccularis), utriculus (macula utricularis), and semicircular canals (cristae ampullaris). These sensory surfaces and the organ of Corti are all bathed in a common space by endolymphatic fluid. The projections of the sensory hair cells are enclosed or covered by an otolithic membrane in which inorganic concretions (otoliths or otoconia) are embedded. The components of the otolithic membrane are illustrated in Figs. 1 and 2 including the otolithic membrane (OM) and the filamentous base (FB) which supports the otoliths (Ot). The otolithic membrane has been partially pulled away to reveal the underlying hair cells (HC) in Fig. 2. In Figs. 3 through 5 the otolithic membrane was manually removed so that the hair cell projections could be more clearly observed.

The surfaces of several sensory hair cells are illustrated in Fig. 3 from the vestibular membranous labyrinth (macula saccularis) of the bullfrog ear. Characteristic structural projections emanate from the surfaces of these cells in the form of finger-like tufts. The lengths of these processes vary in a uniform manner. Studies by transmission electron microscopy coupled with scanning electron microscopy have aided in their identification and have revealed that the projections are of two types: (*1*) a single kinocilium (Ki in Figs. 3 through 5) and (2) numerous stereocilia (St in Figs. 3 through 5). Internally the kinocilia have an axial filament complex similar to that of cilia and flagella. The distal end of the kinocilium is expanded (arrows in Figs. 3 through 5) and is attached at this point to several adjacent stereocilia. The kinocilium of each cell is eccentrically placed (Figs. 3 through 5). The kinocilia are also placed in a similar position on each cell surface comprising about half of the sensory organ, but they are located at opposite poles on the remaining half of the cells in the organ. A portion of both regions is illustrated in Fig. 4. Note that the kinocilia (arrows) associated with the two sensory cells in the top portion of this figure are positioned differently from the kinocilia associated with the remainder of the sensory cells illustrated in this figure.

The stereocilia are variable in length, the longest ones being located adjacent to the kinocilium (Fig. 4). The internal structure of the stereocilia includes a number of thin filaments which extend the length of these processes and terminate in the apical cytoplasm of the sensory cell as a terminal web. This produces a fairly rigid cuticular area (C in Fig. 1) of the cell beneath the stereocilia. The proximal end of the kinocilium is associated with a notch in the cell surface (Fig. 1).

The surface projections from each cell constitute an apparatus sensitive to directional displacement. The base of each sensory hair cell is innervated by two types of nerve endings (Fig. 1). In one case the synaptic vesicles are located inside the hair cell, while in the other case they are located inside the nerve ending. These represent afferent and efferent synapses, respectively. In the case of the semicircular canals, when pressure is applied toward the kinociliary side of the cell, there is an increase in activity in the vestibular nerve. However, when pressure is applied toward the stereociliary side of the sensory cell, there is a decrease in activity in the vestibular nerve.

Movements of the endolymph fluid and the otolithic membrane are translated to the surface projections of the sensory hair cell, and such movements cause a displacement of the kinocilium-stereocilia complex. There is, thus, a mechanical coupling involved in the transformation of the movement or shearing action of the otolithic membrane into a generator potential (cf. HILLMAN, 1969, 1972). In such a system directional movement is distinguished, since the kinocilium is positioned differently on the hair cells comprising each half of the sense organ in the saccule. In contrast, in the semicircular canals in which unidirectional motion is detected, the kinocilia on all receptor cells are aligned in a similar direction.

Important in the overall functioning of the hair cells shown here is the attachment of the kinocilium to the stereocilia and the special arrangement of the attachment of the kinocilium to the cell surface. While the stereocilia attach to the cell via a cuticular arrangement, the kinocilium is inserted over a notch in the cuticle and the base of the kinocilium is in direct continuity with the cell cytoplasm, there being no cuticular modification associated with the kinocilium (Fig. 1). During displacement of the processes, the stereocilia slide in relation to each other, but the displacement of the kinocilium is at its base. When the stereocilia are bent toward the kinocilium and displacement occurs, the base of the kinocilium causes a prominent dimple to form in the cell surface around its attachment point (Fig. 1a). In those cells where the kinocilium and stereocilia are oriented in the opposite arrangement, the directional displacement results in an elevation of the receptor cell membrane at the kinocilium base (Fig. 1b). The deformation of the hair cell plasma membrane at the base of the kinocilium thus appears to be an important event in the functioning of the hair cells. The distension of this membrane region (so that a prominent dimple is produced at the kinociliary base) is thought to lead to changes in the ionic conductance which results in a depolarization of the hair cell and a consequent increase in response in the vestibular nerve which synapses with the base of the hair cell (Fig. 1) (*cf.* HILLMAN and LEWIS, 1971). Conversely, it is believed that a reduction in the amount of dimpling in the cell membrane at the kinocilium base (so that the surface becomes convex) decreases polarization of the cell and there is a consequent reduction in activity in the vestibular nerve (Fig. 1).

There are many structural similarities between the vestibular end organs of the mammalian ear and, for example, the lateral line organs and ampulla of Lorenzini of aquatic fish and amphibia. There appear to be functional similarities in these end organs as well.

(Figs. 1 through 3 and 5 kindly provided by Dr. D. E. HILLMAN; Fig. 4 courtesy of Dr. E. R. LEWIS.)

References

HILLMAN, D. E.: "New ultrastructural findings regarding a vestibular ciliary apparatus and its possible functional significance." Brain Res. **13**, 407—412 (1969).

HILLMAN, D. E., and LEWIS, E. R.: "Morphological basis for a mechanical linkage in otolithic receptor transduction in the frog." Science **174**, 416—419 (1971).

HILLMAN, D. E.: Observations on morphological features and mechanical properties of the peripheral vestibular receptor system." In Basic Aspects of Central Vestibular Mechanisms. A. BRODAL and O. POMPEIANO, eds. Progr. Brain *Res.* **37**, 69—75 (1972).

FB   Filamentous base
HC   Hair cell
Ki   Kinocilium
OM   Otolithic membrane
Ot   Otoliths
St   Stereocilia

Arrows: Expanded ends of kinocilia

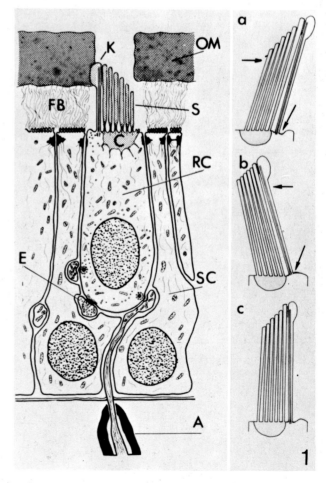

Fig. 1. *Left*: Diagram illustrating the saccular epithelium with its receptor cell (RC). Kinocilium (K), stereocilia (S), cuticle (C), and otolithic membrane (OM) with a filamentous base (FB), which supports the otoliths. Efferent nerve ending (E); afferent nerve ending (A). *Right*: Diagrams illustrate effects of bending the stereocilia toward and away from the kinocilium. The relatively firm cuticular base and the attachment of the kinocilium to adjacent stereocilia causes the pliable receptor cell membrane in the region of the cuticular notch to be thrust up or down with respect to movements a and b. In the vertical position (c) a slight dip is usually noted by both scanning and transmission electron microscopy. (Diagram by permission from Dr. D. E. HILLMAN and E. R. LEWIS, Science **174**, 416—419, October 22, 1971; Copyright 1971 by the American Association for the Advancement of Science); Fig. 2, ×600; Fig. 3, ×3600; Fig. 4, ×10300; Fig. 5, ×11500.

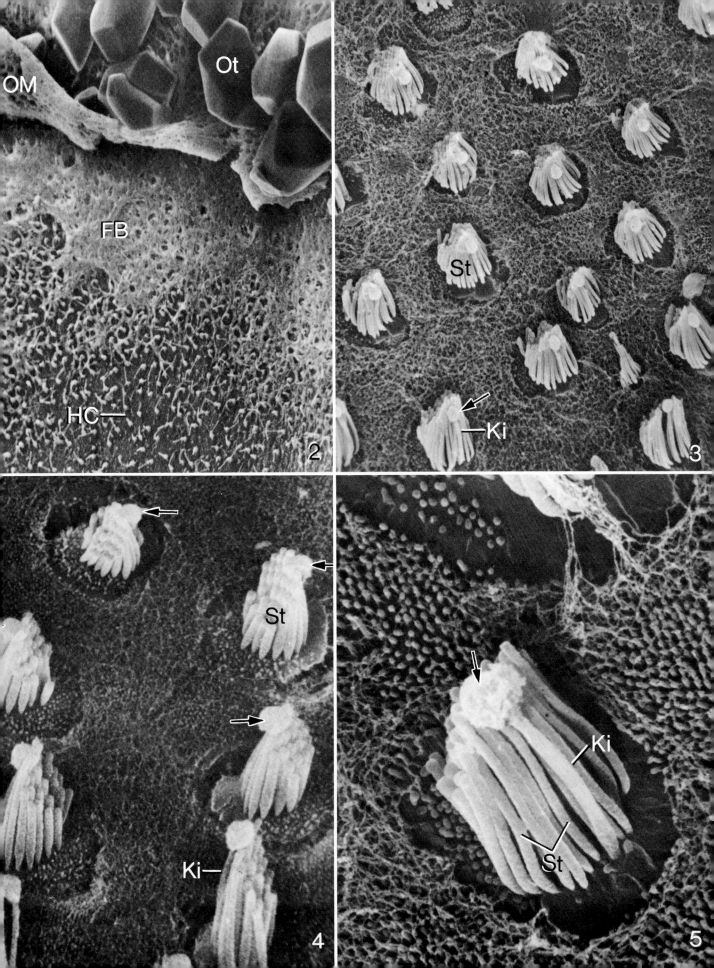

Lateral Line Organ

EC   Epidermal cell surface
Ki   Kinocilia
St   Stereocilia

Arrows: Neuromast

Fig. 1, ×2055; Fig. 2, ×7200.

Because of similarities in innervation, morphology, and embryology, the lateral line organ of teleosts and some amphibians has long been considered as an accessory auditory organ. Thus, a close functional relationship exists between the lateral line and the labyrinth. The lateral line system of fish and aquatic amphibians consists of a series of integumentary mechanoreceptor organs which provide information about water motion in the external environment. The stimulus for organ function depends on a bending or angular displacement of sensory hairs on receptor cells. Thus, the mechanism of sensory transduction in these cells appears to be similar to that of the inner ear.

Rana pipiens larvae develop a lateral line system in specific tracts in the epidermis. While these structures are functional after the embryo hatches and persist through larval stages, they disappear in the adult frog, apparently in response to a more terrestrial environment. The lateral line organ of a stage 24 *Rana pipiens* larva consists of multiple specialized sensory endings positioned in rows on the body surface (arrows in Fig. 1). Each unit is called a neuromast and consists of supporting cells and sensory cells which are innervated by cranial nerves. The sensory cells have a number of surface projections which extend free from the surrounding epidermal cells (EC in Fig. 1). However, in the living condition these processes are embedded in a gelatinous material called a cupula which was removed during preparative procedures used to observe the cell surfaces. An individual neuromast is illustrated at higher magnification in Fig. 2. Two long processes are evident and these are called kinocilia (Ki in Fig. 2). They possess an internal axial filament complex similar to that of functional cilia and flagella. Also present are a number of shorter processes which form a ring around the kinocilia. These projections are called stereocilia (St in Fig. 2) and are much shorter than the kinocilia. These free-standing components of the lateral line organ respond to water movements over the body. When the projections are displaced by water currents, the nerves innervating the cells are stimulated to enhanced or decreased neural activity, since impulses are relayed to the central nervous system even when the structures are not stimulated. The sensory hairs of lateral line organs of other animals may have an organization comparable to that previously illustrated in the labyrinth of the frog ear. That is, there are a number of stereocilia of variable length which increase in height toward an asymetrically placed kinocilium. The function of this system bears many similarities to the sensory cells in the labyrinth as previously described.

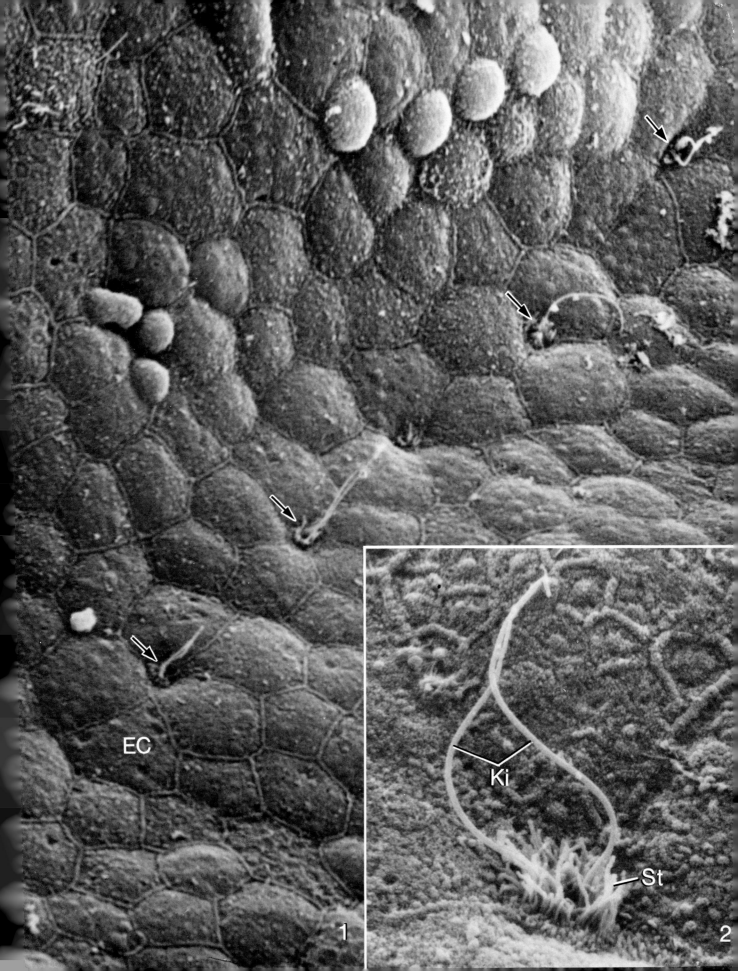

EC

Ki

St

1

2

Chapter 10
Development

Fertilization and the Cortical Reaction in Sea Urchins

Details in the fertilization process and the cortical reaction in the purple sea urchin, *Strongylocentrotus purpuratus*, are illustrated in Figs. 1 through 8. The mature egg is surrounded by a vitelline "membrane," or envelope, which closely adheres to the egg plasma membrane. Finger-like projections, called microvilli, extend from the plasma membrane through the vitelline envelope. Many membrane-bound cortical granules are present around the egg surface in the cortical ooplasm. The interaction of the sperm and egg in fertilization involves a complex of antifertilizin from the sperm and fertilizin from the egg.

Immediately after sea urchin egg and sperm are placed together, many sperm adhere to the egg surface, and the acrosome reaction occurs within a second after insemination. This condition is illustrated in Fig. 1, which shows the egg surface (ES) surrounded by the vitelline envelope to which numerous spermatozoa are attached. The sperm heads (SH) and flagella, or tails (ST), are apparent at this magnification. In Fig. 2, also showing an egg fixed one second after insemination, it is clear that the spermatozoan shown here has undergone the acrosomal reaction and has a visible acrosomal process (AP). The head (SH), midpiece (SM), and flagellum (SF) of the spermatozoan are clearly defined (Fig. 2). By 5 seconds after insemination, the acrosomal processes are no longer associated with the sperm. The egg illustrated in Fig. 3 was fixed 30 seconds after insemination. The tail of the fertilizing sperm (FST) is observed to extend from a small aperture in the forming fertilization membrane.

Immediately upon fertilization by one sperm, the cortical granules in the egg cortex begin to fuse with the egg plasma membrane so as to release their contents into the vitelline envelope. The release of cortical granules from the egg begins at the point of sperm entrance and proceeds in a wave which eventually extends entirely around the egg. The cortical granule material released by the egg interacts with the vitelline envelope material to produce the fertilization membrane.

In the egg illustrated in Fig. 3 the process of cortical granule breakdown and release began to occur 5 seconds prior to fixation. Note that the extent of the release of cortical granules (FM) is clearly defined since no sperm are attached to the egg in this region. Thus, supernumary sperm (SS) are unbound from the egg surface as the fertilization membrane rises by the action of a cortical protease. The wrinkled appearance of the forming fertilization membrane in Fig. 3 is an artifact produced during fixation since for a short time after formation; it is extremely soft until crosslinking has occurred. In Fig. 4 the egg was fixed 45 seconds after insemination. The cortical reaction has covered the top half of this egg; therefore, the fertilization membrane (FM) is about half formed. Supernumary sperm (SS) still adhere to the vitelline membrane around that part of the egg not yet having undergone the cortical reaction.

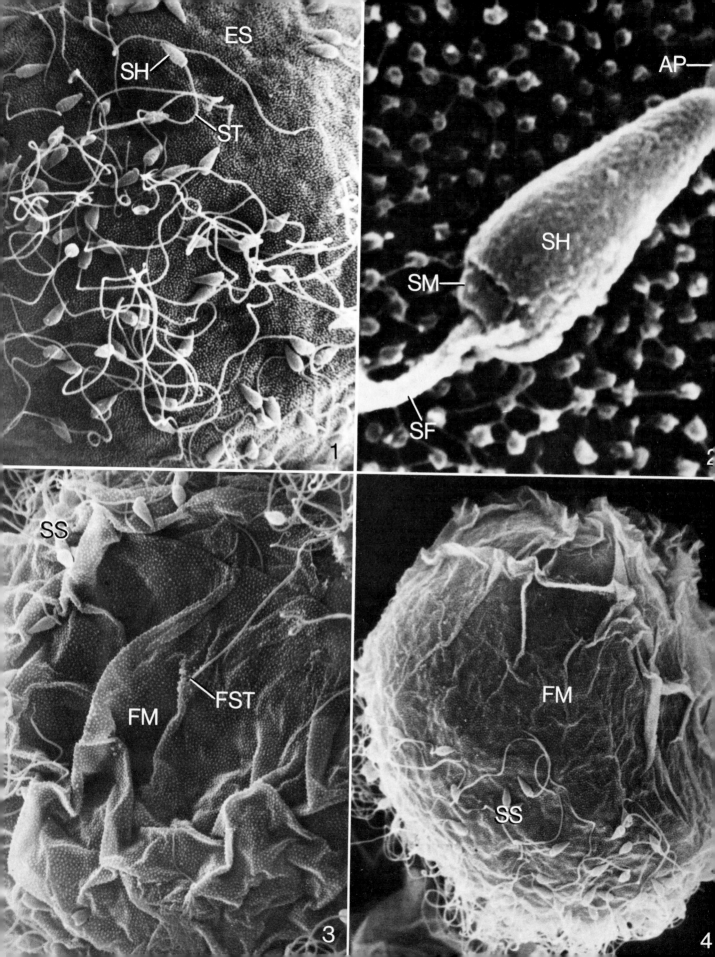

Fertilization
and the Cortical Reaction
in Sea Urchins

(continued)

The egg illustrated in Fig. 5 was fixed 55 seconds after insemination. The cortical reaction is complete and, thus, the fertilization membrane is fully formed. Therefore, all supernumary sperm have been detached. An egg fixed 3 minutes after insemination is illustrated in Fig. 6. This figure also depicts the hardening that occurs in the fertilization membrane so that it now is much smoother in appearance.

The vitelline layer of an unfertilized egg is illustrated in Fig. 7. The projections visible in these figures are impressions of the egg microvilli (Mv) in the vitelline layer. It is apparent in this case that ridges (arrows) extend between the projections. It is thought that these structures represent folds in the vitelline layer and serve as reservoirs of extravitelline material needed for the rapid and marked expansion of this layer at fertilization. A corresponding view of the surface of the fertilization membrane is illustrated in Fig. 8. Note that in this case the impressions of the microvilli (Mv) appear to be shorter and farther apart than those illustrated in Fig. 7. Further, no interconnecting ridges are present in Fig. 8. The changes occurring are consistent with the expansion of the vitelline layer in the formation of the fertilization membrane. (Micrographs kindly provided by Drs. MIA TEGNER and DAVID EPEL; Copyright 1973 by the American Association for the Advancement of Science.)

Reference

TEGNER, M.J., and EPEL, D.: "Sea urchin sperm-egg interactions studied with the scanning electron microscope." Science **179**, 685—688 (1973).

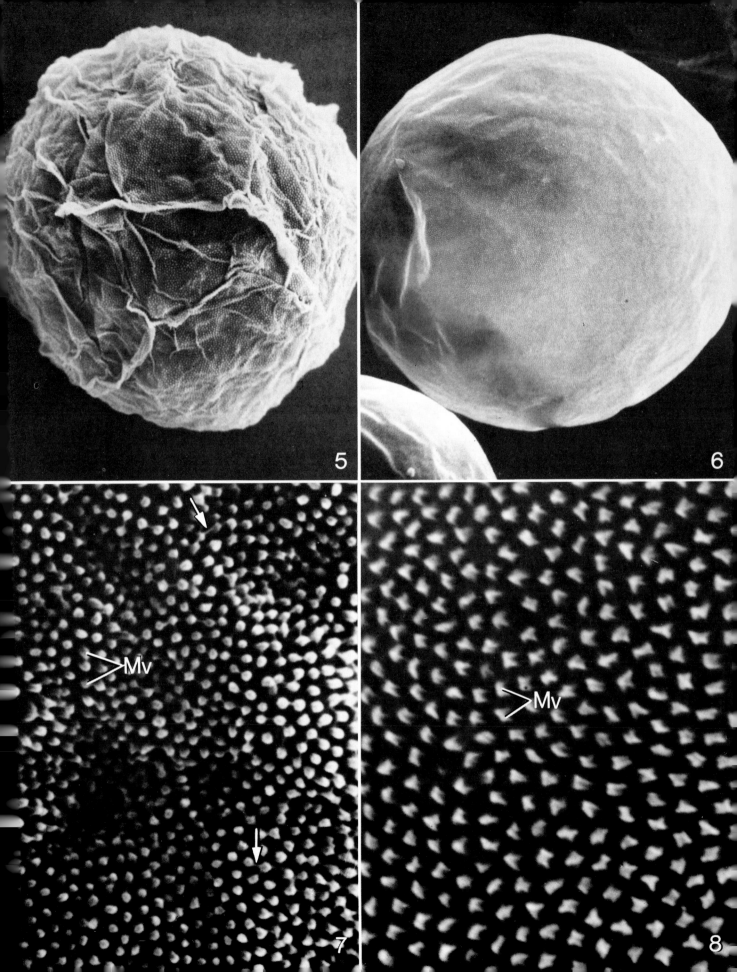

Embryology of the Frog

Fertilized Egg to Late Cleavage

Fig. 1 (unfertilized egg), ×62; Fig. 2 (two-cell stage), ×55; Fig. 3 (four-cell stage), ×65; Fig. 4 (eight-cell stage), ×61; Fig. 5 (16-cell stage), ×60; Fig. 6 (32- to 64-cell stage), ×62.

As ovulated *Rana pipiens* eggs, surrounded by a vitelline envelope, travel through the oviduct, they become closely enveloped by a secondary membrane consisting of a gelatinous layer secreted by the oviduct. This jelly layer swells when the eggs are shed into water. When the egg is fertilized, a fertilization membrane (formed by the interaction of released cortical granule material with the vitelline envelope) elevates and separates from the egg surface (oolema and microvilli) so that the egg is contained within a space separating it from the jelly coat. Thus, the fertilized egg is then capable of freely rotating within the cavity surrounded by the two mentioned envelopes until the time of hatching. The embryo can be observed to rotate within the investing membranes at about stage 15 (i.e., at a time when the neural folds begin to close and when the embryo begins to elongate). This rotation is caused primarily by the action of ciliated cells on the embryo surface. Thus, cleavage, gastrulation, neurulation, and tadpole development all occur within the egg envelopes. Hatching occurs spontaneously late in stage 20, so that the embryo then emerges from the jelly coat which is softened and dissolved by a hatching enzyme secreted by the embryo. At this point, the embryo becomes a free-swimming larva.

Stages of *Rana pipiens* development from the unfertilized egg to the morula are illustrated in Figs. 1 through 6. The jelly layer and the fertilization membrane have been removed from all stages except the fertilized egg (Fig. 1), which has only the jelly layer removed. From the time of fertilization (stage 1, Fig. 1) to the 16-cell stage (stage 6, Fig. 5) a period of about five hours is required at 18° C. During fertilization the sperm enters at the animal hemisphere, and a grey cresent becomes apparent at the opposite side of the egg about one hour later. Approximately 2.5 hours after fertilization, the first cleavage furrow is completed. The first cleavage furrow is illustrated in Fig. 2. It begins at the animal pole and progressively extends toward the vegetal pole, eventually cutting the cell into two. The first cleavage is described as holoblastic since there is a complete division of the egg. It is also equal since each of the resulting cells are similar in size. Further, it is characterized as a meridional cleavage since the cleavage furrow forms along a meridian of the egg. The second cleavage furrow (Fig. 3) forms about one-half hour after the first cleavage furrow is complete and develops at right angles to the first cleavage. The second cleavage furrow is also holoblastic, equal, and meridional. Thus, it also begins at the animal pole and progressively extends toward the vegetal pole (as seen in Fig. 3). The third cleavage furrow is a horizontal one since it pursues a line parallel to the equator of the egg (Fig. 4). The resulting eight-cell stage (Fig. 4) consists of four smaller animal pole blastomeres and four larger vegetal cells. Thus, the third cleavage furrow is holoblastic but unequal. The fourth cleavage furrow of the frog egg extends from animal to vegetal pole and is meridional. As a consequence, in the 16-cell stage there are eight smaller cells in the animal hemisphere compared with the eight larger cells in the vegetal hemisphere. Such a condition is clearly illustrated in the side view of a 16-cell stage illustrated in Fig. 5. At this time a small segmentation cavity would be apparent inside the developing embryo. A late cleavage stage (32 to 64 cells) (stage 7, about 6 hours at 18° C) is illustrated in Fig. 6. The view is from the animal hemisphere and clearly illustrates the smaller size of the animal pole micromeres, but the larger size of the vegetal pole blastomeres cannot be observed from this angle.

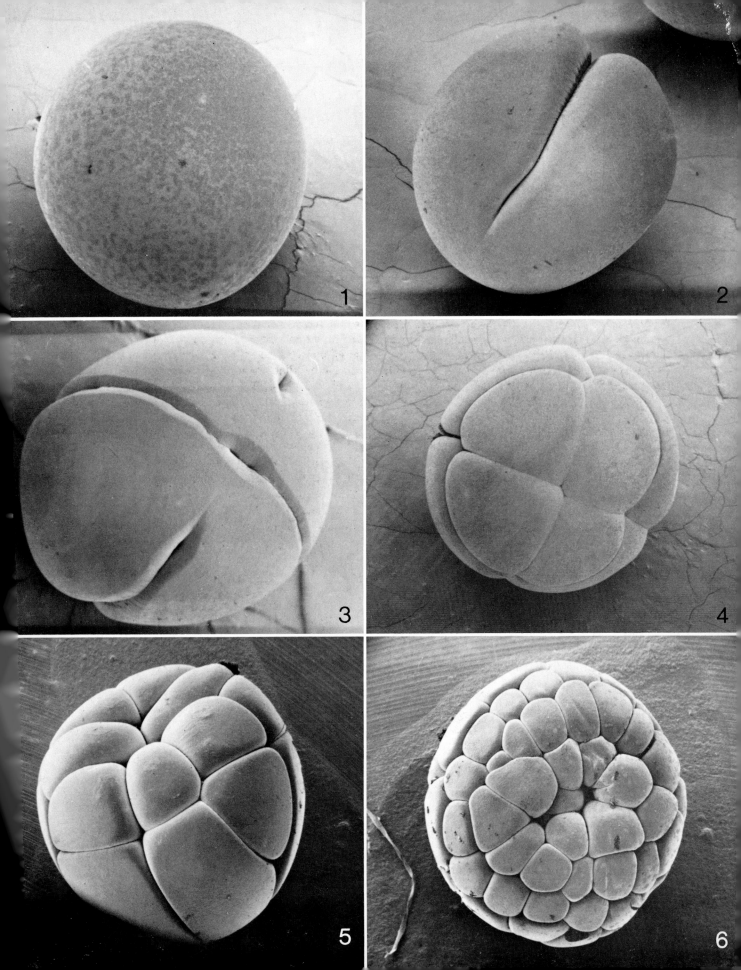

Blastula to Tailbud Development in the Frog

DL	Dorsal lip of	PS	Pronephric swelling
	blastopore	Su	Suckers
Ey	Eye region	TB	Tailbud
GP	Gill plate	VL	Ventral lip of
LL	Lateral lip of		blastopore
	blastopore	YP	Yolk plug
NP	Nasal pit		

Arrow: Blastopore

Fig. 7 (late blastula), ×61; Fig. 8 (gastrula), ×68; Fig. 9 (yolk plug), ×59; Fig. 10 (early neurula), ×48; Fig. 11 (late neurula), ×62; Fig. 12 (tailbud), ×45.

After 12 to 16 hours of development at 18° C following fertilization, the embryo is at a blastula stage (stages 8—9). During this period cleavage continues and the cells at both the animal and vegetal poles become even smaller (Fig. 7). During this period an internal cavity, or blastocoel, develops in the region of the animal hemisphere. By about 19 hours of development, the dorsal lip of the blastopore can be visualized externally in a region just below the equator of the egg. This marks the beginning of cell migration from the animal hemisphere toward the interior through the dorsal lip, and associated with this activity an internal archenteron develops. As more cells migrate internally (gastrulation), the lip of the blastopore extends laterally so that the blastopore assumes the form of a semicircle (Fig. 8, DL). The vegetal pole cells are still larger than the animal pole cells. The lateral lips of the blastopore progress so that eventually they meet to form a ventral lip of the blastopore. As a result, a circular group of vegetal pole cells called a yolk plug (YP) (Fig. 9) is completely surrounded by the blastopore. The regions of the dorsal lip (DL), lateral lip (LL), and ventral lip (VL) of the blastopore are indicated in Fig. 9. During subsequent development to the neurula, the blastopore becomes progressively smaller and squeezes the remains of the yolk plug into the interior. Thus, during gastrulation the animal pole cells (ectoderm) completely cover the entire surface of the embryo. The blastopore then becomes slit-shaped and can be seen at the posterior most region of the developing neural plate and folds. Its position is just out of view in the region of the arrow in Fig. 10. Although previous development has consisted of cell division and cell migration, subsequent development involves the formation of the nervous system from the ectoderm and consists of neural plate, neural fold, and neural tube stages. The sequence of development that occurs from about 38 to 58 hours of development at 18° C is illustrated in Figs. 10 through 12. Fig. 10 illustrates the neural plate and marginal but shallow neural folds of the entire nervous system which represents a thickening of the ectoderm along the dorsal surface of the embryo. The anterior limit of the neural folds (at left in Fig. 10) is opposite the blastopore. The folding process that begins at the lateral margins of the neural plate (Fig. 10) continues so that by stage 15 the neural folds touch in the mid-region and begin to fuse (Fig. 11). As this process continues, a neural tube is formed that is completely enclosed and widens anteriorly as the brain. The complete closure of the neural tube is followed by the tailbud stage (stage 17) (58 hours of development at 18° C), such as illustrated in Fig. 12. In the tailbud illustrated the tail (TB), gill plate region (GP), and suckers (Su) are especially apparent. Other surface markings indicate the internal position of structures and include the pronephric swelling (PS), eye (Ey), and nasal pit (NP). The small tubercles seen to cover the surface of the tailbud are ciliated ectodermal cells.

Reference

KESSEL, R. G., BEAMS, H. W., and SHIH, C. Y.: "Surface structures of the frog embryo as revealed by scanning electron microscopy." Anat. Rec. **175**, 489 (1973).

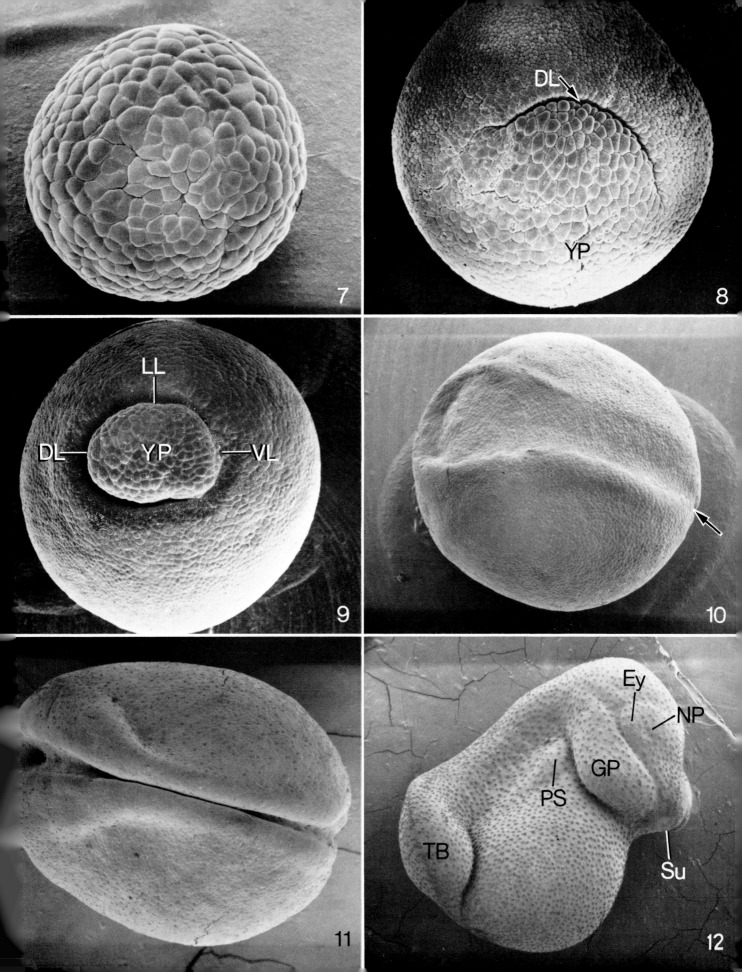

Subject Index

**Edited by
J. K. Koehler**

With Contributions
by S. Bullivant,
J. Frank, K. Hama,
T. L. Hayes, J. H. Luft,
F. A. McHenry,
D. C. Pease,
M. M. Salpeter

Advanced Techniques in Biological Electron Microscopy

Contents

With 108 figures. XII, 304 pages. 1973
Cloth DM 50,—; US $ 20.40   ISBN 3-540-06049-9
Prices are subject to change without notice

The book deals with selected topics of advanced electron optical and preparatory techniques. As such, it is not intended as an elementary guide for beginning workers or students. It contains discussions of new and less well-known embedding media, specimen preparation methods including substitution, and techniques such as inert dehydration, freeze-fracturing and autoradiography, with emphasis on analytical methodology and interpretation. Other chapters deal with image processing using computers, scanning electron microscopy and high-voltage electron microscopy. These papers brought together in one volume are indicative of what constitutes the forefront of research in biological electron microscopy today. No attempt, however, has been made to be exhaustive or all-inclusive with respect to subject matter. In a number of instances, including the chapters on embedding media and computer processing, material not previously available to a general scientific audience is presented in detail. It is hoped that the volume will not only serve as a reference work for scientists already expert in these areas, but will also stimulate biologists to investigate and employ techniques which they may at first consider too exotic or too complex to attempt.

Springer-Verlag Berlin · Heidelberg · New York

München Johannesburg London Madrid New Delhi Paris Rio de Janeiro
Sydney Tokyo Utrecht Wien

Marcel Bessis
Corpuscles
Atlas of Red Blood Cell Shapes

By **Marcel Bessis,** Professor, Faculty of Medicine,
University of Paris, Director, Institut de Pathologie Cellulaire,
Hôpital de Bicêtre, Paris, France

With 121 figures. 147 pages. 1974. Cloth DM 68,—*
ISBN 3-540-06375-7
Distribution rights for Japan: Maruzen Co. Ltd., Tokyo

The hematologist usually examines red blood cells by looking at smears
through a light microscope. This type of examination is likely to remain
the routine technique of blood cytology for a long time to come.
Still, we should not forget that blood smears are artifacts. The smearing
flattens the cells completely, obliterating many of their characteristics and
distorting others.

This atlas reveals the shape of red blood cells as displayed by scanning
electron microscopy. As with the transmission electron microscope,
the cells must be fixed, but they can be observed in their three
dimensions and revolved before our very eyes, thus providing details
of the surface and the shape of cells never before visualized. Having seen
the results of studies with the new techniques, hematologists can now
distinguish nuances in the appearance of red cells in routine smears
which previously could not have been appreciated. Thus routine blood
smears assume a new significance.

Morever, there is an inherent beauty in the pictures displayed here,
a beauty which is enchanting to both the specialist and the layman.

Table of Contents

Springer-Verlag

Berlin · Heidelberg · New York

München Johannesburg London Madrid New Delhi
Paris Rio de Janeiro Sydney Tokyo Utrecht Wien

Living Blood Cells and their Ultrastructure

By **Marcel Bessis**
Translated by Robert I. Weed

With 521 figures and 2 color plates
XXI, 767 pages. 1973
Cloth DM 151,—*
ISBN 3-540-05981-4

Distribution rights for Japan:
Maruzen Co. Ltd., Tokyo

Contents
General Comments on Structure and
Functions of Blood Cells.
The Erythrocytic Series.
The Granulocytic Series.
The Thrombocytic Series.
Lymphocytic Series.
The Reticulo-Histiocytic System
and the Monocyte.
The Plasma Cell Series.
The Mast Cell Series.
Hematologic Malignancies.
Techniques.

Red Cell Shape
Physiology, Pathology, Ultrastructure

Editors: **M. Bessis, R. I. Weed,
P. F. Leblond**

With 147 figures. VIII, 180 pages
1973. Cloth DM 32,40*
ISBN 3-540-06257-2

Distribution rights for Japan:
Maruzen Co. Ltd., Tokyo

Published as a special issue of
Nouvelle Revue Française d'Héma-
tologie, Vol. 12, No. 6, December
1972

Contents
Red Cell Shape and Nomenclature.
Discocyte-Echinocyte and Discocyte-
Stomatocyte Transformations.
Biophysical Studies and Membrane
Models.
Membrane Ultrastructure: Freeze-
Cleave and Freeze-Etch Studies.

---

* Prices are subject to change
without notice